Life on the Edge:
Iain Crawford's Udal, North Uist
The Neolithic and Bronze Age of RUX6

Edited by
Beverley Ballin Smith

with

Judith Finlay Aird, Torben Bjarke Ballin, Julia Beaumont, Solange Bohling, Jo Buckberry, Ruby Ceron Carrasco, Cassandra Hall, Derek Hamilton, Anthony Newton, Susan Ramsay, Catherine Smith, Robert Squair and Caroline Wickham-Jones

Illustrations by

Joanne Bacon, Fiona Jackson, Darmuid O Connor, David Mcnicol, Gillian Sneddon and Leeanne Whitelaw

Archaeopress Publishing Ltd
Summertown Pavilion
18-24 Middle Way
Oxford OX2 7LG
www.archaeopress.com

ISBN 978 1 78491 770 8
ISBN 978 1 78491 771 5(e-Pdf)

© Archaeopress and the individual authors 2018

In memory of Iain A Crawford 1928-2016

For Imogen and Rebecca Crawford

For everyone who worked on the RUX6 Project in 1974 and 1980 to 1984

And

for the people of North Uist.

This publication has been grant-aided by Historic Environment Scotland

HISTORIC ENVIRONMENT SCOTLAND | ÀRAINNEACHD EACHDRAIDHEIL ALBA

All rights reserved. No part of this book may be reproduced, in any form or by any means, electronic, mechanical, photocopying or otherwise, without the prior written permission of the copyright owners.

This book is available direct from Archaeopress or from our website www.archaeopress.com

Funders of the project

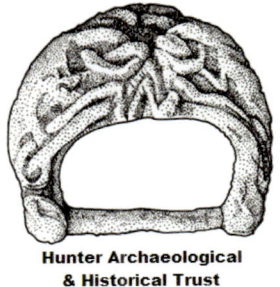

Iain A Crawford
2005

Contents

List of Figures	viii
List of Tables	xvii
Acknowledgements	xx
Foreword - Malcolm Burr	xxiii
Preface - Beverley Ballin Smith	xxiv
Summary	xxvi
Gaelic Summary	xxix
List of Contributors	xxxii

PART 1 Introduction - Beverley Ballin Smith — 1

- Location, topography, vegetation and place-names — 1
- Background to the project and its origins — 4
- The environmental history of the area — 6
- The original research aims 1963 — 9
- Introduction to the excavations — 10
- Methodology — 13
- Post-excavation activities prior to 2008 — 17
- Post-excavation developments since 2008 — 17
- Updated post-excavation research aims and objectives — 18

PART 2 The excavation record - Beverley Ballin Smith 19

    Introduction 19
    Natural events and disturbances 19
    Phase E - natural and anthropogenic deposits 27
    Phase D 29
    Phase C 39
    Phase B 44
    Phase A 57

PART 3 Dating and human remains 64

    Early dating of the site - Beverley Ballin Smith 64
    Radiocarbon dating and Bayesian modelling - Derek Hamilton 64
    The burials on the site - Beverley Ballin Smith 67
    Skeletal analysis - Solange Bohling and Jo Buckberry 70
    Isotopic analysis of the human skeletal remains - Julia Beaumont, Cassie Hall, Derek Hamilton and Jo Buckberry 80

PART 4 The changing natural environment and subsistence farming 86

    Introduction - Beverley Ballin Smith 86
    Botanical remains - Susan Ramsay 88
    Pollen - Beverley Ballin Smith and Keith Bennett 95
    The faunal remains - Judith Finlay Aird 95
    Non-worked marine shell - Catherine Smith 110
    The perforated mollusc shell - Catherine Smith 115
    The crustacea - Catherine Smith 116
    Terrestrial snail assemblages - Ruby Ceron-Carrasco 118

PART 5 Exploitation of natural resources and the uses of artefacts 122

    Introduction - Beverley Ballin Smith 122
    Flaked flint and other fine-grained lithic materials - Caroline Wickham-Jones 123
    The quartz assemblage - Torben Bjarke Ballin 129
    Pumice found at The Udal - Anthony Newton 165

    Worked pumice and pumice artefacts - Beverley Ballin Smith   169
    Stone tools and other items in stone - Beverley Ballin Smith   174
    The prehistoric pottery - Robert Squair and Beverley Ballin Smith   183
    Worked bone artefacts - Beverley Ballin Smith   197

PART 6 Discussion - Beverley Ballin Smith   203

    The origins of settlement on the site   203
    The uses of ritual and domestic structures   204
    Changes to society after the accumulation of blown sand   213
    The people   225

PART 7 Conclusion - Life on the edge - Beverley Ballin Smith   227

Afterword – Iain and Imogen Crawford   229

Appendices   230

    1: Marine shell samples quantified by species   230
    2: Pottery catalogue   231

Bibliography   252

Index   265

List of Figures

Cover illustration: Looking towards Rubha Huilis from the site of RUX6. (Original image © Chris Stewart-Moffitt).

Frontispiece – Iain A Crawford in 2005 - © B. Ballin Smith.

PART 1 — 1

Figure 1.1:	Geographical location of Udal	2
Figure 1.2:	Aerial view of Udal	3
Figure 1.3:	Location of the site	3
Figure 1.4:	The machair vegetation in bloom	4
Figure 1.5:	The dune system and vegetation	8
Figure 1.6:	1964 archive photograph of trial excavations. The RUX6 cairn is under grass at the right hand edge of the image below the beach	11
Figure 1.7:	The section through the cairn revealed in 1974	12
Figure 1.8:	The boulder beach in 1994 after the site had been excavated	13
Figure 1.9:	The section drawing locations across all the features of the site	15
Figure 1.10:	Sieving station on the shore just north of the excavation trench and beyond the protective wall	16
Figure 1.11:	Crawford's Chrystal Palace	16

PART 2 — 19

Figure 2.1:	The gneiss bedrock formations in front of the RUX6 site in 1980. Looking south	20
Figure 2.2:	The accumulation of windblown sand (level X) that brought an end to the late Neolithic settlement. Looking south-east	24
Figure 2.3:	The puddled horizon of mixed material forming the lower part of the first marine inundation, from the lower paler streak in the section to the darker stony layer above. Looking south-east	24

Figure 2.4:	The shingle and stone beach, at the top of the section, deposited as the final part of the marine inundation. Looking west	25
Figure 2.5:	Part of the section seen through the cairn complex revealed by coastal erosion in 1974. Looking south	26
Figure 2.6:	The anthropomorphic layers above natural clays and bedrock. Looking east	29
Figure 2.7:	Location of the Neolithic structures (DJ) west, (DH) centre, and the ritual area to the east	29
Figure 2.8:	The shaft, whale vertibra, platform, stone alignments and Great Auk stone	31
Figure 2.9:	The whale vertibra on the edge of the shaft formed by bedrock. Looking south-east	32
Figure 2.10:	Stonework around the whale bone, with a stone row leading south. The Great Auk stone can be seen on towards the top right of the image. Looking south	32
Figure 2.11:	Stone alignments leading south from the shaft and platform in the north of the excavation, with the Great Auk stone on the left side by the Phase A saw-pit trench through the site. The Othello stone is the black stone at the end of a stone row on the east side of the site. Looking north	32
Figure 2.12:	The curved stone alignment in the foreground (left) with the Great Auk stone (right). Beyond the saw-pit trench are the remains of Neolithic Building 2 (DH). Looking west	33
Figure 2.13:	The Great Auk Stone with packing stones around it (left) the shaft behind it and the Othello stone with a stone alignment (right). The sieving apparatus can be seen towards the top left. Looking north-east	34
Figure 2.14:	Section through the Great Auk stone	32
Figure 2.15:	Detail of the two Neolithic buildings	35
Figure 2.16:	Axe SF 26732 in the wall core of Neolithic Building 1 (DJ). Looking east	34
Figure 2.17:	Neolithic Building 1 (DJ) with central hearth, postholes around it for the roof and the outline of the inner wall (left). Looking south-west	36
Figure 2.18:	Neolithic Building 2 (DH) showing wall outlines, central hearth and other features. Looking south-east	37
Figure 2.19:	Neolithic Building 2 (DH) with excavated postholes with post-pads. Looking south-east	38
Figure 2.20:	Section 16 showing Neolithic levels in red with thicker deposits towards the coast (east)	39
Figure 2.21a and b:	Mattock marks in the infilling sand of Neolithic Building 1 (DJ) Looking north and overhead	40
Figure 2.22:	Plan of Phase C ploughing scars/marks	42
Figure 2.23:	Ploughing scars surviving beneath the cairn complex. Looking north	41

Figure 2.24:	Plan of Phase C pits and linear boundary	42
Figure 2.25:	A bowl-pit exposed and others during excavation. The darker shadows in the lighter sand show others still to be excavated. Looking north-east	43
Figure 2.26:	Phase C bowl pits (blue) in Section 5 cutting through late Neolithic occupation (red)	43
Figure 2.27:	Stone box (CF) that survived the last marine inundation below the cairn complex. Looking north	44
Figure 2.28:	Section 3 through the cairn complex. Burial SF 17436 was found in the pit dug through Phase C into Phase D at the cliff edge (left)	47
Figure 2.29:	Burial pit on the cliff edge immediately the north of the large cist and SF 17642 skeleton	45
Figure 2.30:	Phase B cairn complex with cist, burial, earlier cairn, central domed masonry and the outer kerb (west), and cairn, cist and standing stone (east)	48
Figure 2.31:	The kerb (BA) to the early cairn. Looking north	47
Figure 2.32:	Cist (BB3) and skeleton SF 17642	48
Figure 2.33:	Section 24 through the Phase B cist (BB3).	49
Figure 2.34:	Burial in the cist (left), the cist with levelling stones for the lid (centre), and with its capping (right)	49
Figure 2.35:	Cist with lid on the left of the central core of stone with the foundations for the outer kerb in the foreground. Looking north	50
Figure 2:36:	Construction detail of the inner core of stone over the cist. Looking north-east	50
Figure 2.37:	Construction detail of the outer kerb against the inner core of stone. Looking north	51
Figure 2.38:	Plan of Phase B cairn complex with Phase A disturbances	52
Figure 2.39:	The surviving height of the cairn complex in 1974 prior to deturfing. Looking east	52
Figure 2.40:	The stump of standing stone (BD1) in its socket. Looking north-east	53
Figure 2.41:	Plan of the temporary structure (BB24), postholes, hearth pits and stone edging	54
Figure 2.42:	Floor of temporary structure (BB24) with cist inserted into the floor. Looking west	54
Figure 2.43:	Stone box (BG3) in floor used for quartz raw material cache	55
Figure 2.44:	Plan of cist (BM) with skeleton SF 26319 inserted into the side of the temporary building	56
Figure 2.45:	Cist (BM) with skeleton SF 26319, pot and bones of a calf. Looking north	56
Figure 2.46:	Section 7 through BG24 cist, cairn and temporary structure (BG24) (green)	57

Figure 2.47:	Kerb of cairn (BM) coursing through the temporary building. Looking north-west	57
Figure 2.48:	Section 14 Phases A (brown), B (green), C (blue), D (red) and E (grey)	59
Figure 2.49:	Plan of Phase A features	60

PART 3 — 64

Figure 3.1:	Chronological model for RUX6. Each distribution represents the relative probability that an event occurs at a particular time. The distributions in outline show the calibration of each result by the probability method (Stuiver and Reimer 1993). The solid distributions are *posterior density estimates* derived from the chronological model. This model is exactly defined by the square brackets and OxCal keywords at the left of the diagram	68
Figure 3.2:	Spans for the modelled activity at RUX6. The spans are based on the model shown in Figure 3.1	68
Figure 3.3:	Alternative chronological for the dates from RUX6, with the two charred food residue dates included as deriving from marine-based protein. The model is as described in Figure 3.1	69
Figure 3.4:	Spans for the modelled activity from RUX6, based on the sensitivity analysis. The spans are based on the model shown in Figure 3.3	69
Figure 3.5:	Diffuse smooth but porous compact bone on internal surface of maxillary sinus	72
Figure 3.6:	Seven probable canine and pre-molar teeth, four with pulp cavity exposure.	73
Figure 3.7:	Maxillary molar roots	72
Figure 3.8:	Apical granuloma inferior to left mandibular first pre-molar	72
Figure 3.9:	Possible apical granuloma inferior to right mandibular canine and first pre-molar	73
Figure 3.10:	Linear enamel hypoplasia on lingual surfaces of all four mandibular incisors	73
Figure 3.11:	Possible carious lesions on distal occlusal surface of left mandibular third molar	74
Figure 3.12:	Impacted right mandibular third molar	74
Figure 3.13:	Increased anteversion of right and left femoral heads	74
Figure 3.14:	Varying levels of femoral neck anteversion. Redrawn after Cibulka (2004)	75
Figure 3.15:	Enthesophytes on proximal end of left and right tibiae	75
Figure 3.16:	Rugose gluteal tuberosity on posterior surface of right femur.	76
Figure 3.17:	Articulation of left navicular and intermediate cuneiform from SF 17638 with left lateral cuneiform from SF 17642	76

Figure 3.18:	Associated radial fragments from two different samples (SF 16201 and SF 17436)	78
Figure 3.19:	Comparison between right glenoid fossa from sample SF 24051 and left glenoid fossa from sample SF 17437	79
Figure 3.20:	Comparison of right humerus from sample SF 16201 and left humerus from sample SF 23859	79
Figure 3.21:	A plot of animal dentine and bone collagen $\delta^{15}N$ and $\delta^{13}C$ alongside the human dentine collagen (averaged) and bone collagen. Error bars set ±0.2‰	82
Figure 3.22:	Plot showing bone collagen $\delta^{15}N$ and $\delta^{13}C$ for SF 17437, SF 17642, and SF 26319 with bone collagen data from contemporaneous British sites. Sumburgh (Montgomery *et al.* 2013), Quarterness (Schulting *et al.* 2010), Cladh Hallan (Pearson *et al.* 2005), Yarnton (Lightfoot *et al.* 2009) and Newark Bay (Richards *et al.* 2006). Error bars set at ±0.2‰.	82
Figure 3.23:	Incremental dentine and bone collagen $\delta^{15}N$ and $\delta^{13}C$ profile for SF 26319 plotted against approximate age. Error bars set to 0.2‰	84
Figure 3.24:	Incremental dentine and bone collagen $\delta^{15}N$ and $\delta^{13}C$ profile for SF 17642 plotted against approximate age. Error bars set to 0.2‰	84

PART 4

86

Figure 4.1:	Woody taxa from Phase A (modern)	89
Figure 4.2:	Woody taxa from Phase B (Bronze Age)	89
Figure 4.3:	Woody taxa from Phase D (Neolithic)	91
Figure 4.4:	Woody taxa from other samples	93
Figure 4.5:	Comparative size of RUX6 sheep – 1st phalanges	103
Figure 4.6:	Comparative size of RUX6 sheep – 2nd phalanges	103
Figure 4.7:	Modern day Shetland sheep.	103
Figure 4.8:	Diagrammatic skeleton indicating butchery marks likely to be represented on sheep bones from RUX6 (after Finlay 1981)	104
Figure 4.9:	Diagrammatic skeleton indicating butchery marks likely to be represented on cattle bones from RUX6 (after Finlay 1981)	104
Figure 4.10:	External and internal views of an oyster (*Ostrea edulis*) SF 26800 from Phase D	113

PART 5

124

Figure 5.1a and b:	The two opposed faces of flakes SF 25223 and SF 25743	124

Figure 5.2:	Cores SFs 25020, 23525, 25226a, 26765c and 26778c, thumbnail-scrapers SFs 23206, 23409, 25612c, 26701c, 26765g and 263229, end scrapers SFs 23517 and 23852c, miscellaneous retouch SF 25742 and split pebble SF 26368	125
Figure 5.3:	Length:width ratio of all intact regular flakes; blue: platform flakes, red: bipolar flakes	126
Figure 5.4:	The ratio of bipolar:platform flakes within each phase; hatched bars = platform flakes, cross-hatched bars = bipolar flakes	127
Figure 5.5a and b:	Four refitted orange-segment flakes from the Norwegian site Lundevaagen 21, SW Norway, a) refitted and b) individual flakes	131
Figure 5.6:	The elements of a bipolar orange-segment flake	131
Figure 5.7:	The ratio of sand-blasted pieces by phase (the 'blast ratio') probably indicating the level of erosion the main phases were exposed to	134
Figure 5.8:	The main dimensions of all intact blades/microblades in quartz; hard-hammer pieces (red), and bipolar pieces (blue).	135
Figure 5.9:	Intact blade SF 26004/54	136
Figure 5.10:	The greatest dimension of all intact quartz flakes (2,237 pieces)	136
Figure 5.11:	Crests SFs 26310/37, 23786/58 and 23678/4	137
Figure 5.12a and b:	Split pebble SF 26241/1-4, refitted, a) top view and b) side view	137
Figure 5.13a and b:	The dimensions of the four main core types (intact pieces) recovered at a RUX6 compared with the cores recovered by Elliott's at b) Barabhas: split pebbles (black), single-platform cores (blue), irregular cores (red), and bipolar cores (green)	138
Figure 5.14:	Split pebbles SFs 23332, 25609/55, 23650/28, 23032/11, 23295/7 and 23205/24	138
Figure 5.15:	The dimensions of the split pebbles	139
Figure 5.16:	Main sub-types of split pebbles. Upper row: flakes detached; bottom row: cubic bits detached. 0) Undamaged oval pebble; 1) one half-length flake detached from one terminal, one face; 2) one full-length flake detached from one terminal, one face; 3) two half-length flakes detached to either side of one terminal; 4) two approximately half-length flakes detached, one from either terminal, one face; 5) the pebble split along the central long-axis; 6) diagonal split; 7) two corners split off either side of one terminal; 8) the pebble broke across; and 9) other forms of limited flaking or splitting	139
Figure 5.17:	Single-platform cores (conical) SFs 26020/28, 26327/7, 25091/14 and 25129/7	140

Figure 5.18:	Single-platform/discoidal core SF 26006/1	140
Figure 5.19:	Single-platform cores (broad/flat) SFs 26549/3, 23690/13, 26199/1 and 23470/105	141
Figure 5.20a and b:	The dimensions of the intact single-platform cores. In a), the cores are subdivided according to their morphology (red: conical; green: 'standard' single-platform cores; and blue: single-platform/discoidal cores), and in b) according to the character of their platform (blue: cortical; green: plain; and red: faceted)	141
Figure 5.21a and b:	Single-platform cores sub-types defined amongst the quartz cores from Barabhas (Elliott Collection). a) shows the cross-section of CAT 5167 – these pieces are simply very flat cores with very acute flaking-angles; b) shows the cross-section of CAT 216, 4698, 5526, and 5869 – these cores are based on thick flakes from which smaller flakes were detached along the entire circumference, giving some of them a scraper-like appearance. The thick line indicates the cortical platforms of the pieces	141
Figure 5.22:	The operational schema of the main platform core type identified at RUX6	142
Figure 5.23:	1) Free-hand hard percussion; 2) hard percussion on anvil; and 3) bipolar ('hammer-and-anvil') technique; after Callahan 1987: illus 97	142
Figure 5.24:	The dimensions of the opposed-platform cores (intact pieces). Red: approximately cylindrical pieces; blue: broad unifacial pieces	142
Figure 5.25:	Opposed-platform cores SFs 25204/6, 23470/104, 25724/6, 25148/162, and 26710	143
Figure 5.26:	The dimensions of the irregular cores (intact pieces)	144
Figure 5.27:	Irregular cores SFs 23559/17, 16785/2, 23267/6, 26881/2, 26681/9, 16646/45, 25009/73, 25000/79	144
Figure 5.28:	Bipolar cores SFs 26626/74, 26427/20, 25321/68, 23992/7, 23198/5, 24006/3, 23044/88, 26235/12, 26626/78, 26004/52, 26809/19 and 26468/42	145
Figure 5.29a and b:	The dimensions of the bipolar cores (intact pieces). In a), the cores are subdivided according to whether they are unifacial (blue) or bifacial (red), and in b) according to their number of reduction axes (sets of opposed terminals) (one axis: blue; two axes: red).	146
Figure 5.30:	Knives SF 23803/7 and SF 26146/26	147
Figure 5.31a and b:	Comparison between the short end-scrapers of a) RUX6 (blue) and those of b) Barabhas, the Elliott Collection (red)	147
Figure 5.32:	Blade-scraper SF 23516/28	148
Figure 5.33:	Discoidal scrapers SF 16688/7 (split) and SF 23350/5	148
Figure 5.34:	Short end-scrapers (small) SF 23470/109, 23567/21, 25045/21, 16858/5,	149

	25167/63, 26536/3, 25487/176 and 26487/46	
Figure 5.35:	Short end-scrapers (large) SFs 23141/33, 25025/7, 25148/160, 23470/107, 26020/21, 23023/26, 26209/18, 25336/17	150
Figure 5.36:	Side-scrapers SFs 26020/18, 233650/29 and 25944/9	150
Figure 5.37:	Piercers SFs 26020/19, 26585/4, 26585/5, 25122/38 and 25351/29	151
Figure 5.38:	Piece with edge-retouch with modification along the right lateral side	152
Figure 5.39:	Points SF 26241/14 and SF 25983/8	152
Figure 5.40a and b:	Percussoir/anvil SF 26782/1, a) main face and b) one of the tips	153
Figure 5.41:	Hammerstones SF 26009/1 and SF 26587/1	153
Figure 5.42:	Pounders SF 25669/14 and SF 25110/11	154
Figure 5.43:	Map showing the distribution of archaeological sites in the Western Isles where pumice has been found. Sites mentioned are labelled	165
Figure 5.44:	Graph showing the correlation of the Udal pumice with the other dacitic pumice from archaeological sites in Scotland (Newton 1999a)	167
Figure 5.45:	Graph showing the correlation of the Udal pumice with the SILK tephra layers erupted from Katla, southern Iceland (Newton 1999a Larsen *et al.* 2001)	168
Figure 5.46a: and b:	All pebble sizes, a) Phase B, red – structures, blue – all other contexts, b) Phase D, purple – structures, green – all other contexts	170
Figure 5.47:	Comparison of measurements of worked pumice from all phases	170
Figure 5.48:	Pumice artefacts from Phase A SF 23238a, SF 23788a, SF 23788b, SF 25154 and SF 25174	171
Figure 5.49:	Pumice artefacts from Phase B, SF 23375b, SF 23907, SF 25319d and SF 26223	172
Figure 5.50:	Pumice artefacts from Phase D, SF 23952b, SF 23853b, SF 26621, SF 25497, SF 25943a and SF 26776	173
Figure 5.51:	Axehead SF 23762, stone ball SF 24962 and percussoir/anvil SF 23302	177
Figure 5.52:	Worn piece SF 23302	178
Figure 5.53:	Grinder SF 26211	179
Figure 5.54a and b:	Great Auk stone SF 26904	181
Figure 5.55:	Beaker sherds SFs 23391/2, 23403/8, 23403/10, and 23403/25-27, Grooved Ware SFs 25442/1, 25672/4, 25672/7, 25672/9-10. 25672/15 and 25672/17, early Bronze Age pot SF 26259/2, finger dimple sherds SF 26259x2, and late Neolithic base SF 26817/1	186
Figure 5.56:	Phase D Grooved Ware sherds from Building 2 (DH) all SF 25672	189

Figure 5.57:	Phase D Grooved Ware sherds from Building 2 (DH) all SF 25578	189
Figure 5.58:	Phase D rims from Building 2 (DH) SF 25744 (left), rest SF 25588	190
Figure 5.59:	Phase B shell-edge decorated sherds from the east half of the site, all SF 23403	190
Figure 5.60:	Phase B decorated sherds: SF 23391 (top left) and SF 26143 (bottom left) from the east half of the site, SF 23187 (centre) from the west cist, and SF 23182 (right) from the west half of the site	191
Figure 5.61:	Scoop SF 23908 and pounder SF 25089	199
Figure 5.62:	Modified piece SF 23960, polished piece SF 23991, trimmed pieces SFs 23462 and 23463, point SF 25100, bead segment SF 25226, scoop/plaque SF 25618, points SFs 25618 and 25619, polisher plaque SF 25738, points SFs 25997, 25998, 26041, 26347 and 27205 chisel/polisher SF 27205	200
Figure 5.63:	Unmodified vertebral discs from a possible shark or ray	201

PART 6

		203
Figure 6.1:	Excavation of the bowl-pits above the plough marks. The coastline is immediately beyond the trench with *A' Croig Bheag* inlet between it and *Rubha Huilis* – the headland beyond. Looking north-west	204
Figure 6.2:	Map of sites mentioned in the text	205
Figure 6.3:	The whale vertebra indicating the position of the shaft and platform under a covering mound of stone with a stone alignment leading away from it. The Great Auk stone can be seen on the right opposite the stone alignment. Looking south	207
Figure 6.4:	Map of North Uist sites mentioned in the text	208
Figure 6.5:	Modern ploughing of the machair plain at Udal. Looking north-east	214
Figure 6.6:	Modern stooks of barley cut from the machair plain at Udal	216
Figure 6.7:	Looking south from Huilis across the *A' Croig Bheag* inlet to the site of the excavation, with the exposed gneiss bedrock	218
Figure 6.8:	Removal of the cist capping was achieved using the Landrover winch plus rails from the light railway. Crawford bought rails and a bogie to remove sand from other areas of the Udal project	219
Figure 6.9:	The cairn complex during excavation in 1974. Looking north-east	221
Figure 6.10:	Artist's impression of the temporary structure	223

List of Tables

PART 1		1
Table 1.1:	Excavation dates and days in the field	12
PART 2		19
Table 2.1:	Description of the excavated levels by phase	21 - 22
Table 2.2:	Phases, levels, erosion events, main structures and features of the site	22 - 23
PART 3		64
Table 3.1:	Radiocarbon dates	65
Table 3.2:	Cranial and post cranial indices for SF 17642	71
Table 3.3:	Inter-site comparison of stature	72
Table 3.4:	Inventory of all RUX6 samples; NB: R= right, L= left, MC= metacarpal, MT= metatarsal, C= cervical	77 - 78
Table 3.5:	RUX6 human and faunal samples, age and sex (Bohling and Buckberry 2016) and bulk $\delta^{15}N$ and $\delta^{13}C$	81
Table 3.6:	Strontium isotope ratios for the two individuals, SF 26319 and SF 17642	83
PART 4		86
Table 4.1:	Botanical remains from Phase A (modern)	88
Table 4.2:	Botanical remains from Phase B (Bronze Age cist with kerbed cairn and other structures)	90 - 91
Table 4.3:	Botanical remains from Phase C (Bronze Age pits)	91
Table 4.4:	Botanical remains from Phase D (Neolithic structures and activity)	92
Table 4.5a:	Botanical remains from other samples	93 - 94
Table 4.5b:	Botanical remains from other samples	94

Table 4.6:	Summary of pollen analysis from soil samples	95
Table 4.7:	Species by phase	97
Table 4.8:	Relative frequency of species.	99
Table 4.9a:	Ratios of food mammals	99
Table 4.9b:	Ratios of food mammals	100
Table 4.10:	Sheep epiphyseal ageing data	100
Table 4.11a and 4.11b:	Tooth eruption data (sheep and cattle)	101
Table 4.12:	Cattle epiphyseal ageing data	101 - 102
Table 4.13:	Fragmentation data	103
Table 4.14:	Butchery cuts	104
Table 4.15:	List of identified birds by phase	106
Table 4.16:	List of identified fish by phase	107 - 108
Table 4.17:	Total numbers of marine mollusc shells, with weights and volumes of most abundant species, by phase	112
Table 4.18:	Phasing and identification of crustacea	117
Table 4.19:	Summary of terrestrial mollusc species representation defined by characteristic and habitat	119
Table 4.20:	Terrestrial snails catalogue	120 - 121

PART 5

122

Table 5.1:	Composition of the assemblage by main category and phase	123
Table 5.2:	Raw materials	123
Table 5.3:	Dimensions of the complete cores (mm)	125
Table 5.4:	The various techniques applied to produce the site's flakes	126
Table 5.5:	Retouched pieces by type and phase	126
Table 5.6:	Main artefact categories by phase	127
Table 5.7:	The lithic artefacts from Neolithic Buildings 1 and 2	128
Table 5.8:	General artefact list	132
Table 5.9:	The main artefact categories by phase	133
Table 5.10:	Reduction sequence of all unmodified and modified flakes and blades	133
Table 5.11:	Burnt quartz ratio by phase	134
Table 5.12:	The site's smallest pieces of debitage (chips) and their distribution across the phases	134
Table 5.13:	The composition of the debitage within the main phases (less chips)	135

Table 5.14:	Applied percussion techniques (technologically definable unmodified and modified flakes and blades)	135
Table 5.15:	Flakes and blades with cortical platform remnants (only intact pieces or proximal fragments)	136
Table 5.16:	The distribution of intact bipolar cores across unifacial and bifacial pieces and across pieces with one and two reduction axes	145
Table 5.17:	Distribution of all artefact categories across the various phases	157 - 158
Table 5.18:	Distribution of quartz artefacts across a selection of contexts and features	159
Table 5.19:	The tool ratios of a selection of sites and raw materials. The sites have been sequenced from north to south, with non-Western Isles sites inserted at the bottom of the sequence (Ballin 2000b; 2001; 2002b; 2002c; 2010a-e; 2016; forthcoming a; forthcoming b; Saville and Ballin 2009). When the ratios for the Guinnerso collection was calculated, the finest quartz debris was disregarded, as this may have been produced as temper in Iron Age pottery	162
Table 5.20:	Summary of the main lithic assemblages from the Western Isles	163
Table 5.21:	Details of pumice analysed	166
Table 5.22:	Major element analyses of pumice from RUX6. Total iron is represented as FeO	167 - 168
Table 5.23:	Comparison of worked and unworked pumice by phase	169
Table 5.24:	Number of attributes (types of worn areas) on pumice pieces by phase	170
Table 5.25:	Artefact and samples numbers	175
Table 5.26:	Tool types by stone type, SF number, and phase	182
Table 5.27:	Composition of the pottery assemblage	184
Table 5.28:	Number of vessels with diagnostic sherds	187
Table 5.29:	Number of vessels in each phase	192
Table 5.30:	Number of vessels in each phase with a discernible vessel profile	192
Table 5.31:	Number of bone artefacts considered modified, uncertain or unknown	198
Table 5.32:	Species numbers by phase	198

Acknowledgements

The Excavation

The first acknowledgment is from Iain Crawford who expressed it in the following manner, "It follows that the SDD (Scottish Development Department), in the persons of Patrick Ashmore and Noel Fojut, are to be congratulated on their prescience in supporting this diversion to salvage RUX6 in 1980, 1981 and 1983". In addition, he also thanked crofters and Lord Granville for access to the peninsula.

The Post-excavation

Historic Scotland, with Dr Noel Fojut as head of archaeological funding, provided an initial grant for a full assessment of the finds and samples from the whole of the Udal project. Work on RUX6 continued with funding from Historic Scotland, later Historic Environment Scotland. Roderick McCullagh took over from Dr Fojut till his retirement, Dr Lisa Brown patiently saw the project through to completion and Dr Rebecca Jones provided advice and guidance. I have also had considerable help from Peter McKeague, Dr Iain Fraser and Leslie Ferguson, formerly of the RCAHMS, now HES, on technical matters and archiving.

Comhairle nan Eilean Siar / Western Isles Council were involved with the project from the beginning and have part-funded wider aspects of it. I have worked with a number of people from CnES who have assisted this project in their different ways from its inception to the end: Councillor Uisdean Robertson, Trish Botten (Principal Officer, Libraries and Heritage), Nick Smith (Heritage Manager), Deborah Anderson (Archaeologist), Kevin Murphy, (Archaeologist), Mark Hall (Assistant Archaeologist), and Isabel MacLachlan (Visitor Services Officer).

I have also received specific help from Catriona MacCuish (Museum Development Officer) and Jane Hamill (Conservation Officer) on finds and exhibitions, and Doileag NicLeòid (Policy Officer) for providing the Gaelic translation of the summary.

The Hunter Trust has supported the post-excavation project by specifically funding the digitisation of the 32,000 finds and samples records into a database. Without the creation of this computerised list our work would have been much more difficult.

During the assessment period considerable help was provided by National Museums Scotland, not least by Dr's Alison Sheridan and David Caldwell who visited the collection and gave expert advice on it. Drs Gemma Cruickshanks and Fraser Hunter, also NMS, gave their specialist advice on specific parts of the collections. Others consulted during this period were Dr Daniel Sahlen, Historical Museum, Sweden, Dr Dale Serjeantson, University of Southampton, Dr Jennifer Harland, University of the Highlands and Islands, Orkney, John Stewart, University of York, and Drs Colleen Batley and Richard Jones of the University of Glasgow. To all of them, my grateful thanks.

Dr Euan Mackie re-introduced me to Iain Crawford when he came one day to the University of Glasgow. That was the start of our friendship and the first steps in helping Iain with RUX6. The initial work in 2002 that developed from this meeting was the scanning and digitisation of some of the drawn record from the site, which was funded by Historic Scotland. Kylie Seratis took the lead on the digitisation and she is to be thanked for initiating a high standard of work that we were able to build on later on in the project. Alastair Becket assisted in digitising some of the plans and sections.

None of the post-excavation assessment between 2011 and 2013, or the specialist work since, would have been possible without the kindness and support of my colleagues at GUARD Archaeology Ltd. All have helped me in so many ways to overcome the various problems that affected this project, but special mention must be made of Dr John Atkinson (Director), Pauline MacShannon (Director), Aileen Maule and Bob Will (who are first class problem solvers), and Jen Cochrane and Joan O'Donnell for their invaluable help. Fiona Jackson started the site figures for publication, Dermuid O'Connor carried them on and did wonders with the site grid and 3D modelling, Dave McNicol completed them with the utmost patience and skill, and Gillian Sneddon and Jennifer Simonson provided the final touches. Without them all this book would not have been possible to complete. Gillian took the manuscripts, figures and tables and made them into this book. I cannot adequately express the admiration I have for her as a designer.

I am very appreciative of the very fine drawings produced by Jo Bacon of some of the finds and by Leeanne Whitelaw who skilfully drew the rest. Leeanne also digitised and organised all the finds drawings for inclusion in this publication.

A number of gifted specialists have provided the information at the heart of this project. They are my team and my friends. Some of them I have known a long time and have worked with often. They are Catherine Smith, zooarchaeologist at Alder Archaeology; Dr Susan Ramsay, archaeobotanist, Dr Judith Finlay Aird, zooarchaeologist, Dr Ruby Ceron-Carrasco (zooachaeologist), and Dr Torben Ballin (husband and lithic specialist). During the course of this project a number of others joined us and the team expanded to include Dr Anthony Newton, Department of Geology and Geophysics, Edinburgh University (pumice) and Caroline Wickham-Jones, University of Aberdeen (flint). Both of whom must be relieved to see their work in print after so many years. The team also included for the first time Dr's Jo Buckberry and Julia Beaumont, PhD student Solange Bohling and Cassandra Hall (MSc) all at the Biological Anthropology Research Centre, Archaeological Sciences, University of Bradford; Dr Derek Hamilton and colleagues at the Radiocarbon Dating Laboratory, and Professor David Sanderson, at SUERC, University of Glasgow, East Kilbride. Dr Jim Hansom, Professor of Geography (adjunct) University of Canterbury New Zealand, and University of Glasgow provided much useful information on machair, its development and movement.

Cassie Hall deserves special mention as her original MSc project fell through and Dr Jo Buckberry asked if we could do anything to help her. Jo and Solange were already studying the human skeletal material for RUX6 and it was a delight to provide Cassie with the opportunity to do scientific analysis of it. Not only did the RUX6 material provide Cassie with her degree but we received back a specialist report from her and Dr Julia Beaumont. It informs part of this publication that otherwise would not have been realised. I thank Cassie for her enthusiasm and input into the project.

In addition to the above, a number of others, including friends, have been kind enough to share their thoughts, experiences and their work. They include Robert Squair who undertook his PhD on RUX6 pottery and Ann Clarke who initially examined some of the stone artefacts from the site, Dr Jean Archer, North Uist provided specialist information on the geology of the island, Dr Margaret MacKie, School of Scottish Studies, Edinburgh has provided me with much useful information on Iain Crawford and his work at the School, and Dr Mary MacLeod Rivett from the Isle of Lewis also contributed her memories of Iain Crawford. I have also benefited considerably from discussions with Dr Barbara Crawford, the Strathmartine Trust and University of St. Andrews,

Many people on North Uist have provided valuable information for me on their memories of Crawford's excavations and the impact they had on the island at the time. I am very conscious I that I have forgotten the names of many that have given assistance to this project. Some of them have asked to remain anonymous, others have simply provided good will, information and hospitality, and to all of them I am extremely grateful. However there is one person who deserves special mention.

The exact details are lost in time past, but a couple of emails arrived during the course of a week early on in the post-excavation process. They were followed by telephone calls, which eventually persuaded me that I had better agree to meet Chris Stewart-Moffitt. I did not think I needed help, but Chris persuaded me I did, and he was right. He came to Govan, where some of the Udal collection was stored at that time, and spent the greater part of the next three to four years helping me excavate the samples and artefacts out of their 20 to 40 year old packaging, checking them off the database, re-bagging and boxing them, and moving them about. He is tremendously dedicated, enthusiastic and an inspiration. It was a pleasurable learning experience for both of us to work together on the project and not only is he now a good friend, but the archaeological experience gained him a good undergraduate degree, motivated him to gain a Master's Degree, and he is now undertaking research for his PhD!

Imogen Crawford, Iain's wife, contacted me in 2008 at a difficult time in her life to take away the collection and archive from her house. Without Imogen's generosity, the collection would not have gone through the Treasure Trove system to be allocated to Museum nan Eilean, Western Isles Council. Imogen provided the opportunity for me and others to work on the collection and this book is the first result.

Judith Finlay Aird told me much about her experiences working on the RUX6 excavation and delighted me by putting them down in writing. She also willingly took on the thankless job of reading and commenting on the manuscript along with Dr John Raven, Historic Environment Scotland. I am deeply indebted to both of them. David Davidson and colleagues of Archaeopress are also to be thanked for their patience and for publishing this volume.

My friend Noel Fojut said at the beginning of this project that it was possibly going to be a double-edged sword or a poisoned chalice, or both. He was correct in his assumption. The project has not been an easy one and my life's partner, Torben Ballin, has not only shared the ups and downs of it with me but has supported and nurtured me through it to the end. He single-handedly took on the task to create and populate the database, which is the backbone of all our research on this project and has done much to assist me whenever help was needed. A simple thank you is not enough.

Beverley Ballin Smith
Banknock Cottage
Denny
November 2017

Anthony Newton would like to thank Dr Peter Hill and Simon Burgess for undertaking the microprobe analyses at the Electron Microprobe Unit, Department of Geology and Geophysics at the University of Edinburgh.

Judith Finlay Aird would like to thank the following: For access to reference material and specialist identification, 1981-83 - Dr A S Clark, Mr Lyster, Mr. R McGowan, Mrs I Simpson and Mr J Swinney at Royal Museum Scotland, Edinburgh; Miss S. Colley, Faunal Remains Project, Southampton, Mr D Henderson; and Mr A Wheeler at the Natural History Museum, London and Dr J Herman, Senior Curator, Mammals, National Museums Scotland, Edinburgh for access to reference material and specialist identification.

Foreword

I am delighted and honoured to have been asked to provide a foreword to this monograph which reflects both my professional and personal interest in, and commitment to, deepening and making more publicly available the riches of the Western Isles historical and archaeological heritage. Having had personal experience and knowledge of the rewards of such work in the Orkney Islands, it is essential that the equally rich heritage of our own Outer Hebrides is nurtured and promoted in a similar way, for the benefit of all.

The truth is that while the archaeology of the Western Isles is as rich, diverse and intriguing as that of the rest of Scotland, it is less well known. Comhairle nan Eilean Siar and its partners are working hard to see this position change, and it is therefore a great pleasure to have available this account of the smallest of Iain Crawford's excavations at the Udal site in North Uist, mainly undertaken in the early 1980s, and supported then by the Scottish Development Department and latterly by Historic Environment Scotland. The main purpose of these excavations was preservation of that fragile evidence required to be safeguarded in the face of erosion by sea, storm and simply the ravages of time. The story told by these structures and artefacts, however, reflects the earliest centuries of communities' life experiences on the Udal headland from some six thousand years ago, one of the longest and most fascinating time lines in the archaeology of Scotland. The two Neolithic houses and Bronze Age burial cairns bear testimony to the antiquity and importance of this site.

I wish all readers a happy journey of exploration through this story of a shared past in the knowledge that there are many episodes yet to be told about the archaeology of the Udal peninsula.

Malcolm Burr
Chief Executive
Comhairlie nan Eilean Siar
Stornoway
Isle of Lewis

Preface

By Beverley Ballin Smith

The origin of the book title

The incidence of natural events on RUX6 and the effects they had on the people living there suggested to me that *marginal* and *edge* were words that described the site. Research showed that in 2014 Channel 5 broadcast three documentaries about living on the edge where they explored how people lived 'in the grip of nature at its most ferocious'. This description seemed very apt for people living in the northern part of North Uist during the later Neolithic and early Bronze Age periods. They were living in a landscape that became unstable, where the line between survival and death could have been a knife's edge. Sea inundations perpetuated the image of living on the edge of somewhere unpredictable and dangerous, and finally the site was found on a low cliff, the liminal edge between land and sea.

Iain A Crawford

I had known about the Udal project since the early 1980s, when one of the archaeologists from our excavation on Orkney went to the Western Isles to work for a few weeks. Crawford had advertised for volunteers in the memorable CBA yearly excavations list with its bright orange/red banner, and I must have seen the entry, but by then I was enjoying the archaeology of a different island group. Udal was already becoming an evocative place to dig and Crawford's name was on people's lips. Later, I learnt that not all the stories we heard were complimentary.

I first met Iain Crawford in Liverpool on 19 May 1990 when he was 62 years old at a Viking conference held by the Merseyside Museums in Liverpool. I remember clearly that his lecture did not tell us anything important about the Udal, which was a great disappointment to me. After my experience of working on one multi-period site on Orkney, and having just starting the post-excavation process of another in Shetland with Dr Barbara Crawford from the University of St Andrews, I could see that Iain was in need of some help in sorting out his data. After the lecture I naively approached him to offer my help. I can't remember what he said, if he said anything, but the withering look from a great height said it all.

Ten years later he was brought into my office at the University of Glasgow by Dr Euan MacKie. Iain spilt the papers of his briefcase all over mine on the desk and from that moment we began an interesting working relationship. I visited him at his home in Castle Douglas, Dumfries and Galloway, met his family and the dogs, and was introduced to the basement of archives – an unforgettable experience. The finds and samples were tucked away in three bays between the brick supports in the lower sloping level of the house. They were crammed tightly from floor to ceiling in all manner of containers, from wooden fish boxes, the ubiquitous Haig's red and white cardboard whisky boxes, tea chests by the score, a large wooden chest (including woodworm), wooden apple racks, black plastic bags, and orange carrot or onion sacks. The site records were ordered in his work room on shelves and between tables of various sorts and included an old computer. By 2000 he seemed to have mellowed as a person, but the stories of him falling out with every respected archaeologist in Scotland seemed legion, and what was interesting was that most were true. I have always liked a challenge and difficult personalities seemed to be a speciality, so visiting Iain Crawford at home in his lion's den, was a remarkable experience.

Iain did not fall out with me, perhaps because in some way I was offering him a small step forward in his enormous predicament - the Udal albatross that was forever hanging round his neck. He had obligations to Historic Scotland to write up RUX6 and had presented them with site data that was not in a conventional format. Having worked closely with HS staff for many years might have been in my favour and a communication barrier became unblocked. We began to undertake the digitisation of the sections for RUX6 with HS's hesitant blessing. What was more remarkable was that Iain allowed us to take away his site records, after of course, the signing of papers.

Later, I and a colleague asked a number of researchers to write papers for a festschrift for Dr Euan MacKie who was retiring from the Hunterian Museum in Glasgow, and I invited Iain to contribute too. His paper is an extraordinary contribution, and as we later found out some of the information was inaccurate, but nevertheless we kept it. It had been a long time since he had written anything and his piece on wheelhouses proved to be his last. Many people had warned me that he had threatened them with litigation for one reason or another. He only threatened me once, and that was because I edited his wheelhouse paper and changed some words. His writing style, including his choice of words, was somewhat unusual and like him, was unique and eccentric. With hindsight, I was glad that we published his paper.

Our cooperation on RUX6 ceased when Iain became ill, but the rest of this volume takes up the story from 2010.

Summary

By Beverley Ballin Smith

Iain Crawford began his work on the *Ard a' Mhorrain* peninsula in North Uist in 1963 but it was not until 1974 that the site of RUX6 came to his attention through severe coastal erosion. He and a team excavated the site that year to reveal a kerbed cairn complex with a large cist burial beneath it.

Crawford did not return to the site until 1980 when the rescue excavation was funded by the Scottish Development Department. He completed his excavation there in 1984. One of the main issues for Crawford was to understand the build-up of natural sand layers across the site and the interpretation of erosion events that had also affected it. He was assisted to some extent by a nineteenth century saw-pit, which cut across the site almost down to subsoil. Its sides, together with the eroding section cut through the cairn down to bedrock, revealing the complex stratigraphy of natural and man-made events in this area.

He gradually extended the site to the east to take in archaeological features that were also threatened by continuing coastal erosion, and to build up knowledge of natural events which interweaved with the archaeological remains. The beginnings of human activity on the site started with the contamination of the upper levels of natural deposits with domestic debris and artefacts. This was followed by some evidence of settlement; the fragmentary remains of a possible domestic structure and a large fire pit; but more importantly the conversion of a large slab of protruding bedrock into the side of a formalised shaft within a stone platform. This took place during the late Neolithic, sometime between c. 3000 - c. 2500 BC. The shaft was enlarged for what is assumed to have been the natural collection of ground water flowing off the peninsula. However, its function probably changed over time, and it became a centre of ritual significance. A whale vertebra was positioned besides the shaft to allow access to it and a stone of zoomorphic shape, thought by Crawford to represent a Great Auk, was erected to its west. Two curving rows of stones set on edge, with wooden posts towards their ends, are all that remained of an intentional construction that radiated out from the platform.

Slightly later in date than the creation of the ritual area was the construction and use of a circular domestic dwelling that had been erected towards the west, using driftwood posts to support its roof. Turf walls with stone faces enclosed a space that had a central hearth with a screen in front of a door positioned in the northern circumference of the wall, as well as one or two partitions and a stone platform. Sometime later, a slightly larger structure was built to almost the same plan, which abutted it to the west, and a connecting doorway may have been created between the two. This new building was well-preserved and may have replaced the earlier structure as the main dwelling, with the latter becoming a workshop. Activities recorded from both buildings included the preparation of meat and fish, the collection of shell fish, cooking, as well as bone and stone tool manufacture. The evidence from the older structure was more detailed and included the use of both plain and Grooved Ware vessels and the digging out of the clay subsoil beside the hearth for the making of pots. The building may also have been extended to the east to include a cell or enclosure, perhaps for the stalling of sheep or cattle. Evidence from terrestrial snails recovered from the floors of the buildings indicated that conditions inside them were damp, with drains needed to take seeping groundwater away from the floors. There is no direct evidence for cereal crops being grown or processed at the site at this time, with the economy being based on various strategies, including herding, gathering and possibly some fishing and hunting.

Between the two buildings to the south was a burnt area with ash containing sherds of pottery. This has been interpreted as the remains of a pit used for the firing of pottery. Occupation deposits were noted to the north of the buildings and it is thought that these buildings were all that remained of a larger settlement that had been removed by coastal erosion.

During the occupation of these two structures and the ritual remains to the east, there was a noticeable increase in sand accumulation across the site. The ritual monument of the platform and shaft was covered over with a mound of stone to protect it, but the two buildings had to be abandoned. Strong winds brought sand from coastal dune systems to the west to cover not only the settlement but also its fields to sufficient depths, and to such an extent, that they became unusable. Cattle and sheep would have had to be moved to grazing areas inland, and it is quite likely that people followed to established new dwellings away from the sand accumulation. This event would have been seen by the local community dwelling there as a catastrophe, with notable changes to the landscape and significant changes to their way of life.

The area of the old dwellings was not entirely abandoned, as it still formed part of the community's territory. Once the sand had consolidated and vegetation started to grow, there was an attempt to bring the land under cultivation, firstly by the use of a mattock and secondly by ploughing, as scars of both survived in the sand surface. However, the landscape was fragile and soil development thin. On the eastern side of the site, a linear boundary was constructed of at least three driftwood poles erected in deep pits that were packed with stone, separating the old ritual area from an activity area to the north and west. The activity was the gradual digging of pits in rows that covered an area destined to be a field or fields. The evidence suggests that these pits were likely to be for animal dung or human waste, as a form of structured manuring of the area. They were quickly dug and backfilled before the next pit was dug. Together with the earlier ploughing, these pits indicated that people lived close by and that their land extended further north and west, with the coastline of the time located further out to the west.

The sand accumulation was only one event of a series of natural disasters, including rising sea levels, which quickly followed on from each other and affected the area of the excavation. In the west, a tidal scour reached far inland and removed any archaeological evidence that existed in its path down to near the base of the pits. It deposited in its wake a substantial beach of shingle and rolled stone. Some of the stone was also considered by Crawford to have come from other late Neolithic buildings that had been washed out of the sand and destroyed by the sea. The event would have changed the landscape north of the site beyond all recognition, and a previously marshy area may have been scoured out, leaving the beginnings of the A'Croig Bheag inlet separating the Rubra Huilis from the excavated area. The coastline probably also retreated inland.

We have now moved away from the late Neolithic and into the early Bronze Age with a change not just in architecture but in material culture too. This period on this site is dated roughly to 2100 - 1900 BC. People still lived close by, but away from the coast and the direct influence of sea and shifting sands, and considered the new beach to be part of their ownership, as they used it for the burial of an individual and the construction of a kerbed cairn, of which only an arc of stone survived until 1974. A new burial place was subsequently needed for a young adult male who had died. A rectangular stone cist was built into the ground, close to the kerbed cairn, for the burial of the deceased. However, it took time for a suitable stone for its lid to be found, and the cist remained open to be gradually filled in with sand and other debris. Eventually, a heavy stone was brought to the site and positioned over the cist so that the building of a cairn could commence. The first part to be constructed was a tall core, comprising vertically positioned stones around its edges and flatter stone in the middle that rose like a solid dome of masonry over the cist and abutting the old cairn. The final construction phase of the cairn involved the building of a wide kerb of stone filled in with turf and stone behind, which encircled not just its core but also the remains of the older cairn as well. The new kerbed cairn formed a large mound that made a significant statement in the landscape and would have been the largest man-made structure in the area. Only part of this monument survived subsequent coastal erosion. There were probably activities associated with the cist burial, as fragments of Beaker pottery, typical of the period, were found beneath the cairn.

The story of the activities on the eastern side of the site was different from that on the west. The late Neolithic ritual monuments had not been ploughed or entirely robbed away, and their location was still recognised in the landscape. One post of the early Bronze Age wooden boundary settings was replaced by a standing stone with an earthen and stone plinth. Between this and the old mound of the ritual shaft and platform was a slight dip in the ground and into it a temporary structure was erected, possibly tent-like in appearance, using driftwood posts and probably animal skins, turf and heather in its construction. It may not have been occupied all the time, but the evidence suggests it was well used. Its interior included a number of fire pits, with a significant amount of ash on its floors. The evidence of its used includes the survival of a number of pottery vessels, pits with marine molluscs, knapping waste and raw materials for lithic artefact manufacture, as well as rare carbonised barley grains indicating evidence of the cultivation of cereals close to the site.

This structure seems to have been abandoned prior to the construction of a cist inside its perimeter but against its northern perimeter for the burial of another individual. This older male, who may have been wrapped in an extremely flexed position, was accompanied in the cist by a pottery vessel, a few bone points and a body of a calf. Unusually, there is a disparity of 250-350 years between the age of death of the calf and the individual, suggesting that the human remains could have been curated, possibly mummified, before the burial took place. After the cist lid was placed in position, a small stone and turf kerbed cairn was constructed around it, taking in the remains of the late Neolithic ritual shaft and platform within its circumference. The orientation and position of the standing stone was presumably part of the reason for the location of the kerbed cairn and cist burial, as it marked the position of the latter. As with the ritual monuments on the west side of the site, these too suffered from recent coastal erosion.

This construction was the last of the prehistoric human activities recorded on the site. A further marine incursion on the west took away deposits around the cairn complex and redeposited them without destroying the burial monument. This seemed to be the last major natural event to have affected the site. However, rising sea levels have moved the coastline inland with periods of stability in between erosion events. During the nineteenth century, stone-robbing of the cairn took place for the construction of kelp-drying dykes, a kelp burning kiln and possibly for the building of crofts at Grenitote. The story of the occupation of the site and its structures, from the late Neolithic and through the early and middle Bronze Ages, remained relatively untouched under further sand and turf accumulation until the late twentieth century AD.

Gaelic Summary

Thòisich Iain Crawford obair air ceann-tìre Ard a' Mhorrain an Uibhist a Tuath ann an 1963 ach cha b' ann gu 1974 a thàinig làrach RUX6 gu aire tro bleith-thalmhainn na mara. Chladh e fhèin agus an sgioba an làrach a bhliadhna sin agus lorg iad càrn le ciste mòr fodha.

Cha do thill Crawford chun làrach gu 1980 nuair a chaidh an cladhach a mhaoineachadh le Roinn Leasachadh na h-Alba. Chrìochnaich e a' cladhach an sin ann an 1984. B' e aon de na duilgheadasan bu mhotha a bh' aig Crawford, tuigse fhaighinn air an doimhneachd de ghainmheach air feadh na làraich agus tuigse fhaighinn air an bleith-thalmhainn a thug buaidh air. Fhuair e taic aige gu ìre bho sloc-sàbhaidh a gheàrr tarsainn sìos an làraich, cha mhòr chun fo-ùir. Gheàrr na cliathaichean, an cois an roinn bleith-thalmhainn, sìos tron chàrn chun na clachan, a' dèanamh structairean iol-fhillte de thachartasan nàdarra agus togte san sgìre follaiseach.

Beag air bheag, leudaich e an làrach chun an ear gus pìosan arc-eòlais eile, air an robh cunnart bho bleith-thalmhainn, a ghabhail a-steach agus stòras fiosrachaidh fhaighinn air tachartasan nàdarra air an robh buaidh air làraichean arc-eòlais. Thòisich obair daonna air an làrach le truailleadh air na h-àrd-ìrean le sgudail is innleachdas. Bha an uairsin fianais ann gun do shuidhich daoine an seo; bha pìosan ann a dh'fhaodadh a bhith bho structair leithid taigh agus teine mòr; ach nas cudromaiche pìos chrann air fhighe a-steach. Gheibh seo àite aig deireadh Linn na Cloiche, uaireigin eadar c. 3000 - c. 2500 RC.

Bha a' chrann air a leudachadh, le cuid den bheachd gur ann airson bùrn èirigh a' tighinn bhon a' cheann-tìre a shàbhaladh a bha e. Ach dh'atharraich fheum nuair a thuit ìrean bùirn, agus bha e an uairsin na àite airson comharrachadh cleachdadh. Chaidh cnàimh droma a chur ri taobh a' chrann gus am biodh slighe ann, agus chaidh clach le cruth ainmh-chruthach, a bha Crawford den bheachd a bha a' riochdachadh Great Auk, a thogail chun an iar. Cha robh air fhàgail den dealbhadh cearcall bhon leac, ach dà sreath lùbadh de clachan, le puist fiodh aig am bàrr.

Nas fhadalaiche ann an cruthachadh an àite cleachdaidh, bha togalach ann an cruth cearcall a chaidh a thogail chun an iar, le bhith a' cleachdadh puist airson taic a chumail ris a mhullach. Bha ballaichean monadh le clachan a dìon beàrn san robh teine sa mheadhan le sgàilean air beulaibh doras chun a' bhalla a tuath, an cois pàirteachadh no dhà agus leac àrd. Uaire gin an dèidh sin, chaidh structair beagan na bu mhotha a thogail chun an aon phlana, chun an iar, agus doras a dh'fhaodadh a bhith a ceangal na dhà. Bha an togalach ùr seo air a chumail gu math agus dh'fhaodadh gun do ghabh e àite an togalach na bu thràithe mar a' phrìomh àite-fuirich, leis an togalach eile ga chleachdadh na bhùth-obrach. Bha ag ullachadh iasg agus feòil, cruinneachadh maorach, còcaireachd, is cruthachadh stuthan le cnàmhan is clachan, am measg na nithean a chaidh an dèanamh san dà thogalach. Bha barrachd doimhneachd san fhianais bhon togalach as sine, a' gabhail a-steach innealan Grooved Ware agus cladhach crèadh airson poitean a dhèanamh ri taobh an teine. Dh'fhaodadh gun deach an togalach a leudachadh chun an ear, 's docha airson caoraich no cròdh a chumail. Bha fianais a fhuaireadh bho sheilcheagan air làraichean nan togalach a' sealltainn gu robh iad fliuch, le feum air drèanaichean airson a' bhùrn a ghluasad a-mach. Chan eil fianais dìreach ann airson fàs arbhar air an làrach aig an àm, leis an eaconamaidh stèidhichte air grunn ro-innleachdan, a' gabhail a-steach cruinneachadh bheathaichean, iasgach agus sealg.

Eadar an dà thogalach gu deas tha làrach loisgte le uinnseann le criomagan crèadhadaireachd. Tha cuid den bheachd gur e sloc airson crèadhadaireachd a bh' ann. Bha dùnain gu ceann a tuath nan togalach agus `s iad sin an aon rud a bha air fhàgail bhon t-suidheachaidh a chaidh a ghluasad tro chrìonadh an oirthir.

Fhad 's a bha daoine anns an dà thogalach seo agus na tobhtaichean chun ear, bha e faicsinneach gu robh meudachadh anns na bha de ghainmhich a cruinneachadh air an làraich. Bha tom chlachan air uachdair an ùrlar agus an crann, ach b' fheudar an dà thogalach a bhith air an leigeil seachad. Thug gaoth làidir gainmheach bho shiostaman dhùin air a' chosta an iar is chan e a-mhàin gun deach an tuineachadh a chòmhdach ach cuideachd na h-achaidhean chun ìre `s gu robh iad fo fheum. B' fheudar crodh is chaoraich a ghluasad gu àitean ionaltraidh nas fhaide bhon oirthir, agus tha e glè choltach gun lean na daoine a' stèidheachadh dhachaighean ùra air falbh bho na dùin ghainmhich. Bhiodh a' choimhearsnachd ionadail dha fhaicinn mar chall mòr, le atharrachaidhean bunaiteach air an tìr agus gu sònraichte air an dòigh-beatha.

Cha deach làrach nan seann thoglaichean a thrèigsinn gu buileach, oir bha e fhathast mar phàirt de thalamh na coimhearsnachd. Aon uair `s gun shocraich a' ghainmheach agus gun thoisich planntrais a' fàs, chaidh oidhirp a dhèanamh air an talamh àiteach, an toiseach le caibe agus a rithist le treabhadh, tha làraich iad seo rim faicinn air uachdar na gainmhich. Ach, bha cruth na tìre frionasach agus `s ann tana a bha leasachadh na talmhainn. Air taobh an ear na làraich, bha loidhne chrìoch stèidhichte le co-dhiù trì pòlaichean fiodh-cladaich ann an toill dhomhainn air am bruthadh timcheall le clachan, a' dèanamh sgaradh bhon t-seann àite-cleachdaidh bho na raointean gnìomh gu tuath `s an iar. Beag air bheag bhathas a' cladhach slocan ann an sreathan thairis air àite a bhiodh na achadh no achaidhean. Tha an fhianais sin a' toirt oirnn beachdachadh gur ann airson salchar chon no salchar mac an duine a bha na slocan, mar dhòigh dealbhaichte air an àite a mhathachadh. Tha iad air an cladhach gu sgiobalta agus air an lìonadh air ais mus tèid an ath shloc a chladhach. Còmhla ris an treabhadh, tha na slocan seo a' foillseachadh gu robh daoine a' fuireach faisg air làimh agus gu robh am fearann a' sgaoileadh nas fhaide gu tuath is an iar, agus bhiodh an oirthir aig an àm suidhichte nas fhaide chun iar.

Cha robh ann an càrnadh a' ghainmhich ach aon thachartas ann an sreath de chall nàdarra, nam measg ìre na mara a bhith ag èirigh, a' leantainn fear an dèidh fear agus a thug buaidh air an àite far a bheileas a chladhach. Chun iar, raining am muir fada a-steach dhan fhearann agus sguab e leis fianais arc-eòlais sam bith a choinnich ris sìos gu faisg air bonn na slocan. Dh'fhàg e às dhèidh tràigh mhòr de mhol is clachan-muile. Bha Crawford den bheachd gun tàinig cuid den chloich bho thoglaichean Neolithic eile a chaidh an sguabadh a-mach às a' ghainmhich agus a chaidh am milleadh leis a' mhuir. Bhiodh an tachartas air cruth na tìre tuath air an làraich atharrachadh gu tur, agus far an robh boglach sgùr am muir e, a' fàgail toiseach tòiseachaidh sàilean A' Chroig Bheag a' sgaradh Rubra Huilis bhon sgìre air a chladhach. Bhiodh e coltach cuideachd gum biodh an oirthir air gluasad a-steach an tìr.

Tha sinn a-nis air gluasad air falbh bho dheireadh àm Neolithic agus a-steach don àm thràth anns an Linn an Umha agus chan e a-mhàin gu bheil atharrachadh san ailtireachd ach anns na stuthan a bhathas a cleachdadh. Air an làrach seo, thatar a' tuairmse a bhith eadar 2100 - 1900 ro àm Chrìosd. Bha daoine fhathast a' fuireach faisg air làimh, ach air falbh bhon chosta agus buaidh dhìreach na mara agus gluasad na gainmhich, agus bha iad a' gabhail ris an tràigh ùr mar phàirt den fhearann aca, our chleachd iad e mat àite-adhlaicidh do dh'aon neach le càrn oireach a thogail, agus suas gu 1974 chithear bogha den chloich. Bha àite-adhlaicidh ùr a dhìth do dh'fhir òg a bhàsaich. Chaidh ciste cloiche a dhèanamh san talamh, faisg air a' chàrn oireach, far an deach a chorp a chàireadh. Thug e ùine mus deach clach freagarrach a lorg airson a' mhullaich, agus leis a' chiste fosgailte lìon i le gainmheach agus sprùilleach eile. Mu dheireadh, chaidh clach throm a lorg agus a' cur air uachdar na ciste gus an deidheadh càrn a thogail air a mhuin. Chaidh crann àrd dìreach a thogail an toiseach, le clachan suidhichte inghearach mu na h-oirean agus leac sa mheadhan ag èirigh mar cuach-mhullach chloiche air uachdar na ciste agus ri taobh an t-sean chàrn. Mu dheireadh bhiodh an cabhsair leathann cloiche air a thogail is air a lìonadh le ceapan agus clachan, a' gabhail a-steach tobhtaichean an t-sean chàrn còmhla ris a' chòrr. Bha tom mòr timcheall air a' chàrn ùr a bhiodh na shealladh cudromach san sgìre agus an rud as motha air a thogail le mac an duine anns an sgìre. Ri linn crìonadh an oirthir cha do mhair ach pàirt den charragh seo. `S iongantach mur an robh gnathasan co-cheangailte ri adhlacadh na ciste, oir chaidh bloighean de chrèadha Beaker, nòsach dhan àm, an lorg fon chàrn.

Bha sgeulachd nan gnìomhan air taobh an ear nan togalach eadar-dhealaichte. Cha deach falbh leis, no cladhach, a dhèanamh air càrnan cleachdaidh nua-chreagach agus bha an làrach fhathast ri fhaicinn. Chaidh carragh a chur an àite aon de na crìochan fiodha bho Linn an Umha le bonn chloiche. Eadar seo agus an t-seann chnoc den chrann chleachdaidh, bha sloc beag san talamh far an deach togalach eadar-amail a thogail, rudeigin coltach ri teanta, le bhith a' cleachdadh puist fhiodha agus craiceann bheathaichean, pìosan talamh agus fraoch. Dh'fhaodadh nach robh daoine ann fad ùine, ach tha fianais ann a' sealltainn gu robh e air a chleachdadh tric. Bha grunn slocan teine na bhroinn, is tòrr uinnseann air an làr. Tha fianais ann air cleachdadh an togalaich, le grunn phìosan chrèadha fhathast rim faicinn, slocan le maorach na mara, bun-stuth airson innleachdas nàdarra a chruthachadh, an cois gràinnean eòrna mar fhianais air fàs gràn air an làraich.

Tha e coltach gun deach an structar seo a leigeil seachad mus deach ciste a thogail am broinn a chrìochan, ach ris a' bhogha a tuath tha àite-adhlacaidh neach eile. 'S e fireannach nas sine a tha seo, agus chaidh a phasgadh ann an cruth fìor lùibeach, na chois bha soitheach crèadha, cinn chnàimhean agus closach laoigh. Gu h-annasach, tha suas ri 250-350 bliadhnaichean eadar bàs an laoigh agus an neach, a' cur an aire gur dòcha gun deach an duine air a thasgadh mus deach adhlacadh. An dèidh do leac na ciste a bhith na h-àite, chaidh càrn ceap is chlachan a thogail mu thimcheall, a' gabhail a-steach tobhta an crann is ùrlar bho anmoch san linn Neolithic. Tha e a' coimhead coltach gu robh taobhadh agus suidheachadh an tursa mar phàirt den adhbhar airson làrach a' chàrn agus adhlacadh na ciste, oir tha an tursa a' tomhadh chun chàrn. Coltach ri na carraighean cleachdaidh air taobh siar na làraich, bha iad seo cuideachd air fulang bho bhuaidh crìonadh oirthir.

'S e seo an togalach mu dheireadh de ghnìomhan a' chinne-daoine ro-eachdraidh air an clàradh air an làraich seo. Bhris am muir a-steach a rithist bhon iar a' toirt air falbh an sprùilleach morghain timcheall na carraig adhlacaidh agus ga thilleadh gun call a dhèanamh air, agus tha e coltach gur e sin mòr thachartas nàdarra mu dheireadh a thug buaidh air an làraich. Ach, tha ìrean na mara ag èirigh agus a' gluasad an costa a-steach dhan tìr le amannan socraichte eatarra. Rè an naoidheamh linn deug, chaidh falbh le na clachan a' chàirn gu togail ballaichean tiormachaidh a' cheilp, àth losgaidh a' cheilp agus 's dòcha airson na croitean aig Greinetobht. Bha sgeul tuineachaidh na làraich agus na tobhtaichean bho linn Neolithic agus tro thràth is meadhan Linn an Umha air fhalach bho thuilleadh gainmhich is cheapan gu deireadh an fhicheadamh linn AD.

List of Contributors

Dr Judith Finlay Aird – 15 Dene Grove, Darlington, County Durham, DL3 9LU

Dr Torben Bjarke Ballin - Banknock Cottage, Denny, Stirlingshire, FK6 5NA and Honorary Research Fellow, School of Archaeological and Forensic Sciences, University of Bradford, Bradford, BD7 1DP

Beverley Ballin Smith – GUARD Archaeology Limited, 52 Elderpark Workspace, 100 Elderpark Street, Glasgow, G51 3TR and Research Associate, National Museums Scotland

Dr Julia Beaumont – Biological Anthropology Research Centre, School of Archaeological and Forensic Sciences, University of Bradford, Bradford, BD7 1DP

Solange Bohling – Biological Anthropology Research Centre, School of Archaeological and Forensic Sciences, University of Bradford, Bradford, BD7 1DP

Dr Jo Buckberry – Biological Anthropology Research Centre, School of Archaeological and Forensic Sciences, University of Bradford, Bradford, BD7 1DP

Dr Ruby Ceron Carrasco – Historic Environment Scotland, Longmore House, Salisbury Place, Edinburgh, EH9 1SH

Cassandra Hall – c/o Biological Anthropology Research Centre, School of Archaeological and Forensic Sciences, University of Bradford, Bradford, BD7 1DP

Dr Derek Hamilton – Scottish University's Environmental Research Centre, Scottish Enterprise Technology Park, Rankine Avenue, Glasgow, G75 0QF

Dr Anthony Newton – School of Geosciences, Institute of Geography, Drummond Street, Edinburgh, EH8 9XP

Dr Susan Ramsay – Brownwell Cottage, Standrigg Road, Wallacestone, Falkirk, FK2 0EB

Catherine Smith – Alder Archeology Limited, 55 South Methven Street, Perth, PH1 5NX

Caroline Wickham-Jones – School of Geosciences, University of Aberdeen, King's College, Aberdeen, AB24 3FX

PART 1 Introduction

By Beverley Ballin Smith

Location, topography, vegetation and place-names

The *Ard a' Mhorrain* peninsula, the area where Iain A Crawford spent part of at least 31 years researching, surveying and undertaking archaeological excavation, is situated in the middle of the northern coastline of North Uist. The modern road courses its way north-west from Lochmaddy on the east coast of the island, to follow the line of settlement which divides the predominant fragile and lower lying machair and sands in the northernmost part of the island from the rough grazing and higher land to the south (Figures 1.1 and 1.2). At the township of Grenetote, a track leads north onto the 5 km-long Udal peninsula with its prominent dune hillocks aligned SW/NE, that points towards the small island of Boreray, to which it may have once been attached when sea levels were lower. The geology of the Udal peninsula is predominantly gneiss from the Lewisian Complex, which has been altered by regional metamorphism. Stones derived from igneous granites, rhyolite and other metamorphic rocks, and those of sedimentary origin, would naturally end up as waterworn boulders, cobbles and pebbles on the Udal peninsula, brought by currents from the surrounding coastal outcrops of South Harris, Berneray, other parts of North Uist, and further afield (Geological Survey of Great Britain 2017).

The eastern side of the peninsula merges with the extensive sand beaches that form the coastline leading north-east to Berneray. The exposed western or Atlantic side of this long narrow peninsula is punctuated half way up by the rocky headland of *Rubha Bheilis*, and less than 1 km north-east of the site is the smaller *Rubha Huilis*, which is still attached by a sand beach to Udal. This headland forms the northern limits of Crawford's main project area. The western and southern edges of *Rubha Huilis* are currently under attack from rising sea levels, and it is separated from a larger expanse of solid gneiss bedrock, sand and turf to the immediate south by *A' Croig Bheag*, an ever widening rocky bay. It was on the southern edge of this bay, due south of *Rubha Huilis*, that the rescue excavation of *Rubha an Udail* X6, known as RUX6, was situated at NGR: NF 824 785 (Figure 1.3).

The subsoil layers beneath RUX6 (see PART 2) were complex, acidic, damp, dark and contained low amounts of shell sand and little organic matter (see PART 4 pollen). Members of the Cambridge Quaternary Research Group who visited the excavations in October 1991 (see site archive) examined specimens of what they called *aeolianite*, or *frit* as Crawford termed it. The occurrence of this material indicated that there may have been a freshwater lochan, pond or spring available to the inhabitants of RUX6 in the late Neolithic that calcified the early deposits of blown sand, which then became compressed and lithified during the further accumulation of sand. It is an example of the changing conditions experienced in the landscape and how the past topography was very different to what it is now.

The present day peninsula is impressive, with prominent sand hills covered with machair vegetation, typified by Marram, Bent and beach grasses, with other species such as Red Fescue, Atriplex sp, plantains, Buttercup, Daisy, Galium, Poa sp and Bird's Foot Trefoil forming colourful

Part 1

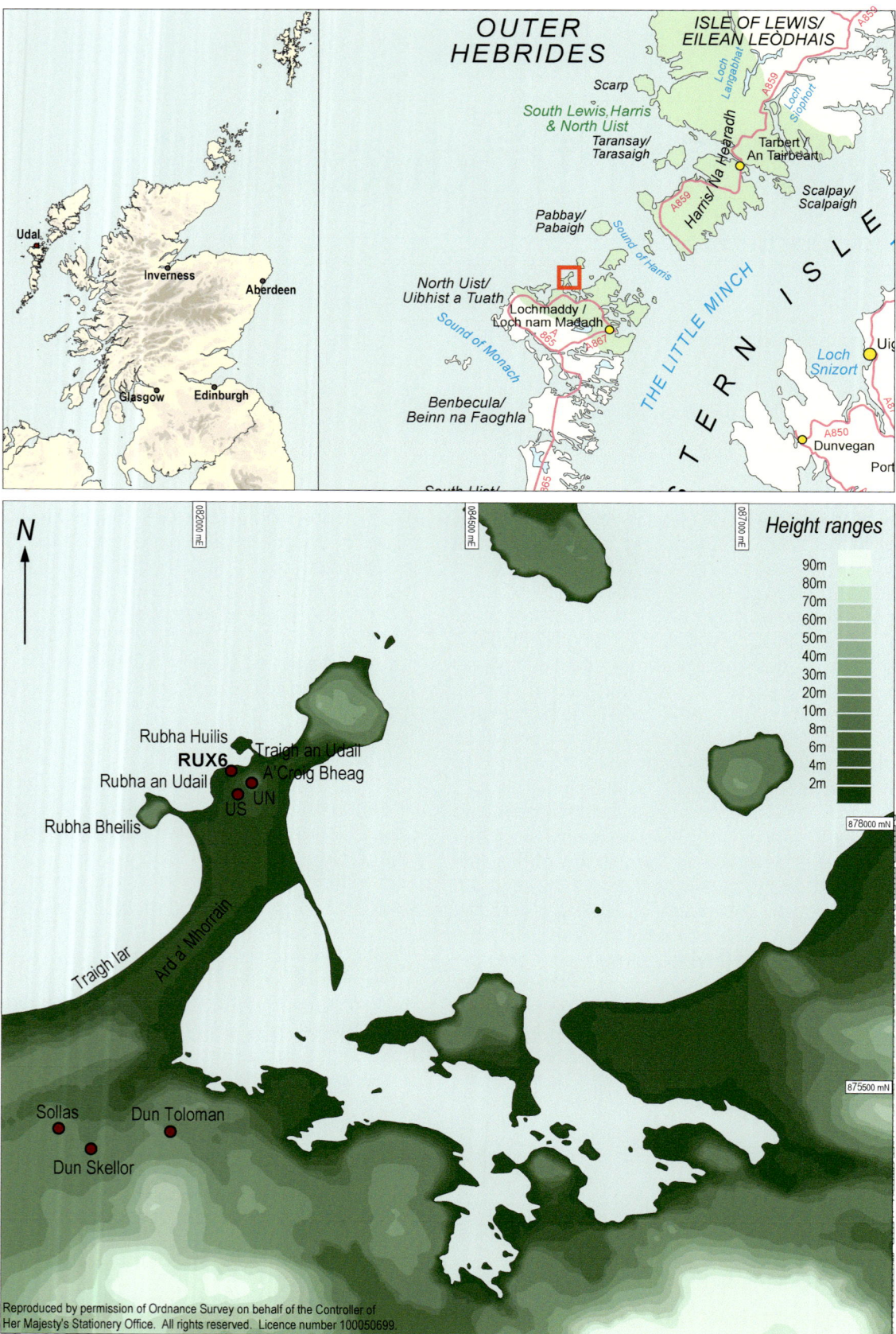

FIGURE 1.1:
Geographical location of Udal

FIGURE 1.2:
Aerial view of Udal - Map data © 2017 Google.

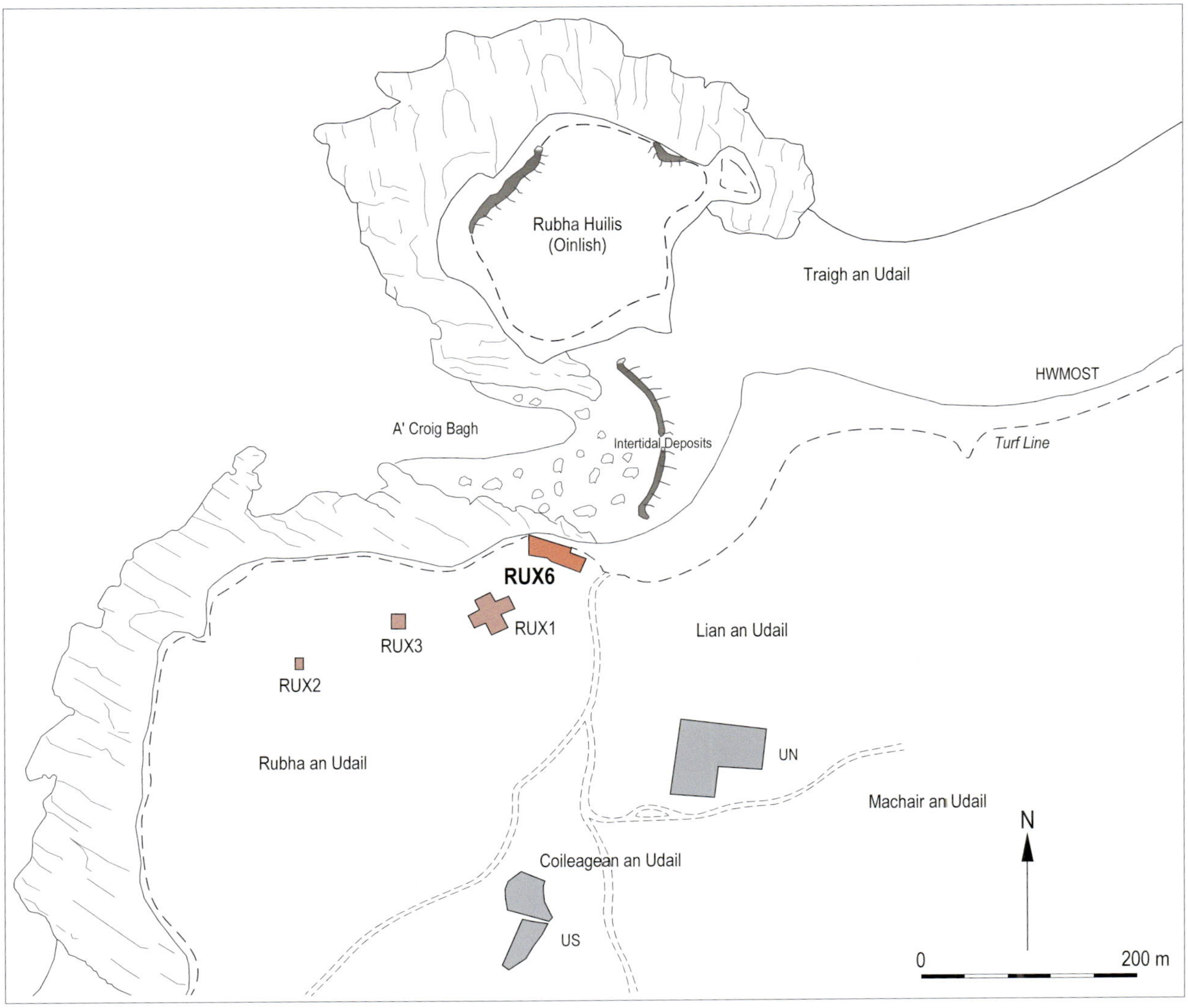

FIGURE 1.3:
Location of the site

meadows on other areas (Figure 1.4). The machair can be both wet and dry, and the flatter parts of the grassland areas have been ploughed and grazed since prehistory (Dickinson and Randal 1979: 271, table 2).

FIGURE 1.4:
The machair vegetation in bloom - © BBS.

Place names

The place name *Udal* is a bit of a conundrum and there is considerable doubt over its meaning. Beveridge (1911, reprint 2001: 95-96) considered the following explanation:

> 'UDAL, close to Oilish at Ard a' Bhorain, is certainly Norse, whatever its meaning. After rejecting several alternatives, *út-dalur* or 'outer valley' appears to be the most suitable, even if this promontory is rather to be described as a plain than a dale. There seems indeed to remain a trace of a 't' in the present local population.'

High sand hills were already present before the appearance of the Norse, as late Iron Age occupation was certainly present on one (the north), and vestiges of occupation might have persisted on the other (south), of the two main settlement mounds of the Udal project (Figures 1.1 and 1.2). The large valley between might be the dale or valley referred to, or another one in the vicinity. The shifting sands of the peninsula make it very difficult to determine the topography of the landscape 1100 to 1200 years ago.[1]

An alternative explanation has been put forward by Graham-Campbell and Batey (2005: 25), 'The term *udal* survives in Scandinavian Scotland from the Old Norse óðal as a technical term for inherited land bound by complex rules.' Odal or Udal Law still survives on Orkney and in Shetland, mainly in connection with property and foreshore rights, and it is still the dominant form of farm landholding in Norway (Linklater 2002: 22 and 25). However, The Treaty of Perth 1266 resolved conflict between the kings of Norway and Scotland by transferring the Outer Hebrides to Scots Law (Beveridge 1911, reprint 2001: 21) thereby removing any connection it had to Udal Law.

Other place names on the peninsula are:

Coileagan an Udail - knoll or dune(s) of the Udal, where Crawford's main sites of Udal North (UN) and Udal South (US) were situated.

Rubha an Udail - Udal headland or promontory, where the sample excavations of RUX1-6 were carried out.

A' Croig Bhàgh / Bheag - Croig Bay (seaweed bay)

Traigh an Udail - Udal beach

Lian an Udail - Udal (wet) meadow

Ard a' Mhorrain - The main or big moraine, east of the large sand dunes, forms the spine of the peninsula, which is cultivated

Rubha an Udail X6 - RUX6 was the 6th sample excavation or exploration on the area north-west of the main sites Udal South and Udal North.

Background to the project and its origins

A short introduction is required to explain Iain A Crawford's interest in the West of Scotland in general and in the Udal, North Uist in particular. He was born in Glasgow, and after military service he studied history at Christ's College, Cambridge, where he gained his BA (Hons) and post-graduate diploma. Between 1952 and 1960, when he became a Research Fellow at the School of Scottish Studies, University of Edinburgh, the details of his life are unclear. It is believed he went to live on the West Coast of Scotland for part of the time where he learnt Gaelic, fished and presumably travelled extensively around the Inner and Outer Hebrides and the West Highlands. It is quite likely that his interest in landscape, archaeology and the settlement

---

[1] I am grateful to Dr Barbara Crawford for further discussion on this matter.

origins of the islands developed then. He must have already been familiar with the Udal peninsula and targeted that area from Edinburgh, when he undertook a parish survey of North Uist, producing in 1965 *Contributions to a History of Domestic Settlement in North Uist*. By that time he had already researched the documentary evidence for Udal and his excavations had been in progress for two years.

The following narrative is mainly the words of Iain A. Crawford, with additional input from Imogen Crawford. The account has been edited from his various interim reports and other unpublished accounts to bring the text up to date. Some of Crawford's unusual words and phrases have been altered to make the text and his meaning clearer.

In general, the work on the *Ard a' Mhorrain* peninsula was based on the background of a general research project into the history of the settlement economy and environment of the north-west highlands of Scotland between Kintyre and Sutherland, an area for which detailed historical information is almost totally lacking prior to the eighteenth century. This is especially true of the approximately 1,500 years which may lie between the many presumed Iron Age sites of the area (few of which were investigated adequately prior to the beginnings of this project) and the end of the medieval period c. 500 AD to 1500 AD. After the collation of the documentary evidence for later periods (Crawford 1965), a collection of oral tradition in the Gaelic of the area was made and more general linguistic material gathered. After field survey, it became quite clear that further progress could only be made as a result of a successful archaeological excavation campaign, preferably beginning on a multi-phase site. The recovery of evidence by archaeological techniques was to be associated with documentary, linguistic and other historical sources for the period in this area. The project may have appeared somewhat ambitious, but as the outstanding problem was the extreme paucity of material, the scope was extensive in a nominal rather than a real sense.[2]

A close appraisal of the situation revealed that by 1968 (5th interim report) the only research of significant calibre had been linguistic. Documentary research had been mainly genealogical and not always of a sufficiently high standard in that field. The archaeological evidence is insignificant to non-existent over protracted periods, and from the mass of evidence for the Iron Age itself, at the time still unassessed satisfactorily, and liable in some aspects to substantial redating.[3] Until the compilation of detailed estate records in the eighteenth to nineteenth century, the settlement evidence is negligible. Physically there exists only the major fortified medieval structures which are largely uninvestigated and in some cases even undocumented, many of the major church settlements are in a similar plight, and additionally, a large number of minor religious foundations are wholly unresearched. The crucial settlement unit, the *baile*,[4] the ancestor of the nineteenth century crofting-township, is virtually unknown in its pre-clearance form (late eighteenth to the nineteenth century), and this ignorance naturally becomes intensified viewed retrogressively backwards from the eighteenth century.

Crawford's immediate tactics were to examine the baile. This was done firstly by intensive field survey to assess surviving surface evidence, to sample particular areas in depth (Crawford 1965), and thereafter to select a series of suitable sites, preferably with at least one or more periods of occupation, especially occupation of the clearance period, or as close to it as possible. His intention was to elicit from these sites a chronology of settlement pattern and of artefacts, and to establish the whole archaeological criteria for the area, and further to build up a framework for economic and ecological studies. The Outer Hebrides were selected for this exercise as being less disturbed by later settlement than the mainland, but also because the machair areas of the Outer Hebrides clearly represent the most propitious settlement area in the West Highlands, especially as regards medieval or earlier settlement.

The Uists and Benbecula were selected as the most favourable settlement zone in the Outer Isles, containing elements of site preservation in the drifting sands of the west coast machair. Ideally, a site was required showing reasonably extensive settlement, covering the maximum possible range of period, and capable of producing evidence of

---

2   This was written in 1968 before the area of excavation and the number of finds and samples grew considerably (Crawford 1968).

3   Euan MacKie had completed his excavation at Dun Mor Vaul by 1964 but the results of his work were not published until 1974, but Crawford would have known of MacKie's work.

4   This is what Crawford considered he had encountered on the top of the Udal North sequence.

buildings and general economy, which could act as a yardstick for the whole area. In fact, the sort of type site of which Jarlshof in Shetland constitutes the classic example for Scotland, and which has hitherto been lacking for the West Coast. A crucial factor in the preservation of such sites is the presence of a natural insulating material. In Western Europe, apart from peat growth, blown sand accumulation is the only common phenomenon of this sort, and Jarlshof is the typical example of sand accretion over successive settlements. In the West Highlands, in particular where soil deposits are very shallow, it seems improbable that well stratified sites will be discovered except in areas of extensive sand deposit. These fertile, alkaline machair areas are most suitable for early agriculture and therefore settlement; and the accuracy of this observation needed to be tested.

The serious disadvantages of sand hills are that they can erode totally, redepositing their contents in disorder in their wake, and they may reform again upon this redeposition. This latter factor constitutes a serious archaeological hazard. Although, as elsewhere, unfortified early settlement sites are rare, there are instances of such in the Outer Hebridean machair, but generally they suffer from the handicaps outlined above.

An intensive field survey of this area (by Crawford prior to 1963) showed Udal to be the most promising candidate for a type site. No other site in the Uists visible to field survey showed small find evidence, surface remains, and the possibility of prolonged occupation similar to Udal. The only remotely comparable site, Sligeanach, Kildonnan (see Parker Pearson and Zvelebil 2014: 5-7) in South Uist, has been almost totally eroded by wind – the material there redeposited in an archaeologically meaningless jumble. Crawford intimated that this fate was gradually overtaking the *Ard a' Mhorrain* peninsula and many other smaller sites in the machair areas.

Previous considerations of the remains at Udal had placed it in the nebulous category of 'earth houses'. A limited excavation carried out by Erskine Beveridge (1911 reprinted 2001, illustrations between pages 128 and 129) at the beginning of the century had reached this conclusion. The RCAHMS on the Outer Hebrides, Skye and the Small Isles (1928, the site was investigated in 1914) comments on the site in some detail commencing as follows.

273. 'Earth houses (ruined) UDAL. Amongst the sand dunes on the western side of the peninsula extending North East from Sollas, at Udal about 2½ miles north of Sollas, are four large sand dunes, the slopes of which are covered with kitchen midden refuse, consisting of shells, animal bones and fragments of handmade pottery. At different places on the slopes are quantities of stones, dislodges by wind denudation, apparently the ruins of earth houses'.

Iain Crawford, with his wife Harriet Crawford, began fieldwork on the Udal peninsula in 1963. He called the area *Coileagan an Udail*, which he described as consisting of two large and two smaller sand dunes, rising to some 12.5 m above the surrounding *machair*[5] level and covering an area of some 11 acres. The following year a number of structures (RUX1-5) were plotted on the *Rubha an Udail* headland westwards of the dune area. It was intended to put trial trenches across many of these but excavations at RUX1 became so increasingly complex that work was confined to that site. He did not return to the area until 1970, when he established that a Beaker occupation phase with possible structures was present in RUX3. Although he was in close proximity to what later became RUX6 situated on the coast, Crawford did not admit until 1974 that a site had been noted among the shore sand hillocks as having a possible stone core.

Although Crawford did not know it at the time, he later indicated that the general area of *Coileagan an Udail* constituted a 'fossil' landscape containing occupation levels, structures and fields. These landscapes had been conserved by sand deposition, insulated and confined by wind, and possibly sea erosion, to the extent that 4,000 years of human occupation and old ground horizons extending much further back, existed undisturbed (1970, 7th interim report).

The environmental history of the area

Since 1963 when Crawford began his field work at Udal, there has been considerable research into the coastal changes of the Uists, the understanding and investigation of machair, and in particular those areas of the northern part of North Uist. It is to his

---

5   Described as low-lying arable or grazing land formed near the coast by the deposition of sand and shell fragments by the wind.

credit that he took great interest in the new study area of the 'environment', a word that only came into general usage from the mid to late 1950s. He may have got to know Professor William Richie at the Department of Geography, University of Aberdeen, who by the early 1960s had already begun to study the machair of South Uist for his PhD. There may well have been an exchange of ideas and points of view that influenced the thinking of both men.

For Crawford, the key to understanding the settlement of any of the Udal sites, but especially that of RUX6, was to understand the natural processes that interleaved with, or affected the survival, of archaeological remains. His new wife, Imogen Crawford, an ecologist, was also interested in the development of machair, and together they may well have developed a deep understanding of the natural processes that affected that site.

It is apparent, reading Crawford's summary reports for each phase at RUX6 that he produced for the Scottish Development Department, that he based his understanding of the cultural remains on a more developed knowledge of the natural process that had been at work on the site from prehistory to the present day. Although he was clearly aware of the different archaeological time periods of occupation on RUX6, his understanding of them was more clearly based on his 'reading' of the natural accumulations of sand, their removal and their redeposition. He must have had much discussion with geomorphological scientists that visited the site and applied the knowledge gained to his better understanding of the natural processes at work.

From an archaeological point of view, his detailed description of the site levels (the natural and anthropomorphic sands and soils) surpasses that of the cultural remains. They interleave in the archaeological story, and the development of the machair had a huge impact on the settlement and use of the site. The following section has been updated to bring in current work and new developments in the understanding of the machair environment and coastal change. Crawford's ideas and words are embedded in the text.

The Outer Hebridean machair, with smaller deposits found elsewhere in Western Scotland, is a 160 km long, mainly calcareous shell sand build-up forming the west littoral of the Outer Isles. It is a light coloured material with a high lime content that produces a fertile soil when covered in vegetation.

The fluvio-glacial sands and marine shell that came inland with rising sea levels and stronger winds is a finite resource derived from a marine platform or the continental shelf off the west coast of the Hebrides (Hansom and Angus 2006: 404). It probably started to be brought onto the land from approximately 3750 BC (Ritchie 1979: 117), Phase E at RUX6, but the major primary deposition took place somewhere after 2400 but before c. 2200 cal BC from the evidence from RUX6 (see PART3, Table 3.1). This indicates the event was a little later than the 2500 BC date proposed by Ritchie. He suggests there was stabilisation c. 1750-1500 BC, but episodes of movement, disturbance and redeposition would continue to occur (ibid).

The stabilisation was probably an indication that the marine deposits were exhausted as there is now no further additional marine resource to add to the dune system. The movement inland of the sand system leaves the coastline exposed to erosion from rising sea levels (Jim Hansom pers. comm. and see also NCCA 2017 for monitoring and mapping of coastal change).

Despite considerable fluctuations in its extent in prehistoric and historic times, machair constituted a crucial and valuable environment for early settlement in the Western Isles, as the many sites of all periods situated there would indicate. It was Crawford's central hypothesis of his Udal research project that the region and the machair were underestimated and undervalued in the history of settlement development when he began work there in 1962. The agricultural use of machair and its link with settlement was part of the cultural context mentioned by Hansom as part of his definition of it (2003: 473). At Udal, a rock escarpment with thick clay deposits and developed soil attracted the earliest settlement discovered at RUX6 – the late Neolithic. Subsequently, massive drifts of shell sand accumulated that covered the settlement and into which later prehistoric settlement developed, only to be subject to the cyclical events of sand envelopment and later deflation.

On the east side of the *Ard a' Mhorrain* escarpment a massive sand plateau or plain, up to 16 metres high built up, which sloped gradually to sea level. This would have been part of the natural cycle of the 'beach-dune-machair' system (see Hansom 2003: figure 9.2). The focus of Bronze Age and later settlement moved into this area, but subsequent erosion only left surviving remnants as isolated hillocks. Extreme sand blow in the seventeenth century AD caused evacuation of the contemporary settlement, and further erosion in 1905 and 1962 continued to threaten the prehistoric and later sites, and created an island from the headland of *Huilis* (*Uilish* or *Oinlish*), or as it was when named in ninth/tenth centuries (see maps generated for National Coastal Change Assessment (NCCA) 2017, using the 1880s first edition OS maps as a base).

Early research by Crawford showed that the extent of the machair plain on the west and north-west coast of North Uist makes the island one of the most suitable of the Outer Hebrides for early agriculture and settlement. In addition, the irregular shore line as compared with the relatively unindented west coasts of South Uist and Benbecula, for example, has made for unusual conditions of sand movement, with important effects on settlement preservation (Figure 1.5).

FIGURE 1.5:
The dune system and vegetation - © BBS.

Crawford's documentary research, together with place-name indications and surviving oral tradition, indicates periodic and considerable mobility of machair sand in the Uists, and whilst in the adjacent islands this has tended to be a steady wind and sea erosion, the more broken coastline of the north-west of North Uist has encouraged redeposition on rocky promontories. In fact, it is clear that on the machair areas of the Sound of Harris, and to a limited extent in the sound between Benbecula and South Uist, the shell sand has spread almost to the east coast before the prevailing westerly winds, and in the process islands have been made and unmade and substantial stretches of arable plains deposited and removed (see processes in Hansom 2003: figure 9.2). It is almost certain that the *Ard a' Mhorrain* peninsula has been much affected by these coastal fluctuations, which have been prominent factors in the disturbance of occupation continuity in the area and in the preservation of evidence. In spite of these issues, Crawford wrote in 1970 that the archaeological investigations at Udal presented an excellent chance of dating the deposition of the machair in that immediate area (7th interim report).

The whole area of north-west machair between Griminish and Port nan Long, some 16 km across, contains a very large number of early historic and prehistoric sites of all kinds. The *Ard a' Mhorrain* peninsula itself (see Figure 1.3) contains on present knowledge (1965) two undated forts (Dun Toloman and Dun Skellor) and three early chapel sites at its landward base, the large wheelhouse at Sollas on the *Machair Leathann* re-excavated by Atkinson in 1957 (Campbell 1991), the *Coileagan an Udail* complex with a nearby Bronze Age cemetery complex partly investigated in 1964[6] (1st interim report), and a fifteenth century cemetery near the tip. Reliable oral tradition states that until the end of the nineteenth century two tidal channels intersected just below the *Ard a' Mhorrain* headland and nineteenth century sailing directions for the west coast of Scotland confirm this. Furthermore, there is strong evidence that people born about 1800 recalled a time when the whole machair plain (the outfield), but in fact the main cultivation area of the Grenetote township on the east side, was a great sweep of sterile sand. Crawford also noted that the same plain had started to diminish again since the mid-1930s. It seems very likely that this peninsula was chosen for medieval settlement, probably for its strong strategic situation across the west end of the Sound of Harris and as the only permanently navigable east-west route between the Butt of Lewis and the Sound of Eriskay, some 240 km to the south, and for defensive isolation which would be accentuated if tidal channels existed at the time.

It seems that there may have been gradual erosion even by 1469, as a ½ pennyland, perhaps as little as 4-8 hectares (10-20 acres) arable at which the charter evidence (SRO C2/13/1) rates the *Ard a' Mhorrain* area, seems very small for the purlieus of a ruling family like the *Siolachadh Ghroaidh* who probably held Vallay as well. The apparent cessation of permanent occupation c. 1666 (SRO CC3/9/30 and GD221/105, Crawford and Switsur 1977: 133)

---

6   Crawford thought he had investigated three cists, but there is no surviving site record or material cultural evidence to suggest burial urns, human remains or a cemetery complex.

may well indicate the destruction by sea and wind of an arable machair which did not reappear until the nineteenth century. Sixteenth century Exchequer rolls state that Uist generally had been diminished by erosion, and there are many minor references to these events.

The prospects for preservation were uniquely favourable for the site, as the two catastrophes (c. 1468 and c. 1666) created major interruptions to its occupation, and insulated it by sand, isolated it by tidal erosion, and rendered it undesirable for later occupation and disturbance by the disappearance of the arable land. When the township of Grenetote was resettled in 1889, despite the distance involved, many houses and walls were raised with tumbled material from the south Udal dune (US) occupation, probably the outward scatter of the Iron Age structures' walls. Otherwise, circumstances have combined to protect this area from the hazards which have destroyed the bulk of West Highland medieval sites. It is improbable that such a fortunate combination of circumstances will be found again in the Outer Hebrides.

During the first excavation of the RUX6 site in 1974 (11th Interim report), Crawford noted that there had been substantial erosion of the surface of the area probably for the most part in the eighteenth and nineteenth centuries – a period of optimum machair destruction generally, and which led to final desertion noted at the Udal North (UN) (see Crawford and Switsur 1977: 133). The narrow line of machair has also been under erosion by wind and sea in the past, especially as a result of rising sea levels, and also currently in exceptional conditions. The effect has been to destroy the machair sand shore-face along most of the Udal headland frontage. Fortunately, the natural rock has just sufficient elevation (up to 5 m OD) to prevent inundation of the hinterland. Only near the head of the relatively sheltered bay, *A' Croig Bheag*, which forms the northern face of the *Rubha an Udail*, has there been the survival of a short length of machair shore face. This remnant of turf shore line appeared to cling to the lee of a small hillock (c. 5 m high), which Crawford identified in the 1960s as a probable cairn. The exceptionally high spring tide of January 1974 (a perihelion event[7] with the moon also in alignment and a following wind) cut open this short face, as most other west-facing coastal sand frontings in the Western Isles, and confirmed the identification of the mound.

Crawford reported in the 17th interim report 1980 that despite a visit of the Machair Research Group in 1978 and the initial doubts expressed by the geomorphologists on the limited exposure of the shingle bed on which the cairn complex stood (level VIII), it was felt that the indications were that it was tidal wash or redeposition of eroded material. The material graded out in size away from the sea, and its surface immediately below was clearly puddled. Other details pointed to this shingle deposit as a tidal wash probably produced by exceptionally high tides, indeed one comparable to that of 1974 but dating between c. 2200 and 1900 BC (see PART 3). Crawford thought that astronomical calculation may possibly date the precise tide in question. During the winter of 1980/81, high tides entered the site area, but the new stone revetment constructed in 1980 prevented any damage as on previous occasions this would have caused severe scour of the shore face (18th interim report 1931).

In spite of the high tide events, the processes of accretion, removal and redeposition of the machair are seen by Hansom and Angus as 'a continuum of essentially similar processes that have operated with only minor variation since at least the middle Holocene' (2006: 407).

The original research aims 1963

When Crawford started his research at Udal, the purposes of the project were diverse, as described above. The primary and over-arching aim was to find the right site, and he had probably considered Udal as a likely candidate in the very early 1960s. His choice was most probably confirmed during his fieldwork for his North Uist parish survey of settlement published in 1965. He possibly knew the area well and had certainly done copious research into the historical background of local settlements, land ownership and investigated unpublished estate records. He also collected copies of pertinent charters and land records going back to the medieval period from the Scottish Record Office (now the National Archives of Scotland). Crawford also researched the published documentation on relevant clans and searched through the map collections of the National Library of Scotland, as well as relevant documents held by the University of Edinburgh. He had also read and digested the 1911 publication of *North Uist* by Erskine Beveridge, which probably helped pinpoint

---

7 The definition of perihelion is the point in the orbit of a planet, asteroid or comet at which it is closest to the sun.

North Uist in general and Udal in particular as a starting point for further work. The implication of all his historical investigation is that Crawford may well have already begun this research before he joined the School of Scottish Studies, University of Edinburgh, but his appointment gave him the opportunity to continue his work, and especially to test his hypotheses through archaeological survey and excavation.

His 1965 publication demonstrates his enthusiasm for this research, which played to his strengths as a historian. It could have also been a period where he was at his most confident. He had completed a post-graduate diploma in archaeology some years previously, and in 1963 he was only 35 years old and at the beginning of what he probably hoped was an exciting future working in the Uist machair.

The purpose of undertaking a trial excavation in 1963 at Udal was for the following reasons:

- the necessity for establishing a type site for the area and period by means of a well-defined stratigraphy with dateable small finds and a pottery sequence
- to establish whether the site fulfilled its surface promise
- to establish whether intact structural plans (building outlines) existed
- to establish whether wind erosion had not proceeded so far as to make more intensive excavation unrewarding
- and to rescue the evidence before wind, sheep and rabbit activity combined to cause complete shifting of the massive dunes containing the site.

These intentions became more specific over time and in 1980 they included for RUX6 the salvaging of further information from the areas damaged by erosion since the emergency rescue operation of 1974, as well as the aim to establish the quality and extent of any further Bronze Age or earlier levels, and to protect them.

Crawford's intention in 1981 was to complete the work of 1980 and 1974, which was to salvage the substantial Neolithic and Bronze Age monuments being eroded at every exceptional high spring tide. In addition, he wanted to secure the excavation area from the severely damaged shore face erosion initiated by the perihelion tide of 1974, extending along the south-east corner of the bay *A' Croig Bheag*. By doing this he would have been better equipped to protect the Bronze Age settlement deposits (RUX1-3) sampled in 1964 and 1970. Further noted in the interim report of that year was the aim to expose the complete Beaker horizon first, and after recording it, to strip it to recover the complete plan of Neolithic Building 2 (DH) (18th interim report).

At the beginning of 1983, only a small area at the eastern side of the site measuring 9 by 9 m of Beaker and Neolithic deposits remained unexcavated. Crawford hoped to excavate this completely down to sterile (natural) levels throughout, but despite good weather conditions this was not achieved, principally due to the complexity of stratigraphy in that area (20th interim report).

By 1984 Crawford had realised that 'it was essential to set up a computer terminal with adequate facilities and a modem connection direct to a mainframe' as 'it was the only way to overcome the associated problems of large quantities of material and shortages of staff and resources' (21th interim report).

Introduction to the excavations

A low mound at RUX6 close to the main Udal complex of sites and adjacent to the coast had been under observation for some time and was thought to be a possible cairn (Figure 1.6). Crawford had examined it and realised that it comprised early deposits that related to a now vanished coastline. These deposits had been at severe risk of marine erosion for most of the 20th century, but the accelerated sea level rise during the last half of the 20th century rendered the situation crucial. The great perihelion tide of January 1974 not only caused severe, widespread and unpredicted damage to the soft coastlines of the Western Isles machairs, but it was probably the highest tide for over 2,000 years.

The 1974 coastal erosion cut through the face of the cairn mound and a fresh vertical section was still exposed in March of the same year (Figure 1.7). Crawford realised that in the now low cliff section, the complex stratigraphy showed elements of not only a damaged cairn, indicating that it was of Bronze Age date, but that it was stratified above earlier settlement horizons. An emergency excavation and improvised shore face conservation were carried out in the summer of 1974 as an addition to the adjacent research programme at *Coileagan*

FIGURE 1.6:
1964 archive photograph of trial excavations. The RUX6 cairn is under grass at the right hand edge of the image below the beach - © Udal project archive.

an Udail. The results of this intervention were more substantial and more important than Crawford could have anticipated. After an interval of six years, the then Scottish Development Department (SDD) agreed to fund the exploration and rescue of the evidence from the exposed and eroding cairn and the associated structural material to its east. The excavation campaign was funded for 1980, 1981 and part of 1983 when the Neolithic dimension was identified. The work continued in 1983 and 1984, and was completed with funding from the Udal Research Project.[8] A stone and concrete shore face dyke was erected in 1980 to protect the immediate machair grazing and the contiguous prehistoric levels previously noted inland (RUX1-3), but this did not long survive the end of the excavation. By 2011 when the site was visited, there was no clear indication of where RUX6 had been as the stone and concrete dyke had been destroyed. By removal of the stone structures and the soft deposits, the excavation had aided the effects of coastal erosion, and all that was left was part of the eroded back section of the site, and an extended boulder-strewn and bedrock beach (Figure 1.8).

Over the five years of excavation of the site, 398 days were spent in the field. Table 1 indicates the amount of time spent on each year in North Uist. In some years, the field season was split into two, with a spring and a summer excavation period.

Year	Fieldwork periods	Days in the field
1974	2 days in the field 27-28 May	2
	the main excavation between 17 July to 5 September	50
1979	site was visited but not recorded or excavated	1
1980	preparation between 12 March - 3 April	23
	the main excavation between 8 June - 16 August	70
1982	spring excavation between 25 March - 14th April	21
	the main excavation between 4 July - 2nd September	60
1983	the excavation between 5 July - 9 September	66
1984	spring excavation between 4 May - 14 May	10
	the main excavation between 14 June – 17 September	95
	Total	398

TABLE 1.1:
Excavation dates and days in the field [9]

FIGURE 1.7:
The section through the cairn revealed in 1974 - © Udal project archive.

---

8   This was largely Crawford's own money.

9   Crawford calculated 378 days were actually worked in the field. The area excavated was 670 sq m.
    The cost of excavation was equivalent in 1992 of £98,000

FIGURE 1.8:
The boulder beach in 1994 after the site had been excavated - © Udal project archive

Site location

The perimeter of RUX6 – the sixth of the *Rubha an Udail* X complex of early prehistoric sites – was restricted to the north by the shore erosion face, to the west by a rock fissure, to the east by a gradual thinning of deposits and a final interruption from a nineteenth/twentieth century cart track. However, to the south deep occupation layers have been shown to run through to RUX1, which can now be seen clearly as part of the same site.[10] To define this salvage excavation an ad hoc boundary was drawn just south of the main structures creating a back section some 15 m behind the pre-existing shore line.

Within these confines extensive settlement and ritual levels were exposed, classified into five phases or periods (PART 2, Table 1).[11]

---

10   Crawford assumed this, although it has never been entirely proven. Occupation deposits were also found in the other RUXs, but we cannot prove they were the same as at RUX6.
11   It was found prudent to continue with Crawford's phasing and his nomenclature, but they are discussed further below.

Methodology

There is no clear or full account of how Crawford set about his excavation of RUX6. The following information has been brought together from investigation of plans and sections, interrogation of the photographic evidence, and his written account of the phases of the site that he wrote in the 1990s.

Site preparations and setting out of the grid

Crawford set out an accurate imperial grid aligned E/W and N/S in 1974 and surveyed in four 12 x 12 foot grid squares across the cairn mound and shore face. These were D, E, probably F, G and H but there is no record of exactly where they were. No sketch or drawing survives of the mound and coastline before the excavation started, and the section (Archive section No 1.) of the eroded face of the cairn has been lost, with again no record of it, apart from that it was included on a list of sections for the site. It may have been redrawn or superseded by another section, but that is not certain.

This early arrangement of excavation grid squares, a method Crawford had used on RUX1-3, did not come to light until the remains of another possible burial had been detected among the disarticulated human remains which were being analysed in 2016 (see PART 3). The question that was asked was where were the human remains located? A copy of a sketch plan present in the 1974 interim report (11th interim report: 6) was found that had been highlighted by Crawford to indicate the likely position of a burial in a pit, situated on the cliff edge, and this was confirmed by photographs. However, there was no mention in his notebook of the finding of this feature, no detailed plan of the pit or its contents, and no detailed photographs.

In 1980, the site grid was maintained but changed to metric and all subsequent plans and sections were drawn using metric measurements. Before each excavation season general maintenance took place such as the repair of fencing, replacement of survey points, and minor excavational (sic) repairs.

The initial excavations in 1980 were confined to the original 1974 area, which was deturfed and stripped of its covering of plastic sheeting. The trench was also extended further east some 9 metres to enable the whole of Neolithic House 2 (DH) to be excavated. In 1981, the projected excavation area was extended probably to the east, deturfed and then machine excavated to the uppermost occupation levels. This produced a cutting 3 m wide by 18 m long and meant the removal of sterile sand deposits to a depth of 2-2.5m.

Plans, sections and photographs

Plans and sections were drawn in 1974 at imperial scales of 1:06, 1:12 and 1:24 and in pencil, ink and felt tip. In 1980 the recording had changed to the metric scales of 1:10, 1:20 and 1:50. The site plans and sections from 1980 onwards are intact, but a lot of reworking and overworking has taken place on the original pencil drawings. Crawford liked to use a variety of inks, felt tips and also Tippex whitener, to enhance or highlight important features or stratigraphy. Unfortunately, the original pencil drawing and details are often lost under layers of later colour and changes of interpretation.

As far as can be ascertained, the black and white photographic record appears to be fairly complete, apart from the beginning of 1974, where there are missing images of the storm damage to the site and the features found there, and 1983, where only half a film of images survives for the 66 days of work on site. It is more than likely that films were lost in the post, were under or over exposed, or destroyed in the processing. The colour transparencies show a fairly complete record, as described for the black and white images, but under exposure was a problem at times.

Written record

The main written record of the site is six of the collection of 31 interim reports produced annually after each year of excavation from across the whole of the Udal project area. Only the 11th, 16th, 17th, 19th, 20th and 21st interim reports are relevant to RUX6. These give a fairly comprehensive account of the features found and excavated, of the main artefacts retrieved, and of the site interpretation at the time.

Crawford used 12 school exercise books for all his day to day sketches and notes for all the work at RUX6. Their contents are largely incoherent and similar to a stream of consciousness, like a shopping list or an *Action Drill* as he termed it. His sketches are largely incomprehensible, as they are usually without context, direction, grid reference, measurements or scale, and in many instances it is difficult to know to what they refer. They could have been drawn to remind him of a detail or a location, but even that is unclear, and it is equally possible that after the passing of several years they made little sense to him.

His written notes are largely restricted to grid issues, what was completed or removed on a given day, and things that needed further attention or clarity. They demonstrate no overall understanding of excavation matters, there are no matrices or stratigraphic accounts of relationships between layers and features, and no measurements or descriptions of shape or depth. It is almost as if Crawford was lost in the minutiae of things but could not stand back and look at the bigger picture. His notes are extremely hard to read, even harder to interpret, and in the end they provide little understanding of what he actually thought about what he was finding. They reveal none of his thought processes or give any guidance to site interpretation.

Crawford's use of day books and their lack of context may explain why he spent a lot of time carefully going over the section drawings and plans, adding notes to them and highlighting details. They may have been his main tools of understanding some of the complex stratigraphy of the site (Figure

1.9). During the post-excavation process it became apparent that Crawford altered his labelling or numbering of some layers, either during fieldwork or afterwards (see PART 2, Tables 1 and 2). He did this as his understanding of the site stratigraphy became more confident, but it has left problems or confusion as to whether the relabelling of organic and sand layers was consistently applied to all records. His reworking of the site drawings gives us his last thoughts on matters of phasing, but there is some doubt concerning the boundaries between phases and whether they reflect reliable environmental and anthropogenic changes or simply Crawford's final interpretation of site data.

At the beginning of the 1974 season, Crawford's then wife Harriet wrote several good descriptions of the day's archaeological activities and what was found and their relationships and possible interpretations. After two weeks, her entries ended. From that point onwards, all the notes in the exercise books were written by Iain Crawford.

On site sieving and post-processing of site finds and samples

The general Udal methodology was that all spade or hand-dug sand and other soils were dry sieved on site. The sieving system at RUX6 was positioned just out of the excavation area and onto the foreshore (Figure 1.10). In addition, six soil flotation samples were processed: Phase B (2), Phase C (1), Phase D (2) and Phase E (1) and it is interesting to note that there were no seeds present in Phase E or the exterior of Phase D. However, Phases B, C and a Phase D floor all produced seeds as expected. In the specialist reporting (see PARTS 3 and 4), of samples and materials, some of which were sieved, it became clear that there were problems. It is suggested that either sieving of building floors was not a consistent operation or that some of the more significant sieved residues were lost, for example, seeds from the floors of the Phase D buildings and the lithic fine material that should have been present. This has resulted in a serious loss of information and interpretation.

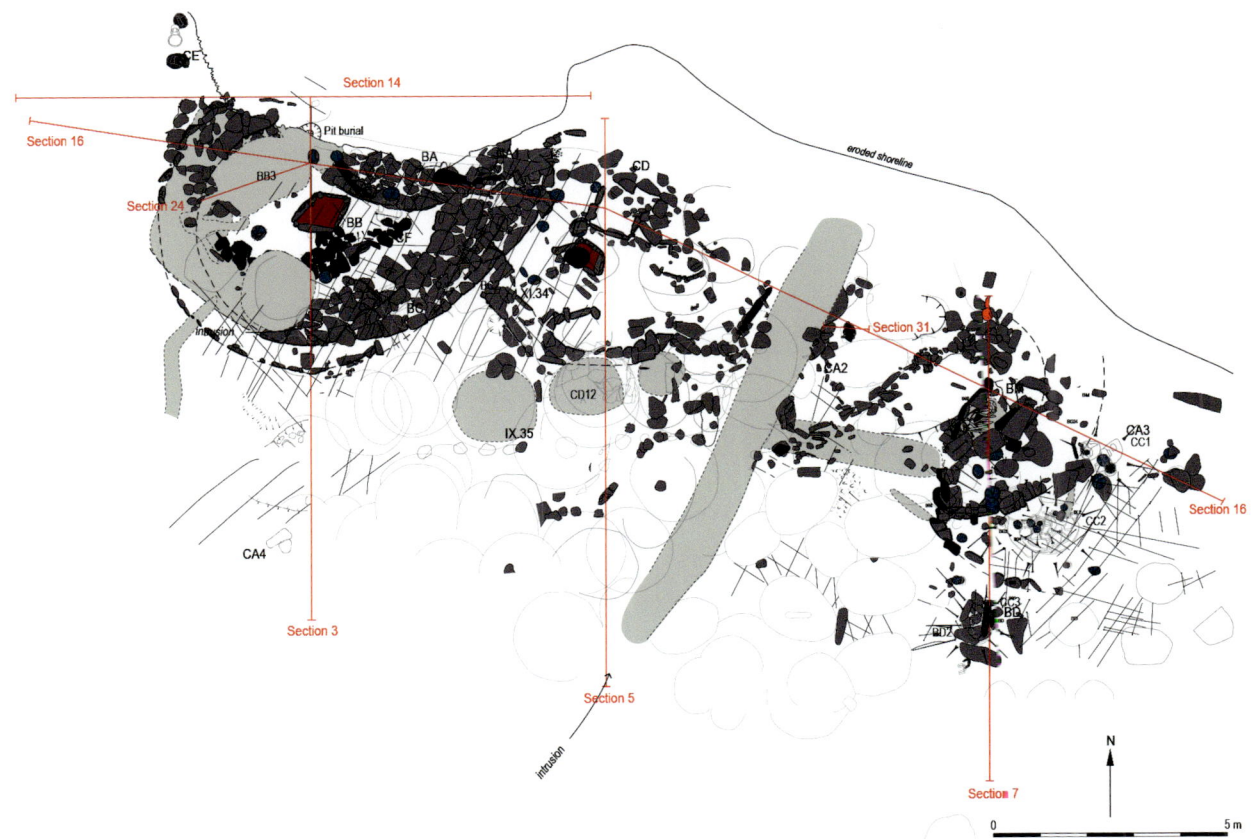

FIGURE 1.9:
The section drawing locations across all the features of the site.

FIGURE 1.10:
Sieving station on the shore just north of the excavation trench and beyond the protective wall - © Udal project archive.

Taken as a whole, the processing of materials and samples from RUX6, accounts for c.10% of the whole Udal project. Greenhouses and sheds with plenty of light were bolted together to create what became known as Crawford's *Chrystal Palace* (Figure 1.11), for the sorting, labelling, registering of finds and writing of finds cards. Usually there was one person on each excavation season that took responsibility for the recording of finds and samples.

FIGURE 1.11:
Crawford's Chrystal Palace - © Judith Finlay Aird and Graham Aird.

Monitoring and protecting the site during the excavation

In 1974, after excavating to the shingle level below the cairn, the site was covered with polythene sheeting and turfed over, and the excavated cairn material used to build a protective wall along the shore face. It is not clear if the site was monitored for continuing erosion in the intervening years, but in 1979 severe sea erosion was noted as having continued at the site, and damage to the underlying 'Food Vessel and Beaker levels had ensued'. The Scottish Development Department agreed to support a limited project to secure the shore face by walling and to excavate the more immediately threatened deposits in 1980. In May 1980 prior to the excavation season, a low cemented stone wall was constructed for c. 10 m along the shore face above the HWMOST.[12] During the excavation that year, the area behind the wall was backfilled and a turf covering placed on the revetted bank. At the end of the excavation the site was stabilised with plastic sheeting, returfed and the totally excavated western area completely back-filled behind the revetment wall.

Further trial trenching took place in 1981 to determine the extent of shore line deposits, which extend along the full extent of the damaged shore face (some 30 m) to the angle it made with the beach of the adjacent bay *A' Croig Bheag* at the point where the main trackway into the area reached the shore. The trenching left a 1 m wide 'tidal' baulk to seaward, and spoil was deposited against it on the shore face and later turfed, forming a grass levee to conserve the junction between the shore and the excavation. Additional spoil was added to it on the seaward face by soil sieving during the excavation. In spite of the attempts to protect the site, the backfill material, the walling defence and the shore face were completely removed in January 1993. The entire area of the excavation to the back section is now beach (Figure 1.8).

Nomenclature of features

The naming of structures and features during excavation was always intended to be informal, descriptive and, most important, relevant to the circumstances of its uncovering. However, what has evolved as on-site communication was not appropriate for formal identification and publication.

---

12   High Water Mark at Ordinary Spring Tides

At RUX6, a parallel system was established which described features using their level (Phase) number and their site grid 6 figure reference number.[13] The 6 figure reference number is derived from 'squaring' the feature and taking the co-ordinates from the bottom left-hand corner. This combination produced unique codes embodying both period and location information.

However, it was felt that a feature code should be used in the reports and lists that was shorter but hierarchically structured to enable easy database manipulation and cross referencing.

1st field - a letter denoting the phase in RUX6 - A, B, C, D or E

2nd field - a letter allocated sequentially A-Z denoting a major, functional unit, a building or a collective group of contexts.

3rd field - a number denoting a sub-feature belonging to the unit of the 2nd field, such as a hearth, floors and pits.

4th field - a number occasionally used to denote plural, collectives, resettings and other minor fine details.

Post-excavation activities prior to 2008

After the final excavation season in 1994, Crawford had an obligation to the SDD to write up the site, as RUX6 had been excavated with their funding. He was required to prepare the site records, write a site narrative, and have specialist analysis undertaken on the finds and samples in order to publish the site. Crawford set about contacting colleagues and others to undertake some of the specialist work, and this is explained in the introduction to PART 5.

Between 1994 and 1995 he produced five documents, the accounts of the levels and features found in the five different phases (A to E) of the site that he considered Data Structure Reports, and a sixth document, *Information and Notes for Specialists: the Udal Collection, with special reference to RUX6*, was compiled in 1996. Each of the phased accounts consisted of a background to the project with a brief site description and a commentary on the levels involved in that phase, with details of the individual levels and the features they contained, with reference to some images, plans and sections, grid references and finds. They included some photographs, interpretive drawings, plan lists, a diagram locating the sections, and lists of sections, features, photographs and a summary finds list. But Crawford had a problem due to his difficulty in separating straight facts from supposition. His descriptive accounts are interesting, but although he provided data, these reports did not include a narrative account of events.

In 2002 Crawford agreed, after a personal approach from the author, to some help in allowing a trial digitisation of the RUX6 plans and sections by GUARD, University of Glasgow, with funding from Historic Scotland (previously the SDD). The aim was to see if it was possible to build on a successful communication break-through to assist him in furthering the publication process. In spite of some hesitancy from Crawford in allowing primary records to leave his study, he may have reached the reluctant conclusion that he needed some external help in bringing his life's work towards publication. The work achieved by GUARD staff (the author as project manager and Kylie Seratis as technical help) was the first major step forward in eight years. The project ended when Crawford could no longer work with GUARD due to a family death which brought on a deep depression. Although communication continued with him over the next two or three years there was no further progress in the project. In 2008, after a long illness, he moved to a care home.

Post-excavation developments since 2008

The transfer of the archive and collection in 2008 from the Crawford family into the author's caretakership was a dramatic story told elsewhere (see online lectures Society of Antiquaries of London and Society of Antiquaries of Scotland). In 2010, the Comhairle nan Eilean Siar was updated on the availability of the collection for study, and together with Historic Scotland part-funded a two year assessment of the full Udal collection until 2013. This work resulted in the 're-excavation' and examination of the entire collection, the rebagging and boxing of all the artefacts, and the production of a detailed assessment report (Ballin Smith 2013). Running in tandem with this was the initiation and production of a database created by the digitisation

---

13 This is not a context number but is derived from the sand layers identified on Udal North and applied to RUX6. The level is identified by Roman numerals.

of over 32,000 record cards generated from the Udal project finds and samples. Most of the database work has been privately funded, but its maintenance and updating has also been funded through Historic Scotland (now Historic Environment Scotland).

In 2013 the collection was declared to the Treasure Trove Unit and the ownership of it was legally transferred to the Museum nan Eilean and the Comhairle nan Eilean Siar in 2014. The post-excavation programme for RUX6 funded by Historic Scotland, Historic Environment Scotland and the Comhairle nan Eilean Siar, was initiated in 2015, but was mostly carried out in 2016, ending on 31 March 2017. The results of that work are found in this volume.

Updated post-excavation research aims and objectives

The final research aims were written as an addendum to the 2013 Udal post-excavation research design (Ballin Smith 2015). They included the following:

Specific research aims and objectives

The aims of undertaking research at this site were to understand:

- the relevance of the site to the understanding of the earliest settlement in North Uist.
- the natural landscape around the site and the environmental and man-made changes that impacted on it, including the development of the machair.
- how sea level changes affected the landscape and the viability of settlement at this site.

The research objectives were:

- to undertake the collation and analysis of the stratigraphic evidence to produce the stratigraphic narrative and identify the sequence of events at the site.
- to undertake specialist analysis and research of all surviving artefacts in order to investigate their manufacture, and the exploitation and procurement of resources by the inhabitants of the site.
- to undertake specialist analysis and research of all surviving environmental samples in order to understand the nature of the natural environment, the exploitation and procurement of wild resources, the practice of agriculture and animal husbandry, the availability of food stuffs for the human diet, and changes in the landscape.
- to obtain taphonomically secure samples for radiocarbon dating specific structures and features across the sequence of events at the site, and to establish the date, type and function of structures and related deposits.
- to include recent research into sea level and coastal change using Lidar data and other techniques.

PART 2 The excavation record

By Beverley Ballin Smith

Introduction

The excavated evidence is presented in two sections. The first deals with the major natural events that affected the site during the late Neolithic, into the Bronze Age, and to the present day. The second section comprises Iain Crawford's detailed understanding of the site, and how he put the excavated information together. Much of it is in his words, but an attempt has been made to clarify them through editing. The site history is divided into five phases, with most beginning with a description of the sands/soils and other contexts that formed a major part of Crawford's understanding of how the site developed over time. This is followed by details of the features found within them.

Crawford's choice of words and his descriptive names for features are sometimes very difficult to understand and often misleading. An example of this is the word *niche*, normally referring to a cavity in a wall. However, as the structure in which this particular niche was found did not have a wall, it eventually became clear that Crawford had referred to a stone box set into the floor. There are many instances of site interpretation, prior to full excavation or understanding that led to misleading terms for features, and to some extent this restricted his wider thinking of what these features or their purposes could actually be.

Crawford's existing written work is sparse and largely unstructured, which has made difficulties for our understanding of his interpretation of what he excavated. Separating fact from interpretation and sometimes from supplementary but useful historical information has been very time consuming. This issue has been compounded by Crawford's change of mind on many stratigraphic details. His reworking of sections is just one example, but also the renumbering of levels and the relationship of structures to evels became complex. The narratives below take into account Crawford's views but they have had also to be ordered logically and stratigraphically, and to some extent simplified. Sometimes this has worked well, but with other features it is not so satisfactory, and therefore the stratigraphic story is a best fit scenario. This is discussed further in PART 6 in the interpretation of structures and their phasing.

This account is not the conventional way the evidence for an archaeological excavation is normally presented, but as this is a legacy project where our understanding is based on sometimes inadequate information from the past, there has been little alternative but to proceed with what there is but in a more structured way than when it was compiled by Crawford. In this account the descriptions link directly with the site archive.

Natural events and disturbances

A number of natural occurrences played a significant role in the story of the site. These environmental factors are described in order to set the scene of how their occurrence affected the late Neolithic settlement and also the use of the site during the early Bronze Age. They are important events that would have affected a wider geographic area than RUX6 or the *Ard a' Mhorrain* peninsula, and have significant implications for our understanding of the beginning of machair development and of rising sea levels.

Part 2

FIGURE 2.1:
The gneiss bedrock formations in front of the RUX6 site in 1980. Looking south. © Udal project archive.

Due to the erosion of the site in 1974, Crawford was presented with a vertical picture, a low cliff section (Figure 1.7). running across the site, down through time and through both natural and anthropogenic materials to bedrock (Figure 2.1). In his first intervention on the site he used this information to examine the build-up, not just of structures but of clearly defined sand and organic horizons between and around them (Table 2.1). In examining these, and with his knowledge from other areas of the Udal project, he began to investigate the environmental factors that had affected the site since the Bronze Age, portrayed by the cairn cut through by wind and water. He noticed there was nothing of any antiquity present in the section above the cairn, and the accounts below are predominantly his understanding of events (see also Tables 2.1 and 2.2).

The deposition of windblown sand (level X)

The earliest recorded natural event on the site was the accumulation of wind-blown sand (level X). It marks a significant stratigraphic division between the last part of the late Neolithic (Phase D, level XI) and the early Bronze Age horizons above (Phase C, level IX). It represents the initial deposition of a large amount of shell sand along the western edge of the *Ard a' Mhorrain* peninsula, as well as other western coastal areas of North Uist. The accumulation of this sand to a depth of c. 200 mm marks a complete environmental change, from natural acid gley tills with high siliceous content (levels XIII and XIV), to alkaline coarse shell sands with a moderate percentage of quartz sand.

Level X comprised a layer of white shell sand which probably extended throughout the *A' Croig Bheag* bay area and beyond. It derived from a marine deposit of shell and sand west of the island when the shoreline was an estimated 7-10 m lower than now.[14] The bottom of the shell sand deposit developed on top of predominantly wet or damp soils, whereas its upper levels displayed more aeolian characteristics and were less directly influenced by the hydrology of the immediate environment. Crawford considered this to be the earliest dated example of machair development in the Western Isles. Although difficult to ascertain with precision from the radiocarbon dates (see Hamilton in PART 3 and Table 3.1), it is likely that that level X could have started as early as 2465 cal BC, but probably more likely around 2300 cal BC.[15] The colour and texture of the sand indicates that it was deposited in a period of rapid accretion with minimal humic development (Figure 2.2).

---

14 Crawford's estimation of sea level change.
15 All dates are expressed at 95% probability unless otherwise stated.

Level	Phase A - AD 1925-75	Depth
I	Stable machair grassland ground surface with dark grey sand and organic content. Banked grassland to shore edge prior to perihelion tide of January 1974 when it was sliced vertically through Levels I to XV. Deposition commenced prior to 1951 (cf. DoH aerial photos). Extended over 350 m^2 across the excavated area and beyond.	0.1-0.2 m
Level	Phase A - AD 1900-25	Depth
	Kelping horizons with absence of significant organic content, fast accumulation but reduced by erosion.	
II.1	Pale grey sand across most of excavated area.	0.35 m
II.2	Dark grey sand with humic content, across most of excavated area.	0.03 m
II.3	Brown sand, across most of excavated area.	0.05 m
II.4	Pale grey sand, across most of excavated area.	0.05 m
Level	Phase A - 1910-25	Depth
	Identified as very distinctive banding in the section,	
III.1	Dark organic sand with ash from kelp kiln (AJ)? over the southerly 2/3 of the excavated area up to the 730 section.	0.05 m
III.2	White sand drift. Extent as III.1	0.10 m
III.3	Dark organic sand with ash. Extent as III.1.	0.05 m
Level	Phase A - 1880-1910	Depth
IV1	Grey/pale grey sand over two-thirds of the excavated area.	0.3 m
IV2	Pale brown sand with clear indications of ploughing. Extent as IV1.	0.6 m
IV3	Pale grey sand, drifted and very variable depth. Extent as IV1.	0-0.22 m
IV4	Very variable sand which partly derived from earlier levels and partly pale-grey and brown infill and collapse. Its extent was more restricted than IV1-3. It was identified by (restricted to?) the major disturbances rather than as a detectable horizon across the site.	0.80 m
V	Level V was a characteristic erosion-enriched matrix sealing the early deposits. It represented the final truncation horizon of perhaps a number of different episodes which eroded a now unascertainable depth of material, the final being late nineteenth century. After it, substantial early twentieth century machair horizons built up.	
Level	Phase B - early/middle Bronze Age	Depth
VI	Below the cairn complex and the eroded vestiges of VI, is another mauve level above.	
VII	Against the back section, leading into the unexcavated areas, a fine red ash band appeared as level VII.	
VIII	Paler mauve / light brown sand capped the upper-most structure (BM), and is the top surviving level in most of this sector.	
VIII	IX.2 is a suite of consolidated horizons whose distribution and depth is conditioned by later erosion periods.	
IX IX.1	Pale to deep mauve permeated with red and black organic material and ash. Not readily distinguishable from each other over appreciable distances. Distributed across the site according to erosion survival or feature consolidation. These soils were ploughed.	Slight to 0.5m
IX.2	The compacted ash floor horizons at (BG24) are probably the source of patchy ash spill through the east strata in IX.2/.1.	
IX.31	Shingle (B1/AC). The lower coarse pebble and stony single horizon of a single natural phenomenon. Covers an arced area of 80 m2, thinning out towards the edges of the arc.	<150 mm
IX.32	The upper horizon of a single natural phenomenon. Covers an area of 150 m2, thinning out towards the edges of the arc. This is a patchy, dark and pale, mottled deposit of shell sand showing clear evidence of wash and puddling.	<100 mm
IX.33	Deep mauve sandy soil indicative of humic development in damp, low-lying machair extending across the excavation area. Stratified between late Neolithic and early Bronze Age. Possibly mixed.	Slight where discernible

TABLE 2.1:
Description of the excavated levels by phase

Level	Phase C Late Neolitic/early Bronze Age Phase C	Depth
IX.34	Early bowl-pit horizon with post settings. Darker soil, ploughed.	
	This was the most extensive ploughed land on the site, characterised by turf build-up. Ploughing took place through the turf and into the white sand below. Plough marks showed in the white sand surface level X as dark humic furrow tips.	
X	Creamy white shell sand: generally sterile and represented the initial and probably rapid sand accretion. Its lower portion contain more orange/ginger till with lithogenic content and pale grey shell sand with a hard, dense, almost metallic texture. First significant machair deposit. Contains mattock marks that went down to the top of the building floors	c. 0.3 m
Level	**Phase D Late Neolithic**	**Depth**
XI	Settlement horizon including occupation debris and two buildings. Pale mauve to light brown in colour with organic enrichment	
XI.1, XI.2, XI.31, XI.32	Includes levels exterior to buildings and X1.2, black and red ash tamped floors, within the buildings	0.3 m
XI.21 & XI.22	East half of site between 349 - 358 grid lines. Pale mauve shelly sand with charcoal concretions. Contains the stone settings and shaft with stone platform/cairn facade. Cut by shore erosion face. Excavated 1984	100 mm
Levels	**Phase E Anthropogenic and Natural**	**Depth**
X1.3 - XI.6	Anthropogenic: banded sediments with animal bone, flint and pumice	80 mm
XII and XIII	Sterile weathered glacial till with traces of windblown shell sand	10 mm
XIV	Glacial till, olive-green in colour	0.4 m
XV	Gneiss bedrock	

TABLE 2.1 continued:
Description of the excavated levels by phase

Phase	Level	Events	Main structures and features
A	I	A - Perihelion shore face 1974	
Recent		E - Rabbit burrow system collapse	
	II.1	B - Pre-1974 shore face	
	II.2		Rabbit burrows
	II.3		Saw-pit (AA)
	II.4		Kelp drying dykes (AD-AI, AZ)
	III.1		Kelp kiln/pit (AJ)
	III.2		Robber pits
	III.3		
	IV.1	F - Fissure?	Robber pit
	IV.2		
	IV.3		
	IV.4		
	V	C - Eastern and N/S surface creep Deposition of an enriched horizon D - 2nd Marine incursion with eroded organic levels	
B	VI		
Early Bronze Age	VII		

TABLE 2.2:
Phases, levels, erosion events, main structures and features of the site

Phase	Level	Events	Main structures and features
	VIII		Cairn and cist burial (BM in BG24)
	IX.1		Kerbed cairn with cist (BA, BB, BC) Structure and pits (BG24)
	IX.2	Shingle in the east, pebble beach in the north	Late bowl-pit (BE)
		Sand deposition	Standing stone (BD)
	IX.31	Eroded organic levels with water inundation and erosion	
		1st Marine incursion	
BC interface	IX.32		
C			Box (CF) and pit with pot (CE)
			Bow pits (CD) Beaker?
Early Bronze Age	IX.33/.34/.35	Homogenous matrix	Post settings (CC)
			Ploughing (CA)
			Mattock marks (CB)
	X	Blown sand accretion	
D	XI.01		Neolithic House 1 (DJ)
	XI.02		Neolithic House 2 (DH)
Late Neolithic	XI.2		Pre-Neo House 1 (DA)
	XI.21		Platform (DB), shaft (DC), Gt Auk stone (DD) and other stone settings (DE)
	XI.22/.23		Ash mound (DG, DL)
	XI.31		
	XI.32		
E	XI.3		Subsoil with animal bone and artefacts intrusions
Natural/semi-natural	XI.4		
	XI.5		
	XI.6		Mineral soil
	XII		
	XIII.1		
	XIII.2 XIV		Glacial till
	XV		Bedrock

Dates:
Blown sand accretion from c.2465 cal BC, but probably more likely around 2300 to 2200 cal BC
1st marine incursion sometime before 2150 cal BC
2nd marine incursion between 1560 and 1215 cal BC

TABLE 2.2 continued:
Phases, levels, erosion events, main structures and features of the site

Part 2

FIGURE 2.2:
The accumulation of windblown sand (level X) that brought an end to the late Neolithic settlement. Looking south-east. © Udal project archive.

The first marine incursion (levels IX.31/.32)

The large amounts of blown sand and subsequent machair development indicated a turbulent time at the beginning of the Bronze Age. This was most probably demonstrated by increased wind speeds and strength (bringing in blown sand) and higher tides (causing erosion and deposition), most likely due to sea level rise. Phase C is a division of Crawford's making in the story of the site that artificially breaks up the natural events of this dramatically changing landscape. The events perhaps should be seen as a continuum of significant weather patterns that affected the human use of the landscape.

The interface between Phase C and the succeeding Phase B is marked by unusual and relatively complex stratigraphic evidence. This complexity is a product both of the survival of information from anthropogenic and natural deposition, and of erosion due to geomorphological processes, which are intimately related. A major marine water incursion or flooding is indicated during levels IX.31/.32 that washed out the top of Phase C levels IX.33/.34 while leaving traces of the bowl pits that were dug into it and scars from ploughing etched on the underlying horizons (IX.34 and IX.35). In the western part of the site only relict features survived but with their associated soil horizons scoured away.

Layers IX.33/.34/.35 represented a block of stratigraphy identified in the north-facing section of the southern part of excavation. These layers were distinguished by a clear structural sequence within a homogenous matrix, which survived undisturbed in the eastern part of the site, but were largely missing in the west where they were replaced by deposits of sandy wash and shingle.

The lower part of this horizon was an extensive fan of shallow, and puddled mixed sandy sediment (IX.32), with patches of humic material (possibly turf) from level IX.34, mingled with stained white sand, to give it a deep mauve colour and a mottled appearance. It had a deeply dimpled surface, the result of the swirling wash and puddling and was recorded over an area of c. 150 m² with a variable depth between 10–100 mm thinning towards its edges (Figure 2.3).

FIGURE 2.3:
The puddled horizon of mixed material forming the lower part of the first marine inundation, from the lower paler streak in the section to the darker stony layer above. Looking south-east. © Udal project archive.

It provided a bedding layer for an upper horizon of shingle (IX.31) that was deposited in a single event as a fan or arc across the northern part of the western half of the excavated area to a depth of 0.15 m, but which thinned out towards the north. It was recorded over an area of c. 80 m² in extent but was considered less widespread but much heavier in composition than the sand wash horizon IX.32 below it. It comprised water-rolled beach pebbles c. 0.20 m in diameter with occasional much larger stones but contained no organic matter (Figure 2.4). Some of the larger stones are not water-rolled, and are indicative of redeposited building material from nearby structures perhaps further to the north.[16] These are likely to have been concealed in sand, then exposed, and later destroyed by the flooding. A Neolithic wall exposed within Phase C (level X) sand-blow was composed of similar stone and a building is a possible source of this material.

---

16 This is a rare piece of evidence that demonstrates that the RUX6 Neolithic settlement extended further north into what has become the *A' Croig Bheag*.

FIGURE 2.4:
The shingle and stone beach, at the top of the section, deposited as the final part of the marine inundation. Looking west. © Udal project archive.

Both the sand layer (IX.32) and the shingle deposit (IX.31) are likely to been brought to the site by sea drift, possibly from a shore exposure or storm beach at some distance to the north or the north-west, or from marine deposits to the west. The evidence suggests that these two episodes might have been part of the same event that would appear to be the major inundation of the past 4,000 years, and indeed may well be the highest of the group of marine transgressions recorded for the 3rd millennium BC. The RUX6 deposits showed no other comparable event down to glacial till, therefore indicating a rise in sea level during this later period. From the archaeological evidence, the cairn complex had not been built before this event took place and the radiocarbon dates (see Hamilton in PART 3 and Table 3.1, SUERC-69665) indicate that it is most likely to have occurred sometime between 2270 to 1920 cal BC.

This marine event (levels IX.31/.32) removed and replaced earlier soil horizons to a greater or lesser degree with the washed and damaged condition of the surviving IX.33/.34 features being crucial confirmation of this interpretation. It is worth reflecting that the damage to surviving Neolithic (and earlier) coastal settlements, including that of RUX6, throughout the west coast islands by this event may have been severe. Record of this incursion may have been noted elsewhere, such as c. 50 m north of RUX6 across the A' Croig Bheag and on the opposite shore of Huilis, there is a shingle exposure at a similar elevation to those at RUX6. In other, but occasional sheltered, locations e.g. Baleshare and Ceardagh Ruad further examples can be found. A shingle horizon is also visible in a shore exposure east of RUX6 at a similar elevation to IX.31. It, and other such nearby horizons, may represent the same phenomena suggesting that it was more than just local.

The second marine incursion (D) and deposition (levels V and VI)

Phase B activities were terminated by a marine

incursion (D) of the area that stretched inland beyond the Phase B cairn complex. It was similar to the earlier event but was not quite as dramatic. During the flooding all organic and sand layers down to horizon level VII, and even the top of that level, were eroded away but the stone structures of the cairn were not destroyed by this activity.

Following the erosion was a deposition (level VI) of an enriched horizon of mixed materials and fine residues, most likely derived from the eroded organic levels. What the interval was between erosion and deposition is not clear but the lack of a turf line suggests it was a continuum with little interlude between the two. In Tables 2.1 and 2.2 it is quite clear that the event occurred sometime during or after the Bronze Age, which separated its surviving structures from all later activities. It appears to have been the last major incursion of the sea until 1974 and is likely that it occurred sometime between c.1560 and 1215 cal BC (see Hamilton in PART 3 and Table 3.1).

Modern erosion (A)

Prior to 1974 the hillock at RUX6 presented a grass covered slope behind the rocky shore (Figure 1.6). The combination of tides and strong north-west winds in January 1974 eroded the shore-face back vertically to reveal a series of prehistoric deposits that were exposed to further destruction in the absence of any seaward protection. The deposits exposed a stratigraphic sequence from topsoil through to bedrock (levels I-XV) over a fresh section about 8.5 m long (Figure 2.5).[17]

The destructive tide produced a shingle deposit, which later was shown to lie at approximately the same vertical height as the shingle of Phase B/C, level IX.31 (see above) from some 4,000 years earlier. This suggested that no similar high tide inundation and deposition had occurred as high or so far inland between these two events.

---

17 The section drawing no longer exists and does not form part of the site archive.

FIGURE 2.5: Part of the section seen through the cairn complex revealed by coastal erosion in 1974. Looking south. © Udal project archive.

Modern erosion (B)

A c.15 m long stretch of the shore face survived the 1974 erosion and its turfed edge remained undisturbed at this time. Crawford in his examination of it realised that there must have been other comparable high tides earlier in the 20th century, followed by a period of stability, as the remains of deposits (levels III and IV) from the 19th and early 20th century kelping industry were revealed to be partly eroded beneath the now stabilised turf edge that developed along the whole seaward face of the site. The roots of the plants had prevented both wind and sea erosion to the machair and therefore hindered coastal retreat.

Soil creep (C)

Crawford provided little information about the soil creep (C) that occurred in Phase A and this could perhaps be interpreted as the destabilisation of turf and topsoil after the earlier episode of erosion and deposition. Whether this was due to soil creeping downslope to the shore or the effects of the prevailing winds, and therefore a natural event or whether there were other causes is not known. In the overall picture of events at the site, these might have been minor landscape adjustments.

Modern disturbance (E) – rodent burrowing

This event was a typical sub-circular maze of old and more recent rodent burrows (bioturbation) that existed south of the cairn and into its stonework. The disturbed area measured c. 6 m wide and 0.5 m deep below the level I surface. Collapse of the burrow tunnels was identified as a depression filled with dark mauve sand derived partly from the topsoil and turf. The burrows were clearly associated with levels I and II. The rabbit was only introduced into North Uist between 1880 and the 1890s, and therefore the burrow complex probably dates to sometime from the end of the nineteenth century to about 1975.

Modern disturbance (F) – probably a robber pit

Crawford originally thought that this event was a discrete incursion of the sea causing deep erosion located beneath level I that cut into the soft horizons immediately to the west of the Phase B cairn complex (BA, BB and BC) down to natural deposits (level XV). It lay c. 12 m south of the 1974 shore face and tapered inland from a width of c. 10 m and Crawford likened it to a fissure. It eroded the west face of the cairn by undermining the stonework and in doing so also removing the whole western segment of the outer kerb (BC), the adjacent stone filling and packing behind it, and much of the inner kerb (BB). It led to some cairn collapse with also redeposition of material that accumulated down slope and filled in the hollow caused by the event.

Crawford later argued that it was highly unlikely that the sea was responsible for the creation of the feature and the destruction of the west side of the cairn. He considered that in all probability that it was more likely to be the result of human interference, possibly a failed attempt to construct a saw-pit. He had been told by members of the local community in Sollas and Grenetote that there were a number of saw-pits were dug in the area wherever wrecks were accessible. However, the precise location of these is long forgotten as was the location of saw-pit (AA, below). The disturbed area provided an opportunity for exposed stone from the cairns to be removed and reused by the kelp drying dyke (*sgeirean*) builders.

PHASE E - natural and anthropogenic deposits

Although the presence of the many Phase E levels was apparent, if indistinctly, in the shore face exposures of 1974 their dating was unknown. No serious investigation of them was possible before the closing season of 1984. They were then more clearly perceived and investigated in deep sections and horizontal exposures as overall bands of fine grain sediment distinguishable by their individual colours. Lying above the gneiss bedrock and heavily coated with glacial till, the upper levels showed some slight evidence of human occupation but there were no primary features, except for the penetration downwards of stone settings from later periods, including the Phase A saw-pit trench (AA).

Pollen was reported to be present (see PART 4) in the lower natural horizons (levels XII - XV), which were sealed by the earliest anthropogenic levels XI.3/.6. There is no indication throughout this sedimentation sequence of any significant natural disturbance or denudation. The undisturbed presence of glacial till (level XIV) at the shore edge, which is early Holocene in date, demonstrated conclusively that this was the post-glacial optimum HWMOST at this point.

Bedrock (level XV)

There are significant factors concerning the structure of the bedrock that underlay the site. A broken fold in the Lewisian gneiss shows all along the inter-tidal exposure on the shore as massive denticulated (toothed) slabs up to 2 m in length (Figure 2.1). These rock-fast slabs have an inclination of c. 30° from the horizontal and the interstices between them form deep crevices, which hold tidal deposits of shingle and sand. This rock formation continued uniformly through the site area, except where its interstices contained a substantial drift of glacial till or gleyey clay (level XIV).

The implication of the bedrock formation is that the size of the foliation and the good quality of the hard gneiss slabs made the exposure highly suitable for quarrying for building materials (Professor R Graham, Swansea University, pers. comm.). Professor Graham specialised in the micro-geology of North Uist, and considered it the best available source of building stone in the island, and suggested that this could have been a major factor in the choice of the location for the late Neolithic settlement. This might be overstated, but the easily available stone could have been an advantage. Most of the late Neolithic structures on the site indicated that this advantage was used, but the resource was not exploited for some of the structures in the early Bronze Age. Due to machair movement, stone as a raw material may not have always been exposed, and therefore the opportunity to exploit it was not always available. The examples of one of the standing stones, the side slabs of a large cist and most Neolithic building walls would all appear to be derived from the bedrock.

In addition to quarrying the bedrock, there was also direct in situ use in Phase D (level XI.3/.4) of one deep crevice between particularly long denticulate slabs, which broke through the surface of level XII. The crevice was most likely emptied of its deep filling of till and later post-glacial sediments (see features DB and DC below).

Another feature allied to the gneiss faulting is the presence of a freshwater spring 20 m to the west of the site edge (see Phase A, feature AY, below). Investigation showed that this was the focus of natural water drainage from the whole north face of the *Rubha an Udail* moraine. There is no means of confirming that this fresh water resource was available millennia ago. It may have seeped unobserved on to the natural and served to feed the possible marshy area that seems likely to have occupied much of what is now the *A' Croig Bheag*.

Glacial till (level XIV)

Covering the bedrock on *Rhuba an Udail* is olive-green clay with a red coloured granular lower portion derived from degraded gneiss bedrock. Its recorded maximum depth was c. 0.4 m and it is post-glacial in derivation. Its sediment is particularly fine-grained, and as a natural horizon it was not disturbed except below the shore face perimeter where it had been subject to tidal scour. At that point it represented the Holocean maximum tidal advance in this region. This is a distinctive and significant feature of the Western Isles littoral as it contrasts with the mainland and even other machair islands such as Tiree.

The till was dug into by the construction of Neolithic and later structures and it was also dug out of the ground as an important natural resource. A shaft or pit in the bedrock in the centre of Neolithic Building 2 (DH) was emptied of its clay till for the manufacture of pottery at the site. This convenient resource, and others, would have been utilised around the settlement, and perhaps further afield as there are significant thicknesses of this material exposed in stream beds in the hills.

Subsoils (levels XII and XIII)

The first recorded soils on the site were coloured and c. 10 mm in depth (XII was ginger/red, and XIII was pale cream in colour), and were composed of weathered glacial till (XIV) with an increasing component of windblown silicious material from degraded gneiss and some shell sand. The latter indicated the first deposit of shell sand from its off-shore reservoir due to rising sea levels.

Anthropogenic layers (levels XI.3 - XI.6)

These finely banded sediments varied from cream coloured (XI.3) and brown (XI.4) across the site, to steel grey (XI.5) and brown/khaki colours (XI.6) on the west shore side only. Their maximum depth was c. 80 mm in total and they overlay levels XII and XIII, the earliest soils (Figure 2.6). Their colours as well as the presence of some animal bone, flint and pumice suggested they had been subject to/or derived from human disturbance and mixing.

FIGURE 2.6:
The anthropomorphic layers above natural clays and bedrock. Looking east. © Udal project archive.

Phase D

This phase represents the last vestiges of the late Neolithic settlement of this locality. It consists of three successive Neolithic horizons with possible evidence of earlier ones constructed on, and into, the upper natural layers of Phase E (Figure 2.7). The earliest structures of the Neolithic sequence are regarded as ritual, as there is no apparent other function for the stone settings (DD, DE, DF1 and DF2, level XI.22). The central focus of this complex was a well-positioned whale bone over a natural shaft in the bedrock (DB and DC). This sequence is succeeded by two settlement horizons (XI.02 and XI.01).

These Neolithic structures were damaged by the shore face erosion of January 1974. The Neolithic buildings 1 (DJ, level XI.1) and 2 (DH, level XI.2) lost a similar amount of the area of their circumferences, which included their doorways and adjacent walling. Any associated occupation deposits to the north of the buildings were also lost at this time. Neolithic Building 2 (DH) re-used the remains of the earlier ritual facade (DB) to the shaft in the bedrock (DC) as a revetment, suggesting the latter had been out of use for some time.

Some levelling took place prior to construction of Building 2 (DH), and further levelling westwards took place prior to the construction of Building 1 (DJ). Building 2 (DH) continued in use but with perhaps a change of function as an outhouse, byre or workshop. The levelling for Building 1 (DJ) almost wiped out the extremely slight remains of a possible third Neolithic structure extending westward. Due to the levelling and subsequent coastal erosion its identification and precise relationship cannot be determined, but it was considered by Crawford to have been the vestiges of an earlier structure (DA).

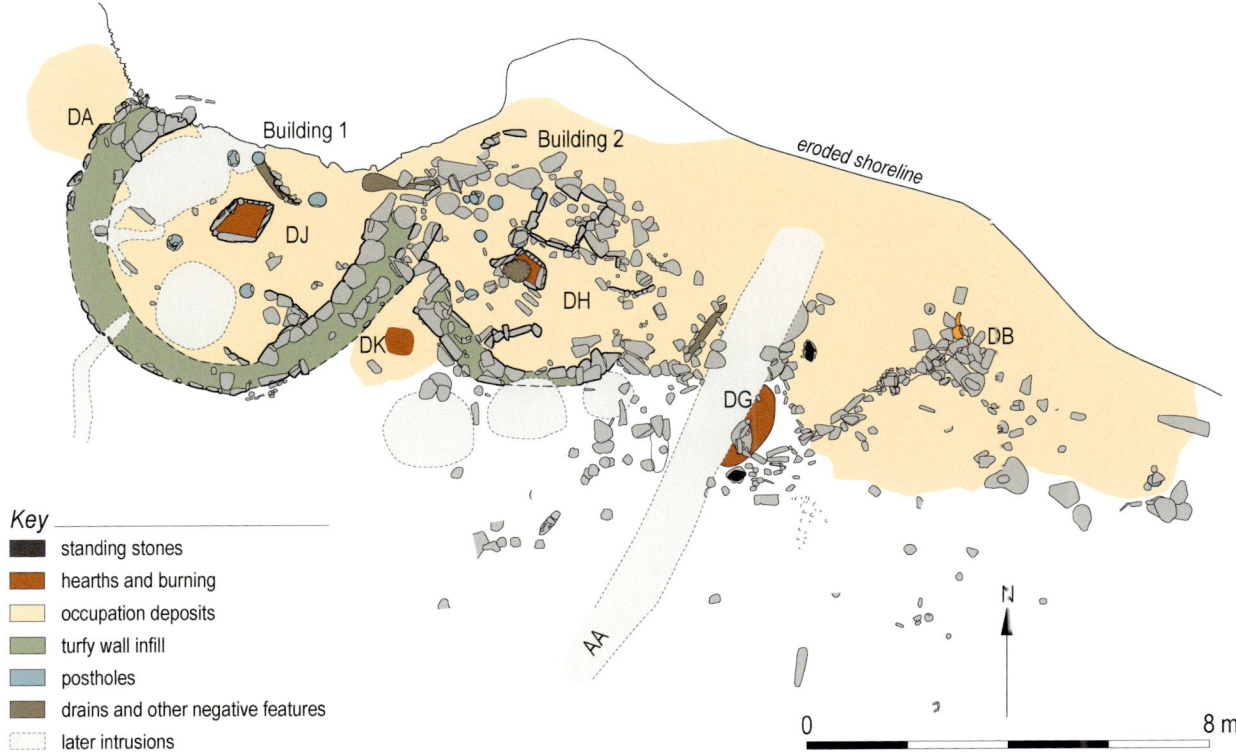

FIGURE 2.7:
Location of the Neolithic structures (DJ) west, (DH) centre, and the ritual area to the east.

Building 1 (DJ) was robbed down to a single course and was identified only by its floor edge for much of its perimeter. Ploughing and the digging of pits in the succeeding Phases C and B period were partly responsible for its destruction.

Soils descriptions

The soils horizons or levels descriptions are to be found in Tables 2.1 and 2.2, and mostly relate to the buildings. All the late Neolithic features fit into this sequence of soils which developed on top of the anthropogenic layers of Phase E through to a termination caused by a massive sand blow (level X).

Feature descriptions

Earliest structure (DA)

The scant remains of a possible structure (level XI.2) were noted to the immediate north-west of Building 1 (DJ). There appeared to have been a c. 3 m^2 area of occupation deposits or floor (XI.1.22), which partly lay under the west wall of Building 1. Crawford considered that the deposits may have been deliberately spread over a wet patch from natural drainage coming off the *Rhuba an Udail* (see Phase A feature AY). It was considered that if this had been a structure, most of it lay further north, into what was the tidal zone. An area of tumbled stone overlay the occupation deposits, which was refaced to form part of the west wall of Building 1.

Two possible postholes with stone packing and a pit with the base of a pot were allocated to Phase C (see Phase C, feature CE), but could be related to these deposits but their exact relationship is uncertain.

The following complex of shallow features on the east side of the site was not particularly well-recorded by Crawford or clearly understood by him. His inconsistent use of descriptive terms and the lack of factual detail make it difficult to comprehend or interpret the evidence. There were other features on the west side of the Phase A saw-pit that may have been part of the same complex, but again, the lack of information makes it impossible to identify them with any certainty in the site record.

Ash mound (DG)

A large mound of red and black ash (DG) measuring 2 by 1 m lay to the south-west of the Gt Auk stone and was located on either side of the saw-pit which cut through its middle (level XI.22/.23). It is the earliest occupation evidence on the site and predates the platform (DG) and its associated features.

The mound was oval in shape and seemed to have filled a shallow fire pit or hollow in the underlying deposits. Crawford considered that there may have been an earlier building here that was only survived by its hearth.

The shaft, platform, stone settings and alignments (DB, DC, DE and DF)

Level XI.22 was an extensive deposit of pale mauve shelly sand with charcoal flecks or concretions, and appears to the earliest occupation horizon of any substantial scale into which the stone settings are dug and on which the platform rests. It occupied the area east of the saw-pit, approximately c.1/3 of the excavation area, and was bordered in the north by the eroding shore face. The Phase A saw-pit trench (AA) split the buildings (DH and DJ) from activities to the east and destroyed the stratigraphic connections with features there and our clearer understanding of them.

In this area of occupation deposits were the surviving fragments of a stone platform (DB) with a radiating arm of small vertical stones. The stone platform surrounded a protruding but sloping denticulated piece of white gneiss (level XV, see Phase E bedrock description), which was part of a broken fold of bedrock that lay beneath the site (Figure 2.8).

A shallow pit (DL) was dug to enlarge a hollow beside the protuberant slab of bedrock to create a shaft from one of the crevices between it and a less protuberant adjacent slab. Crawford thought this had been deliberately filled in with a dense black deposit accompanied by stone and bone artefacts, but it could equally have naturally filled in over time. Three low stones (DC), perhaps a wall fragment, but eroded by the shore face, edged the east perimeter of the pit (DL).

A large intact whale vertebra (SF 26888, Figure 2.9) was placed upright on top the low stones (DC) and in front of the protruding slab of bedrock with its neural canal positioned above the shaft. The effect was accentuated by the prominent lamina projections either side of the canal which implied the bone was a spout.[18] A light accumulation of sand in the shaft indicated that it had been open when the whale bone was positioned above it. The area around the shaft was capped with stone, resting on low vertical stones, some considered to be pillars that were held in place with packing stones (1984 Calendar), to form a slight mound or platform (DB). By creating this feature, both the shaft and much of the whale bone were hidden in the stonework (Figure 2.10). Crawford thought the platform may have been larger before coastal erosion took some of it away.

An arrangement of stones was considered to be a contemporary feature with the use of the platform (Figure 2.11). However, it is impossible to deduce the complete lay-out of this complex due to the lack of evidence. One of the stone alignments (DF) comprised low vertical positioned stones in a line radiating towards the west and away from the platform and ending in a prominent black stone - the 'Othello' (DE). The alignment also contained two to three posthole or post-sockets, towards the beginning and end of the line. Crawford implied there were additional radiating low, hook-shaped off-shoots of stones curving northwards from the platform with each ending in a wooden post, and their plan gave him the impression of a series of small enclosures constructed of attractively shaped or coloured beach pebbles with vertical posts positioned at their ends. He also argued (1984 interim report) that some of these alignments had been demolished with debris left on the ground, suggested that some rebuilding may have been considered, and that the feature running due south from the platform was probably a dismantled linear stone setting that contained two small postholes.

---

18 Crawford called this a libation orifice and the shaft a libation pit.

FIGURE 2.8:
The shaft, whale vertibra, platform, stone alignments and Great Auk stone.

Part 2

FIGURE 2.9:
The whale vertibra on the edge of the shaft formed by bedrock. Looking south-west. © Udal project archive.

FIGURE 2.10:
Stonework around the whale bone, with a stone row leading south-west.. The Great Auk stone can be seen towards the top right of the image. Looking south-west. © Udal project archive.

FIGURE 2.11:
Stone alignments leading south-west from the shaft and platform in the north of the excavation, with the Great Auk stone on the left side by the Phase A saw-pit trench. The Othello stone is the black stone at the end of the stone row. Looking north. © Udal project archive.

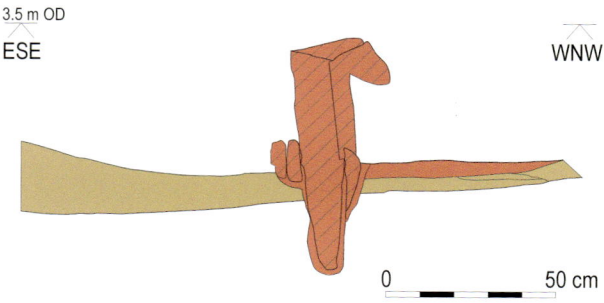

FIGURE 2.14:
Section through the Great Auk stone with some of its packing stones visible.

Positioned half way between the platform and Building 2 (DH) to the west was a standing stone identified as the Great Auk stone (DD), which may have been an integral part of the pattern of features lying to the east. This stone was positioned to face the shaft of bedrock and the platform but stood between a line of possibly three wooden posts with packing stones, and the arms of a stone alignment (Figure 2.12). Its shaped head and the small amount of shaft protruding from the occupation deposits gave it a zoomorphic appearance. Crawford considered that the shaft of the stone had been deliberately shaped to terminate in a curved keel, but none of this was visible as it was buried deep in the subsoil. The stone was secured by nine packing stones, some of which were reused artefacts (see Stone Tools PART 5, Figure 2.13 and 2.14). Crawford thought the stone was a small version of a statue menhir.

FIGURE 2.12:
The curved stone alignment in the foreground (left) with the Great Auk stone (right). Beyond the saw-pit trench are the remains of Neolithic Building 2 (DH). Looking west. © Udal project archive.

FIGURE 2.13:
The Great Auk Stone with packing stones around it (left), the shaft behind it and the Othello stone with a stone alignment (right). The sieving apparatus can be seen towards the top left. Looking north-east. © Udal project archive.

Neolithic Building 1 (DJ)

The Neolithic Building 1 (DJ) was one of the most informative structures on RUX6. It was an almost circular building measuring c. 7 m (E/W) by c. 5 m (N/S) with an almost intact ground plan and with surviving interior detail (Figure 2.15). The northern arc of the building, including the position of the doorway had been removed during erosion by shore advance. However, this area was probably still present during the first observation of the exposure of the settlement in 1974, but was lost before further excavation commenced in 1980.

After excavation of the building, it was noted that levelling had taken place before its construction, and that this may have removed some of the remains of a possible earlier structure (DA) to the west. During construction the circuit of the walls would have been laid out first. The building had a double-faced wall c. 0.7 m wide, but only half of the exterior wall face survived in situ. The latter was built of substantial long slabs set edge down with small flat stones organised in a V-formation paced in front, which may have prevented the walling from slipping outwards and stopped it being undermined by water runoff from the building's roof. A shallow drip gully (DJ7) was noted running parallel with the wall on the east side of the building.

The inner face was built of large flat horizontally laid stones that survived in places to four courses high (c. 30 m). Between the two wall-faces the cavity was packed with mauve coloured sand, probably interpreted as decayed turf. The polished stone axe (SF 23762, Figure 2.16) was recovered from the wall core.

FIGURE 2.16:
Axe SF 26732 in the wall core of Neolithic Building 1 (DJ). Looking east. © Judith Finlay Aird and Graham Aird.

FIGURE 2.15:
Detail of the two Neolithic buildings.

The roof of the building was supported by wooden posts inserted into in four large postholes arranged around the central hearth but extending 1 m beyond it in each direction. The postholes (DJ3) were circular roughly 0.5 m in diameter and deep (Figure 2.17), and some contained packing stones when they were excavated. The hearth (DJ) was sub-rectangular and measured c. 1 by 1 m. It was constructed of stones on edge with a cobble in at least two of its corners. North of the hearth, presumably to shelter it from the door was the fragmentary remains of what Crawford termed a fire screen (DJ2), positioned at an angle between the two northernmost roof posts. Other internal features included part of a stone-built partition (DJ4) in the SE sector of the building with a shorter length of presumably a similar partition (DJ5) to the south. Flat stones on the floor of the building near to the inner wall in the west may also have indicated the position of partitions.

Where the inner wall-face was missing the edges of ash deposits indicated the extent of the floors. It appeared to have three floors (XI.1), the first was c. 10 mm thick pale grey sand with a hard almost metallic texture, possibly derived from the subsoil. Patches of orange/ginger coloured floor deposits also survived, perhaps of a second floor, but the uppermost and best preserved comprised a layer of thin black and red tamped ash more than 50 mm in thickness with clay and stone content. The limited thickness of the floor deposits suggested they were regularly cleaned out.

As this structure was built partly over the remains of a possible earlier building to the north-west, which had a damp patch in its floor, there may have been a corresponding issue with ground water seeping through into this building due to impervious midden build-up. One drain or soakaway (DJ6) was dug

FIGURE 2.17:
Neolithic Building 1 (DJ) with central hearth, postholes around it for the roof and the outline of the inner wall (left). Looking south-west. © Udal project archive.

through the floor and into the underlying occupation deposits to deal with the problem in the north-west corner of the building close to its eroded edge. Another corresponding drain was excavated in the east close to the junction of Building 1 and 2.

The exterior wall of this house was built over the north-west corner of House 2, indicating that the latter was the older structure and that it might have had a secondary rather than a primary domestic function at this time.

Building 1 seemed to have had only one period of use, with little evidence of alteration. It was larger than Building 2 and was better preserved. Crawford felt that there may have been a blocked interconnecting doorway between the two buildings where their walls joined but the evidence is unclear. In spite of its better preservation, it had its fair share of disturbances. The pit for the Phase B large cist (BB3) was dug down to bedrock through the floors of the building and removed a large proportion of the floor north-west of the hearth. In addition, a large Phase C bowl-pit removed a significant part of the floor in the southern part of the building to the immediate west of the two partition fragments, and a stone-robbing hole disturbed another portion of the north-west part of the building. Burrowing rodents were also an issue in the south-west and western parts of the structure and subsequent ploughing and mattocking (CB1 and CB2, Phase C) was probably a contributory factor in the removal of wall stones in the southern half of the building.

According to Crawford, at the time of excavation this Neolithic building provided the only genuine datable house plan in the Western Isles.

Neolithic Building 2 (DH)

This building was probably the earliest of the two but it was less well preserved due to coastal erosion, possibly disturbance from the saw-pit and alterations on its eastern perimeter. It occupied almost the whole of the northern half of the central area and was probably circular like Building 1. The outline of the house as it survived was predominantly traced by the presence of compressed ash floors (Figure 2.18).

FIGURE 2.18:
Neolithic Building 2 (DH) showing wall outlines, central hearth and other features. Looking south-east. © Udal project archive.

The internal width of the building was 4.5 to 5 m (NW/SE). It was constructed with double-faced stone walls 0.7 m in width similar to Building 1, and presumably packed with occupation deposits and turf. Its stone-lined hearth (DH3) was irregular in shape but four-sided, measuring c. 1 m by 1 m, and located south of the centre of the building. North of the hearth was what is interpreted as part of a three-sided screen or recess, which would have sheltering the hearth from the door. Adjoining it was a platform (DH1) built of upright and flat stones to its east that measured 2 m long by 1 m wide, which was situated in the north-eastern part of the building. A small partition built of upright stones positioned against the internal wall south of the hearth completed the internal arrangements. The floors were thin and indicating that they were probably regular cleaned.

The remains of three postholes (DH2), smaller in diameter than Building 1, formed an arc west of the hearth, with an additional large posthole with a stone pad between the hearth and the recess to the north. Basal stones in postholes were used for stability or for preventing the rotting of the post if conditions were wet or damp (Figure 2.19), and the use of such stones suggests the building had problems with water ingress. Corresponding postholes on the east side were not found, but it is possible that this structure had an inner ring of six posts to support its roof. Additional postholes, one with stone packing, in the north-west of its interior suggest repair, alteration, or extra support when Building 1 was constructed. Like Building 1, this structure also had drains, which were covered and either were external to the building or were positioned by the north wall.

The walls of this structure were overlain in the west by those of Building 1, indicating Building 2 was earlier, but it is possible, but not proven that they were accessible to each other by a doorway at that juncture. The relationship of the two structures indicates that Building 1 replaced Building 2, which became an ancillary structure.

The interior of the building was also much disturbed by secondary use during the late Neolithic, when its function could have altered to that of a workshop or outhouse. The activities of this later use include the digging of a shaft immediately south of the hearth into subsoil deposits for the removal of grey-green coloured glacial till for pottery manufacture. Some of this clay was deposited inside the building, possibly in preparation for use.

The eastern side of the structure has been much damaged and Crawford suggested the wall had been taken down to create a rectangular cell (DI) or enclosure, which may have required extensive levelling up to 0.5 m in depth that spread out and covered the platform (DB) and the Great Auk stone (DD). Crawford intimated that there were floor and building elements east of the saw-pit that may have resulted from this building but the surviving evidence does not support this interpretation.

Burnt area (DK)

There is little information concerning this feature which was located outside (to the south of) the junction of the two buildings. Its presence indicates the possibility of a dump of hearth waste, a bonfire or other deposits.

Occupation deposits (D?)

Crawford suggested that there were thick occupation deposits noted on Section 16 (Figure 2.20), north of the surviving building remnants, with shell and sizeable stone that tapered from the west to the east. Some may have been floor deposits derived from both buildings, but as the distance between the buildings and the eroding excavation edge was very narrow Crawford may not have entirely understood what he saw or the meaning of it.

The late Neolithic occupation of these two structures and the site was brought to an end by the deposition of windblown sand (level X), which is discussed above.

FIGURE 2.19:
Neolithic Building 2 (DH) with excavated postholes with post-pads. Looking south-east. © Udal project archive.

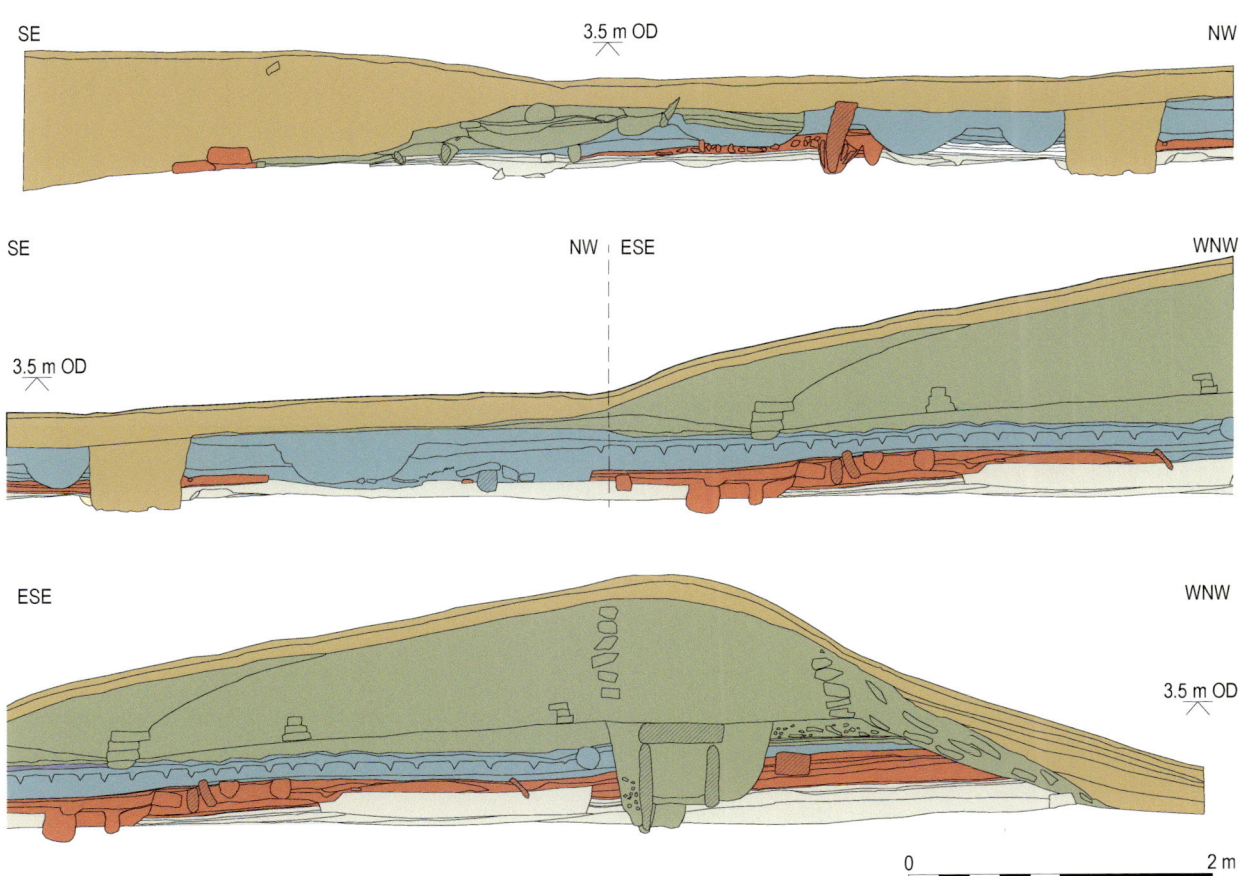

FIGURE 2.20:
Section 16 showing Neolithic levels in red with thicker deposits towards the coast (east)

PHASE C

This phase sees the beginning of the development of machair at RUX6 with the establishment of the basal sand matrix level X (Table 2.2), which was most likely a rapid accretion of blown sand, which lead to the probable hasty abandonment of the Phase D settlement and any fields that were under cultivation (see above). The geomorphological and archaeological records survived much more extensively in this phase than in Phase D, but the consequence of environmental and geomorphological factors are of greater importance. While Phases D and E can be seen as significant dating indicators regarding sea level rise, Phase C provided the crucial date of initial quantities of shell sand deposition on the land and the great physical changes to the landscape that took place because of it. It also became a period of landscape consolidation with machair development. The colour of the lowest horizon (IX.35) was indicative of turf growth from damp meadows which became established on the machair. This extensive area later became a spade or mattock dug area and even later an intensively ploughed field or fields, which may have belonging to a nearby inland settlement.

Archaeologically, Phase C was an environment in which evidence survived in greater detail than in the later phases. In terms of the excavation, there were stark colour contrasts of plough marks and single colour infills of pits against the white sand matrix. The Phase D to Phase C transition is defined by a succession from the Neolithic settlement (Phase D) to a ritual complex of posts and related pits and ploughing (Phase C) in a substantially changed environment. This change had significant consequences not just for archaeological retrieval but also understanding the excavation.

Phase C was totally excavated, but while a significant proportion was lost to marine advance and erosion, substantial uneroded elements survived in the stratigraphy extending inland beyond the excavation. Apart from the ploughing, Phase C was considered to be a dominated by 'ritual' activity centred on level IX.34. It had a boundary comprising three or more upright posts, with exceptionally deep settings, which formed an effective limit to an extensive area of 73 recorded (but probably several hundred) pits of a standard bowl-shaped profile (Crawford 1986:7). This boundary appears to have had a prolonged existence and survived into Phase B. Two small features (CE and CF) survive on the west side of the site in areas where there was no ploughing or pit digging. These are possibly the only evidence of domestic structures that were removed by the flood (see Natural events, above).

It is significant that the most easterly portion of Phase C activities was superimposed over a Neolithic ritual monument (Phase D) and was succeeded by an early Bronze Age ritual setting (Phase B), which share common factors, indicating a long period of continuous, or intermittent, practices in a single location.

Soils development

Ploughed soil (IX.35)

Level IX.35 followed as a period of landscape consolidation. It was a 100-150 mm deep mauve-coloured sandy soil that covered most of the excavation area but was eroded to both west and east. It was indicative of humic development in the damp, low-lying machair and was visually indistinguishable from the IX.34 horizon which overlay it. This soil is the first instance of fully developed machair meadows which have persisted to the present on the peninsula. It was subsequently developed as an extensive and intensive plough-land, probably belonging to settlement inland to the south. It also contained plough-marks (CA1-4, divided according to location (west, central, east and back).

Sandy soil (IX.34)

This horizon only partly survived subsequent flood erosion (IX.32/.33) (see above). In the western half of the site it was eroded away. In the east and inland it was undamaged but is visually indistinguishable from the IX.35 plough-land that preceded it.

Feature descriptions

Mattock marks south (CB1) and north (CB2)

A total of 22 wedge-shaped marks were noted in two slight mounds of dark material infilling Neolithic Building 1 (DJ) below the accumulation of blown sand (level X) (Figure 2.21). They were uniform in size measuring 150 mm in width by 0.5 mm in depth, with a draw-back of c. 80 mm at an angle of c. 45 degrees. They represented the use of an implement such as a mattock or draw hoe. The marks were made from the top of the white sand but cut through to the underlying darker material. As they were filled in with sand when the tool was withdrawn they presented a sharp colour contrast with the dark colour of the building infill.

FIGURE 2.21a and b:
Mattock marks in the infilling sand of Neolithic Building 1 (DJ). Looking north and overhead. © Udal project archive.

These randomly distributed marks indicate cultivation and appeared either in full plan with a square outline if the stroke was true or as right-angled triangles if skewed.

Plough marks (CA) - west (CA1), centre (CA2), east (CA3) and back (CA4)

The top of the accumulated and consolidated white sand (layer X), shows the surviving basal scars of furrow marks made by ploughing with an ard that took place once the area had accumulated turf and topsoil (Figure 2.22). The scars have survived because the topsoil was thin and the plough was drawn into the sand below. The contrasting colour of the marks is due to the darker topsoil (level IX.35) dropping into the furrow to fill it as the plough moved forward (Figure 2.23).

The entire excavated area was intensively ploughed with criss-crossed furrows. They were only absent where erosion in the west and the thinning of levels in the east left no plough marks or they were indiscernible, or removed by bowl-pit digging. Many of the furrows ran directly into the present *A' Croig Bheag* inlet and some towards its mouth to the sea. They showed beyond doubt that the early Bronze Age shore-face lay a long way north of the present one. They are one of the main indicators that no sea inlet existed here some 4,000 years ago and that there were machair grasslands with a dune system well beyond the present inlet mouth when sea levels were appreciably lower.

The plough-land (CA) itself is perforated throughout by some 70 large ritual pits of the succeeding IX.34 horizon.

Boundary post-settings (CC1, CC2.1, CC2.2 and CC3)

An area in the eastern side of the site was defined as a linear barrier between, or most likely containing, the bowl-pits. Three 0.5-0.9 m deep post-settings (CC1-3) were distributed in a line 5 to 6 m long, through this area. They were aligned NE/SW but they were not equidistantly spaced (Figure 2.24). The central post was reset (CC2.1 and CC2.2) and in doing so it skewed the alignment of the posts. The inter-post gaps originally were 1.9 m between CC1 and CC2 and 2.5m between CC2 and CC3. When the centre post was reset, the gaps became 2 m and 2.4 m respectively

A notable feature of the settings was that long packing stones were used around the posts, rather than the more ubiquitous shorter stones used elsewhere on the site, and they were positioned in a double tier. This strongly suggests that the posts were either very tall or included a superstructure, such as cross-members for example. Both factors would have required enhanced rigidity and strength. Each post was also supported by the addition of flat stones 1 m across and c. 0.3 m in height laid on the ground and capped with turf.

FIGURE 2.23: Ploughing scars surviving beneath the cairn complex.

Part 2

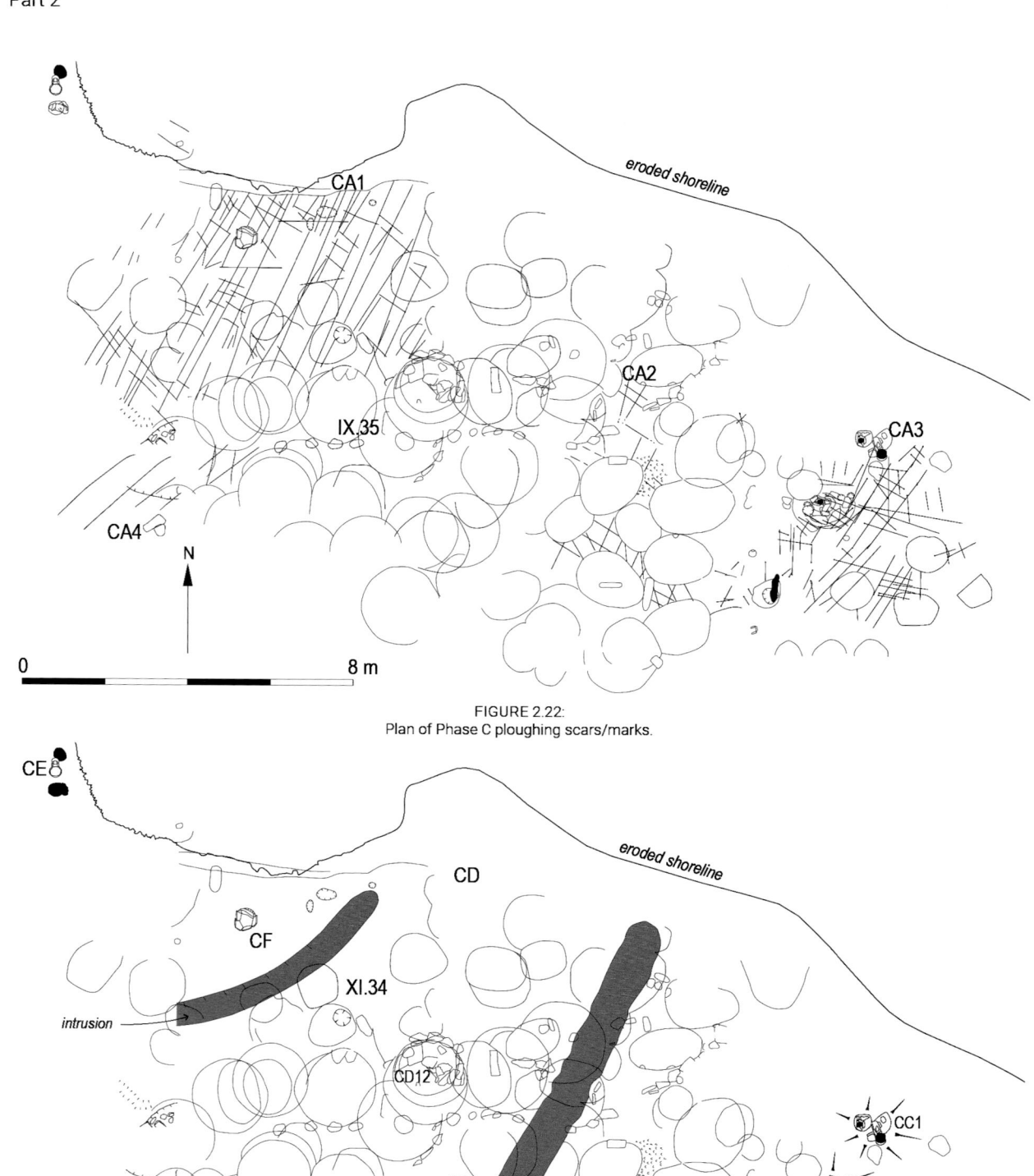

FIGURE 2.22:
Plan of Phase C ploughing scars/marks.

FIGURE 2.24:
Plan of Phase C pits and linear boundary.

There are indications that the middle post (CC2) used an isolated bowl-pit (CD29) that was dug in this area when it was reset. The stone setting of CC3 was also reused when the post was replaced by a substantial standing stone (BD) in Phase B (see below). This was the primary arrangement of this feature, with what may be the earliest bowl-pit (CD29). Crawford recorded (1980 interim report) that the line of posts may have extended north-east where two additional large postholes were found in close alignment. There is, however, no further information in the archive concerning these or their location.

Bowl-pits (CD 1-73)

This horizon contained approximately 73 massive pits dug into the sandy soil (IX.34), only one of which was stone-lined, In general, their fills contrasted only slightly in colour from the surrounding matrix (Figure 2.25). They were distributed throughout the excavated area and were considered to be focussed on the alignment of post-settings (CC1-3). In all excavated cases where a full profile survived, their usual dimensions were approximately 1.4 - 1.6 m in diameter and up to 0.6 m deep. Occasionally there were also ovoid pits which measured 1.2 by 1.6 m.

FIGURE 2.25:
A bowl-pit exposed during excavation. The darker shadows in the lighter sand show others still to be excavated. Looking north-east. © Udal project archive.

The pits, which generally had no discernible content and rarely any finds, were deliberately backfilled with layers of turf soon after being dug. Occasionally, stone from earlier stone walls were encountered in their bases when they were dug, but with CD12, Neolithic walling was reused to form an almost complete lining of the pit (Figure 2.26). The digging of the pits was remarkable in that very few were disturbed through later pit digging, suggesting that the pit diggers knew the exact location of previous pits. If this was so, there could have been surface markers of some kind, such as stones or low mounds, but the latter were undetected.

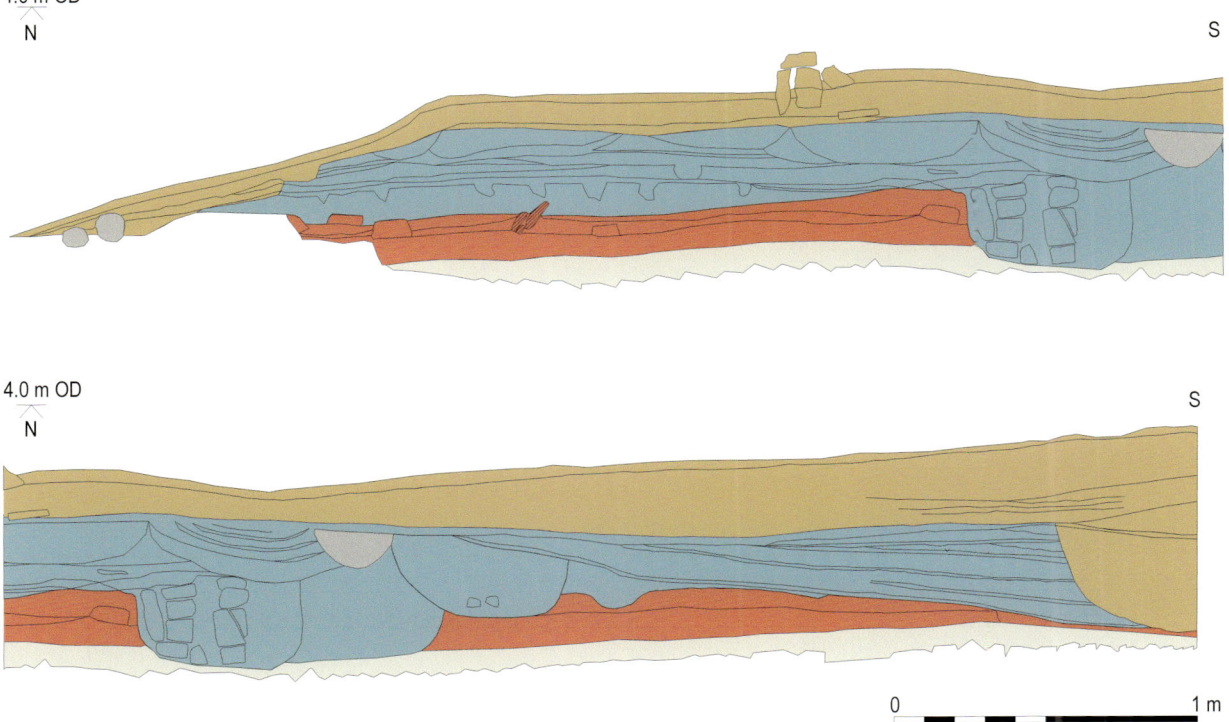

FIGURE 2.26:
Phase C bowl pits (blue) in Section 5 cutting through late Neolithic occupation (red).

At least five of the bowl pits (BE 37-39 and 71-5) are seen to be stratified in Phase B and they relate to the continued use of the area for pit digging.

Pot pit (CE)

The lower portion of a pit, 0.3 m diameter, dug into layer IX.34 contained the base of a pot (SF 23294). The upper portion of the vessel had been removed either during preparation for the construction of the Phase B cairn complex or it was removed by the severe marine erosion at the Phase B/C transition. On the north side of the pit was a 0.2 m deep posthole with packing stones (Figure 2.24).

Like CF (below), this feature is a survivor of the pre-flood horizons and activities. It is later than the level IX.34 ploughing but prior to the construction of the IX.1 cairn and is therefore a contemporary feature with the small square box (CF). The exact stratigraphic placing of these features is in doubt, as the pot pit (CE) was also considered by Crawford to belong in Phase D as part of the possible third structure (DA).

Square feature (CF)

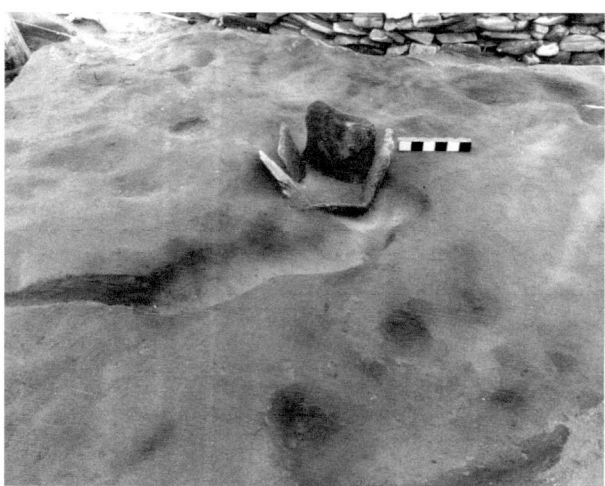

FIGURE 2.27:
Stone box (CF) that survived the last marine inundation below the cairn complex. Looking north. © Udal project archive.

A square stone box measuring 0.32 by 0.32 by 0.28 m in depth lay beneath the cairn complex approximately 1.5 m east of the large cist (BB). It was formed of fine slabs of gneiss positioned vertically, but it is not known whether it had a slab in it base. There appeared to be no top, but this could have been washed away during the scouring caused by the marine incursion (IX.32). Some of the slabs were barely supported by the IX.33 material on which it was constructed, indicating the soil matrix it was dug into had also been washed away. Its upper half was completely buried in shingle IX.2. Although Crawford thought this feature was likely to have been a small cist or cremation burial and associated with the cairn complex, it is more likely to have been a floor-set box within a domestic dwelling (Figure 2.27). Similar small boxes of this type are known from Houses 1 and 7 in the late Neolithic village of Skara Brae, Orkney (Childe 1931: Plan of Skara Brae) and Catpund, Shetland of middle to late Bronze Age date (Ballin Smith 2005: illus 5).

PHASE B

The interface between Phases B and C is marked by dramatic and relatively complex stratigraphic evidence from the earliest marine incursion (see Natural events above). The structural sequence for these events survived intact, unlike its associated soft stratigraphy. Ritual use of the eastern part of the area was uninterrupted throughout Phase C and B but the boundary between the phases was placed at the point where three post settings (CC1-3) were replaced by a standing stone (BD). Phase C was largely characterised by ephemeral monuments while Phase B started with the standing stone and continued with the kerb cairn complex, displaying stone-built and enduring monuments. The persistent nature of ritual activities at this time is underlined by the continuation of the digging of bowl pits (BE and CD) during the succeeding levels to Level VI, which reached down to the IX.34 horizon (Table 2.2).

Phase B is unusual within the whole Udal complex as it was a time of active landscape creation interacting with intensive human activity and the construction of structures. It consists of features originating in the early part of the phase that persisted as dominant features thereafter. Beneath the cairn complex, deep organic deposits of levels VIII, IX.1 and IX.2, had developed but they had been eroded away. The organic horizons ran inland to the south into unexcavated areas where they strongly suggest the presence of related settlement in that direction.[19]

The Phase A saw-pit (AA), which bisected the site, had a particularly divisive effect during this phase. The cairn complex was segregated in the west by

---

19 The RUX1 site, which was close by RUX6, was thought by Crawford to be an unbroken extension or continuum of the same events inland but this was not proven archaeologically.

FIGURE 2.29:
Burial pit on the cliff edge immediately the north of the large cist and SF 17642 skeleton. © Udal project archive.

the saw-pit from its contemporary or earlier ritual activities on the eastern side of the site. These consisted of a standing stone (BD) with a plinth, which may have been erected late in Phase C, and seemed to be the focus of the pit digging, and of ritual activities on the site throughout its existence. A small ovoid building (BG24) abutted the standing stone (BD), and had a succession of ash floors indicated its protracted use. However, its functions were terminated by the construction of a small kerbed barrow (BM) with a cist over part of it. A low dip in the landscape and lack of obvious stone between the two Bronze Age foci may well have influenced the siting of the saw-pit.

In the western half of the site levels VI and VII, included a substantial stone monument of a kerb cairn complex (BA, BB and BC) early Bronze Age date. The earliest (BA) stonework overlies the shingle bed (level IX.2) while the outermost stonework (BC) was large enough to cover the other two parts of it. However, these structures, including one cist inhumation and one robbed central pit, only partially survived modern coastal erosion but their related stratigraphic levels did not.

The upper limits of this phase are defined as level VI or what was left of it after the severe erosion event D (see above), the end of which is marked by a deposition of concentrated sifted residues which are assumed to have formerly sealed the horizon to level VII. Levels VI, VII were organic deposits which must have accumulated to a considerable depth although they subsequently were eroded off most of the excavated area (see Table 2.1 for level details).

The presence of the cairn complex (BA, BB and BC) must be regarded as the principal factor in the conservation of the earlier suite of levels at RUX6 from wind and tidal erosion Once the cairn was removed it resulted in the site becoming a tidal storm beach. This process was well under way when excavation commenced and has accelerated since 1974 by sea level rise and increased turbulence.

Soil descriptions

The soil horizons in this phase stretch from level VI at the top of the sequence to level IX.33 at its base. They include both natural depositions from erosion events (see above), the development of turf on top of soil and the build-up of soils associated with the use of structures. Their full descriptions can be found in Table 2.1.

Feature descriptions

The implication from the stratigraphy, as Crawford understood it, was that the structures on the east side of the site were slightly earlier than those on the west. However, the radiocarbon dates of the two cist burials indicate that the reverse is more accurate. The large cist burial and cairn complex on the west was the earlier of the two with a date range of 2119–1892 cal BC. The burial in the BM cist returned a date range of 1877–1658 cal BC, but the cattle skeleton was earlier at c. 1518–1409 cal BC (Table 3.1). The story is complex and is discussed in PART 6.

Features west of the saw-pit

The presence of the compact shingle deposit (IX.2) may have influenced the location and construction of the kerb cairn complex (BA, BB and BC) in level IX.I, which was built directly on top of it. It is not entirely clear from the evidence what the sequence of events was from the excavation, but the interpretation of those events produced here is different from Crawford's understanding of the structures.

The kerbed cairn complex

First burial

During Crawford's initial recording and excavation in 1974 he found a pit, which appeared in the edge of the eroded section through the cairn complex, but situated below the cairn stonework. It was dug from above the shingle (level IX.2) down through eroded and deposited layers into the blown sand of Phase C (level X) and deeper. The upper part of the pit was wide and measured at least 0.80 to 1 m in diameter as it survived, and it was c. 0.5 m deep. The pit was dug from the north into the largely natural horizons including that of level X, which formed a straight back edge to it. Close to the southern end the pit narrowed to an oval-shaped shaft of 0.30-0.40 m in diameter and c. 0.22 m deep, which had been excavated deep enough to disturb the late Neolithic levels of Building 1 (DJ) (Figure 2.28 and 2.29). The lower portion of the pit was then used for the reception of a human burial SF 17436, without the formality of a cist. When excavated in 1974 the burial place had been disturbed by the tidal erosion and the human bone that remained was disarticulated and in a poor condition (see PART 3). The backfilling material, any grave goods and any structure above the burial had been removed by the sea. Crawford found a large

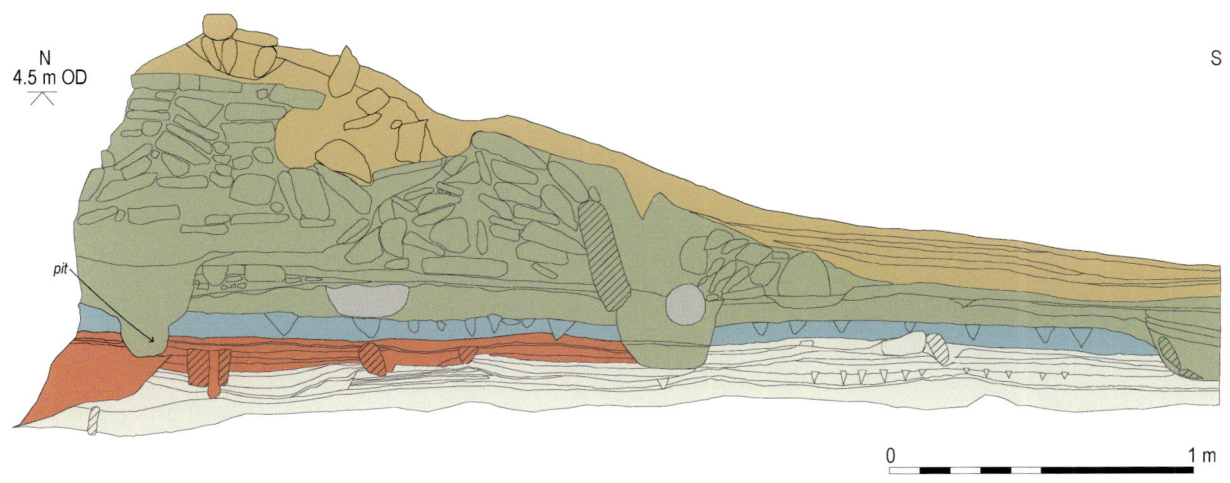

FIGURE 2.28:
Section 3 through the cairn complex. Burial SF 17436 was found in the pit dug through Phase C into Phase D at the cliff edge (left).

slab measuring 1 by 1.5 m lying on the newly eroded beach below the remains of this cairn. He thought it resembling a cist side-slab (as in BB) implying that the first cairn may have had a central cist burial.

The pit burial (SF 17436) may have been the focus for the next series of events – the construction of the cairn complex.

First cairn

FIGURE 2.31:
The kerb (BA) to the early cairn. Looking north. © Udal project archive.

A quarter arc of a kerbed cairn (BA), comprising a c. 1 m wide stone kerb with its central fill of dark sand/turf remained after the marine erosion of 1974, where it was exposed in the cliff face. The well-built kerb survived four courses high of compact dry-stone walling filled in behind with dark sand, indicating turf in its composition (Figure 2.30 and 2.31). The cairn, if it was circular, would have been approximately 6.5 m in diameter but 75-80% of it was lost to coastal erosion, and to pit digging (AB) in Phase A for retrieval of stone at its remaining west end. When constructed, its western arm would have continued over the pit containing burial SF 17436, and either the latter affected the position of the cairn, or other factors lost to the sea may have done so.

Second burial

Sometime after the first cairn (BA) had been constructed, the death of an individual (SF 17642) required burial and a new construction – a large cist and a new and larger cairn. A pit was dug 1 m deep through the floor of Neolithic Building 1 (DJ), down almost to glacial till. A rectangular cist (BB3) which measured 1.25 by 0.7 m was built inside the pit, with the gap between the pit sides and the cist backfilled to support the stones. The sides and ends of the cist were constructed of large single slabs of gneiss that were supported at the corners by smaller wedges resting on the base of the cist, with with smaller stones filling in where the larger slabs did not meet (Figure 2.32). The section drawing (Figure 2.33) indicates that the cist did not have a basal slab and the body was placed on sand in the bottom of the cist in a crouched position. It rested on its right side, with its head to the north-west and its hands resting in front of its face. There were no accompanying grave goods. A radiocarbon dated sample produced a date range of 2119–1892 cal BC for the death of the individual, indicating the person died in the early Bronze Age (see PART 3, Table 3.1).

Part 2

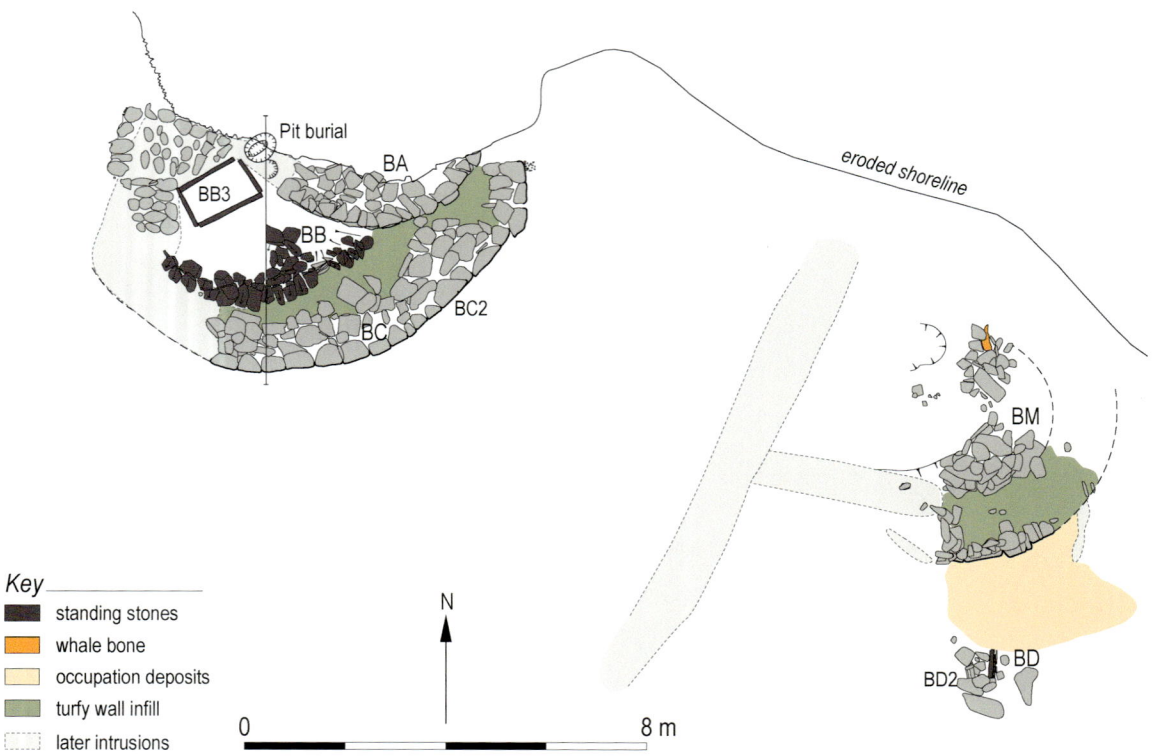

FIGURE 2.30:
Phase B cairn complex with cist, burial, earlier cairn, central domed masonry and the outer kerb (west), and cairn, cist and standing stone (east).

The backfilling of the pit around the cist was raised to meet the top of its side and end slabs. The feature was also levelled off by horizontal stones placed on top of the cist sides. A large dense stone up to 20 mm in thickness and of the right size may have already been selected for capping the cist. Such stones may have been difficult to find, and therefore its dimensions of 1.8 by 1.4 m may have determined those of the cist (Figure 2.34). Another layer of levelling stones were placed around the cist sides ready for the reception of the lid, but analysis of the fill of the feature above the body suggests that the cist may have been left open some considerable time before it was eventually capped. It was filled with different layers, colours and grades of sand and there were rodent bones, marine shell, crustacea fragments and a probable gull's nest before the capping stone was positioned to seal the cist (Figure 2.33) (see PART 4: Faunal Remains, Marine Shell and Crustacea).[20]

---

[20] Such was the weight of the lid that Crawford, during the excavation of the cist, used two rails from his light railway employed on the Udal North to remove large quantities of sand from the site, and attached a rope around the stone and dragged it to the shore using the winch on his Landrover (see PART 6, Figure 6.6).

FIGURE 2.32:
Cist (BB3) and skeleton SF 17642. © Udal project archive.

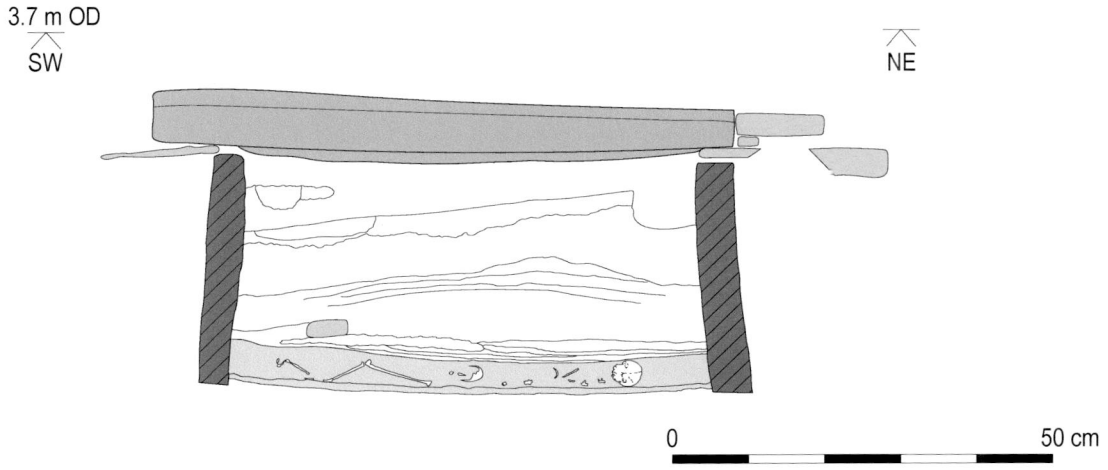

FIGURE 2.33:
Section 24 through the Phase B cist (BB3).

The inner core of stone

One the lid was in place the next stage of construction took place. It would appear from the photographic evidence but also partly from Section 3 (Figure 2.28) that a shallow trench was dug in an arc around the cist to lower the ground level in front of it (to Phase C and D levels). This enabled a row of large, 1-1.5 m long upright stones or orthostats (BB) to be placed against its inner edge in a vertical position. The orthostats curved from the kerb of the earliest cairn (BA) (Figure 2.35), which they abutted and ran alongside for 2 m, around the cist to the west, where the evidence disappeared. The orthostats formed a defined edge to the inner core of the new cairn, and behind them large and heavy water-worn stones were placed flat over the cist (Figure 2.36), but as more were used to fill in the gap to the orthostats, they were gradually angled to fill in behind the upright stones. This construction of well-pack boulders achieved a diameter of c. 5 m and a height of well in excess of 1 m to create a domed central core for the next and last phase of construction.

The final kerb and the completion of the cairn

The whole west side of this structure had been badly damaged by Phase A robber pits but enough survived to indicate that the completion of the cairn (BC) was a massive undertaking. Approximately one quarter of the kerb (BC2) to the cairn survived the marine erosion of 1974, but it was constructed to include the stonework of the first cairn (BA) within its circumference, which would have been c. 11 m in diameter.

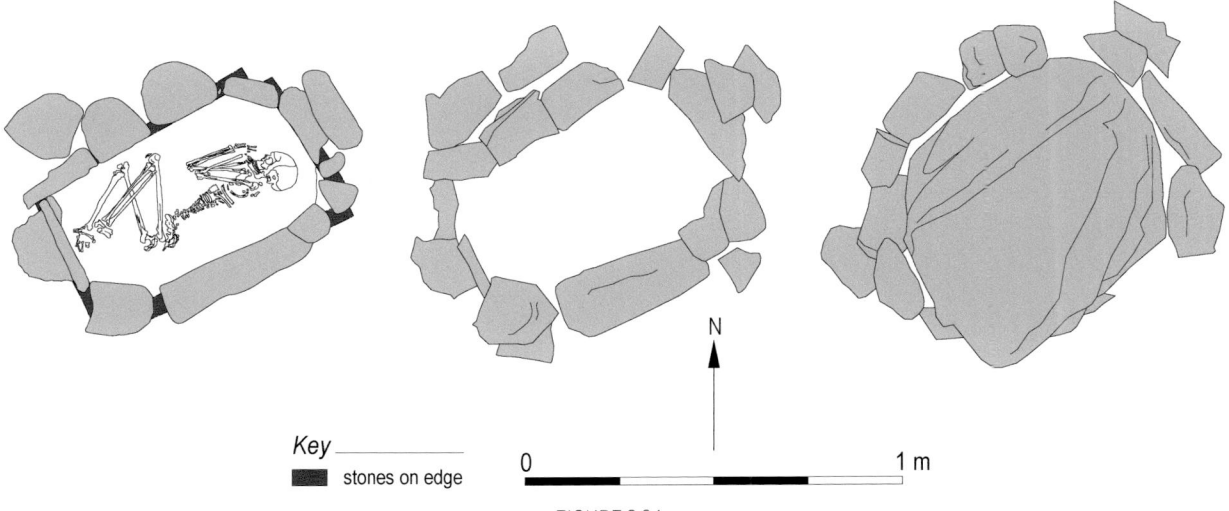

FIGURE 2.34:
Burial in the cist (left), the cist with levelling stones for the lid (centre), and with its capping (right).

FIGURE 2.35:
Cist with lid on the left of the central core of stone with the foundations for the outer kerb in the foreground. Looking north. © Udal project archive.

FIGURE 2.36:
Construction detail of the inner core of stone over the cist. Looking north-east. © Udal project archive.

The kerb (BC2) was of drystone with walling surviving up to four courses in height (0.46 m) (Figure 2.37) with the lowest course placed in a shallow foundation trench packed with turf. The c. 2 m gap between the kerb and the core of orthostats (BB) was filled in with both horizontal stones and turf, requiring an enormous amount of stone (Figure 2.38). The new cairn would have gradually sloped upwards to attain the height of its internal core and also to cover the remains of the first cairn, which may not have reached a significant height. When the cairn complex was encountered in 1974 it still measured 1.5 m in height, although significant amounts of stone had been robbed for building the kelp drying dykes of Phase A (Figure 2.39).

Late bowl-pit (BE43)

This is a bowl-pit identical to the Phase C series but unlike all others in that it is very precisely stratified within IX.2 as it underlay the cairn complex (BB and BC) (level IX.1). The pit IX.31, and its backfill included layered turf and the pebble wash of IX.31 not illustrated.

Late bowl-pits (BE70-5)

These later versions of the continuous sequence of bowl-pits are identical to the Phase C pits but far less numerous. Their scarcity may result from removal by erosion, their concentration being to the south in the unexcavated area, or to fewer of them being dug. The six examples (not illustrated) survived in the deeper deposits running inland but they lost their upper profile due to surface erosion.

Although they are technically unstratified, Crawford considered that they must relate to the last stages of level VIII, or more probably VI. However, he did not question his assumption that they were a continuously dug feature from Phase C through to the latter part of Phase B. The pits puzzled him enormously and he had no idea what they were used for.

FIGURE 2.37: Construction detail of the outer kerb against the inner core of stone. Looking north. © Udal project archive

FIGURE 2.38:
Plan of Phase B cairn complex with Phase A disturbances.

FIGURE 2.39:
The surviving height of the cairn complex in 1974 prior to deturfing. Looking east. © Udal project archive.

Features east of saw-pit

Standing stone (BD1) and plinth (BD2)

The stump of a badly damaged standing stone (BD1) and its stone and turf plinth (BD2) had been a significant monument. The standing stone had been broken at the nineteenth century ground level, probably by a mason's hammer during the excavation of the saw-pit. The remaining portion was a piece of foliated gneiss, characteristic of rock exposures on beaches in the vicinity. Its measurements were 0.6 m from base to its broken top, 0.5 m across its width, and its thickness was 0.10 - 0.15 m. It was orientated very closely to the present magnetic north.

A pit was dug some 0.4 m deep through a pre-existing posthole (CC3) to enlarge the pit, suggesting a deliberate choice as to the location and positioning of the feature. The standing stone was inserted and packed tightly with a stone of similar proportions but shorter. Two rows of chock stones were wedged in tightly to support the upright stone, which when positioned and secured were capped by flat stones around its base. The rest of the pit was packed with turf, and more turves were used to form a turf mound only 0.25-0.4 m in height but extending 3 by 2.5 m around the stone. Two large slabs, 1 m in length, were laid on the mound to the south and east to finalise the undertaking. Crawford considered that the standing stone (BD1) must have stood to some height given its intricate packing and depth and that the plinth (BD2) emphasised its striking appearance (Figure 2.24 and 2.40).

FIGURE 2.40:
The stump of standing stone (BD1) in its socket. Looking north-east. © Udal project archive.

This monument remained a dominant feature throughout the phase. Structure (BG24) and kerb barrow (BM) succeed the erection of the standing stone but are constructed in association with it.

Bowl-pits (BE5, 37, 38 and 39)

These examples of bowl-pits survived in the deeper deposits running inland in the southern part of the excavated area (not illustrated). Crawford thought they may relate to the last surviving stages of levels IX, VIII or more probably VI, and therefore they were later in date than the Phase C examples.

Structure (BG24)

An oval structure located immediately north of the standing stone (BD) was constructed in a hollow in level IX.2. It was faced by a single thin row of pointed upright stones (BG2), c. 0.3 m high on its west side, with other small stones denoting its southern, eastern and northern extremities. The structure did not have a broad or coursed wall and appeared to be light and possibly temporary in construction (see below). During the excavation approximately seven postholes (BJ) were noted in the floor, three of them were deeply angled. This suggests that the structure's roof supports may have been replaced or moved on a number of occasions. Some of the smaller postholes lay beyond the edge of the floor, one to the north, one to the south-west and one to the south-east indicating additional external supports for a roof. Three larger postholes were situated in the centre of the structure and a fourth towards the south-west edge of the floor (Figure 2.41 and 2.42).

Its measurements indicated the structure was c. 5 m in length by c. 3 m in width. Some of the stones bordering the floor were repositioned a number of times indicating the outline of the structure changed over time. They outlined an area of compact ash floors (BG4.1-3) that diminished in extent with each use. The east and north sides of the structure were damaged by erosion, although the eastern floor edges were well-defined as they curled almost vertically up the inner face of the stones. This profile indicated that the floor in centre of the structure was cleaned out at intervals.

The southern extent of this feature was identified by a pile of flat stones and turves (BG1), which directly overlay the plinth of the standing stone (BD), along with its latest floor (BG4.1), indicating it was stratigraphically later.

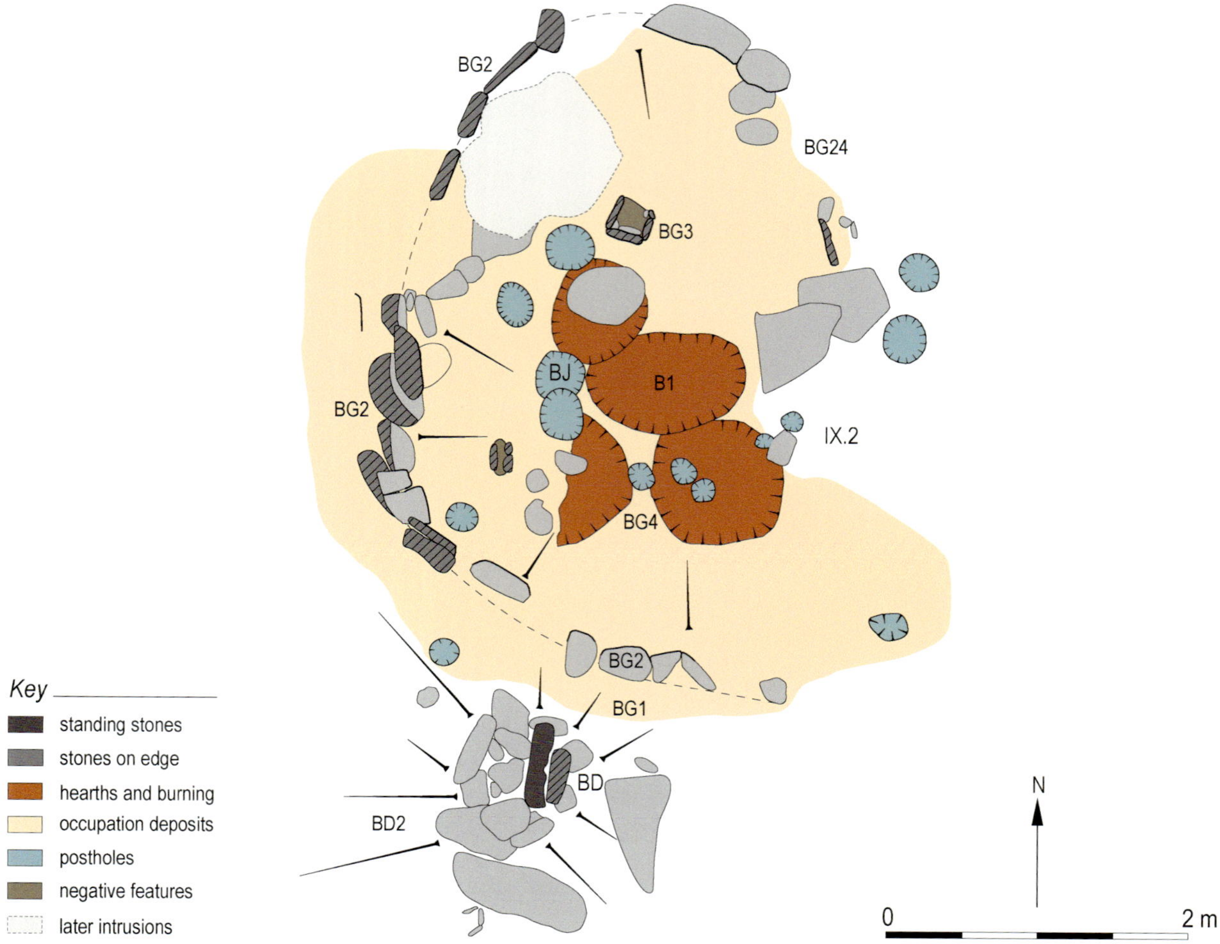
BG2
BG24
BG3
BJ
B1
BG2
IX.2
BG4
BG2
BG1
BD
BD2
N
0
2 m
Key
standing stones
stones on edge
hearths and burning
occupation deposits
postholes
negative features
later intrusions

FIGURE 2.41:
Plan of the temporary structure (BB24), postholes, hearth pits and stone edging.

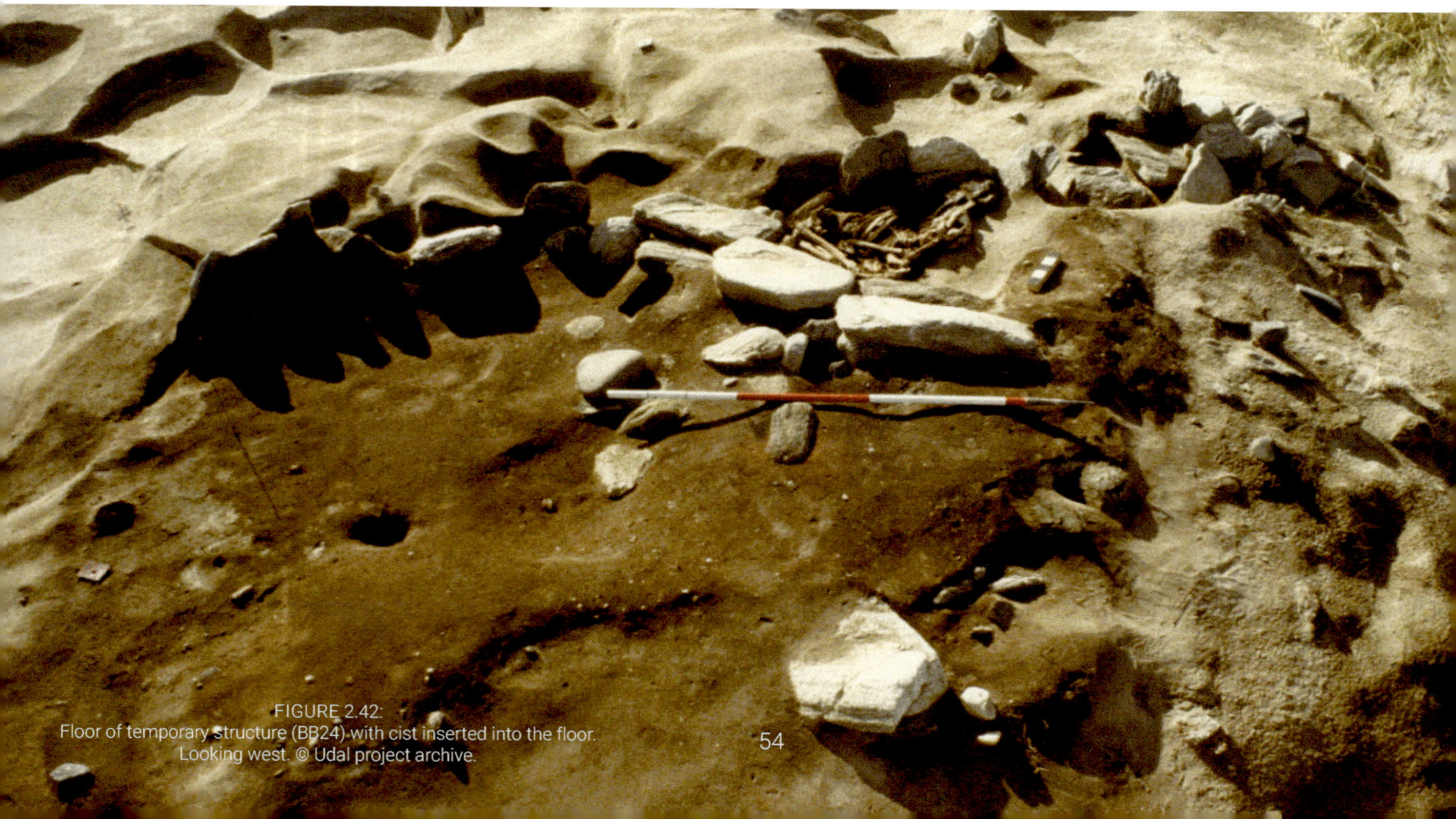

FIGURE 2.42:
Floor of temporary structure (BB24) with cist inserted into the floor.
Looking west. © Udal project archive.

FIGURE 2.43:
Stone box (BG3) in floor used for quartz raw material cache. © Udal project archive.

The structure contained at least three successive floors of ash, which were maroon, red and black in colour (BG4.1-3) and appeared to contain little occupation debris, except for a number of fire-cracked cobbles. The middle floor spilled outside the structure through a possible entrance and towards the standing stone (BD) to the south. There was no defined hearth slab but beneath the compacted floors there was evidence of up to five fire pits packed with ash that measured c. 1 m in diameter (BJ1), obviously as one filled up a new one was dug. Small negative depressions in the floor may have been stake-holes possibly used with the fire pits. Also positioned in the floor towards the northern end of the structure was a roughly 0.3 m square box (BG3) constructed of cobbles and flat stones, filled with stored quartz (a raw material cache, SF 26241) (Figure 2.41 and 2.43).

Kerbed cairn with cist (BM)

Against the north-western perimeter of structure BG24, a shallow rectangular cist (BM) that was constructed of upright stones, which measured 1.2 by 2 m was inserted into the floor (Figure 2.44). A crouched inhumation (SF 26319) was buried within it, along with a calf (SF 26321) placed in front of the east side slab and partly over the skeleton, and a smashed but complete pot (SF 26259) laid besides the west side stone (Figure 2.44 and 2.45). The burial lay in a thick dark chocolate-brown deposit with particles of crushed limpet shell, together with three small bone points. Several stones capped the cist in order to seal it and a saddle quern (SF 26508) presumably from the BG24 structure was reused to support the cist sides. More stone was laid on top of the cist but that was not sufficient to prevent recent disturbance. Unfortunately, this small burial chamber suffered considerable damage from marine erosion after 1974. Due to scouring, the north end slab disappeared, the roof collapsed and the side slabs moved out of position partially crushing the skeleton inside.

The cist was stratigraphically later than structure BG24. The hollow where the structure had been, plus the availability of some loose stone from it and other features in the area, may have provided a fortuitous location for a burial. The cist was placed in the centre of a relatively small cairn of c. 3 m diameter.

A kerb (BM3), formed of four courses of drystone walling, was laid through the middle of the old structure (BG24) (Figure 2.46 and 2.47), and presumably the construction carried on in a circuit to form a ring of stone around the cist (Figure 2.48).

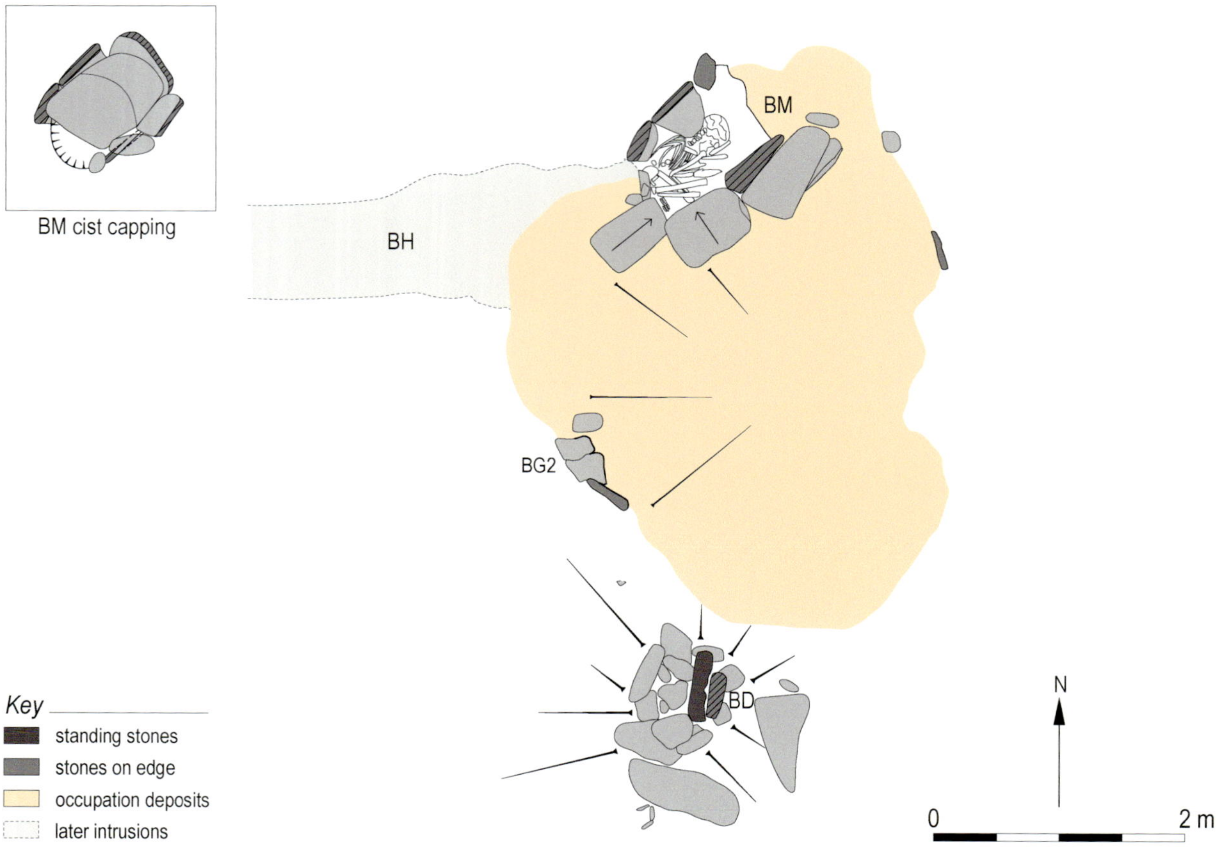
BM cist capping
BM
BH
BG2
BD
N
Key
standing stones
stones on edge
occupation deposits
later intrusions
0
2 m

FIGURE 2.44:
Plan of cist (BM) with skeleton SF 26319 inserted into the side of the temporary building.

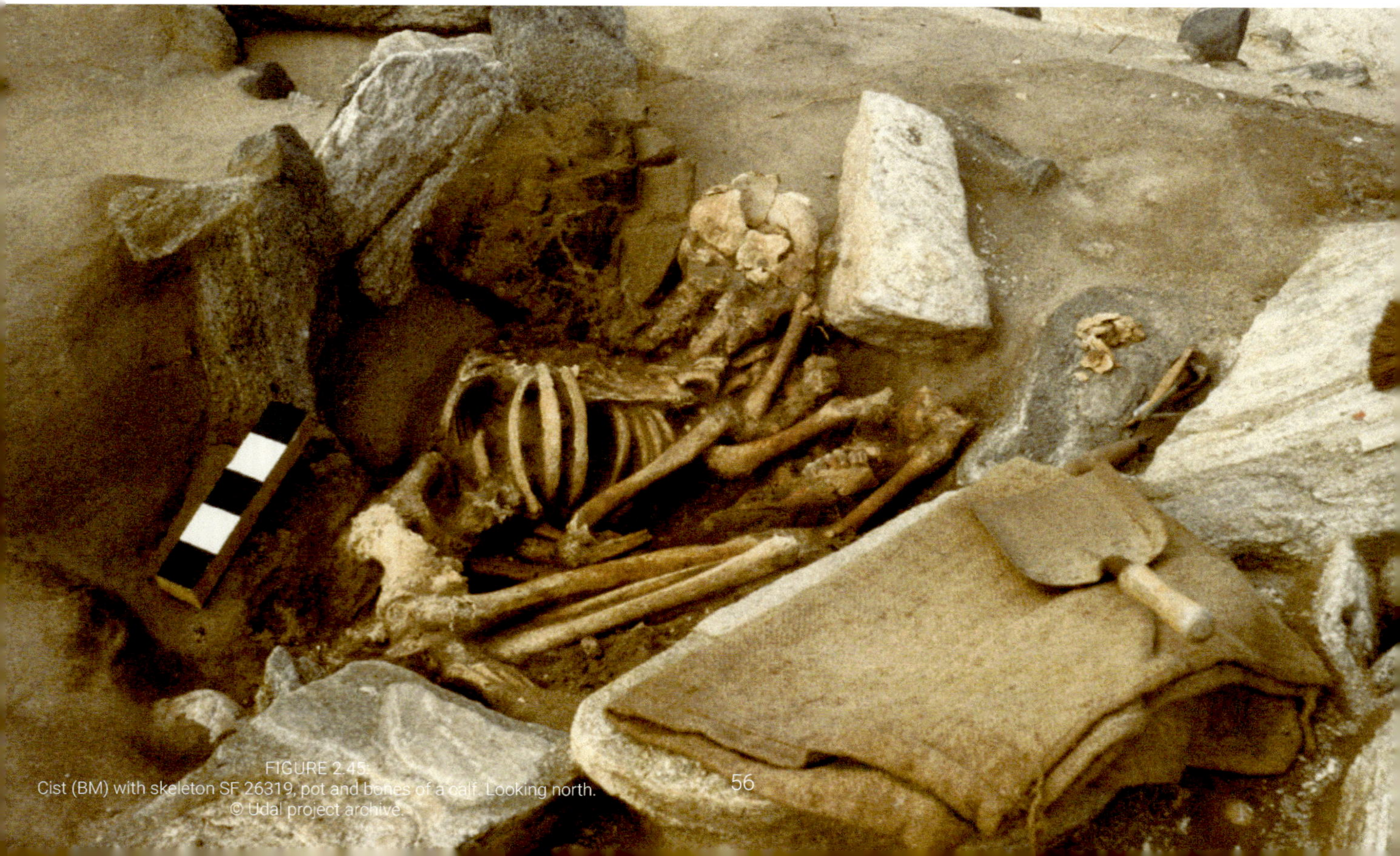

FIGURE 2.45:
Cist (BM) with skeleton SF 26319, pot and bones of a calf. Looking north.
© Udal project archive.

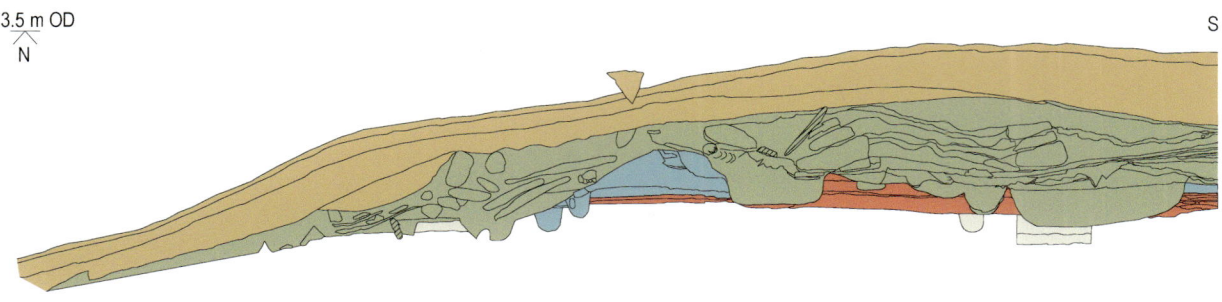

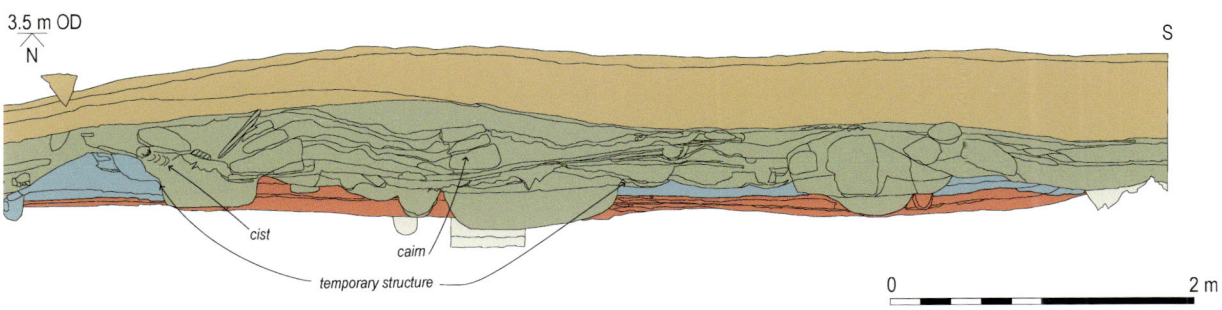

FIGURE 2.46:
Section 7 through BG24 cist, cairn and temporary structure (BG24) (green).

The area between the kerb and the stone over the cist was filled with turves and sand to form a low barrow-like mound (BM2) edged with stone. This might imply that there was a shortage of available stone for a complete cairn. Most of the structure suffered damage from either coastal erosion, or subsequent stone robbing, possibly in Phase A, and only the lowest section of the kerb survived in the hollow of BG24. This burial complex was built in level IX.1 and covered by level VIII deposits.

FIGURE 2.47:
Kerb of cairn (BM) coursing through the temporary building. Looking north-west. © Udal project archive.

Linear trench (BH)

An ephemeral linear feature aligned E/W, which measured c. 0.5 m deep by 0.5 m wide, was traced for over 10 m from the north-west side of the BM cairn, due west (Figure 2.44). Dark circular blotches along the bottom of the feature may represent the ephemeral remains of post or stake-holes. Its function is unknown, but given the presence of the later saw-pit, this may have been an earlier one, or even a robber trench to remove stone from the cairn.

PHASE A

Summary

Phase A differs from earlier phases in that the natural processes of erosion, accretion and deposition became dominant. The build-up of substantial nineteenth/early twentieth century AD machair horizons, levels I to V (Table 2.2), comprised all deposits found in situ above the uppermost surviving horizons (level V) of the degraded surfaces of the early Bronze Age. This material also included cliff-face tumble, erosion, rodent damage and crofting excavation including the excavation of a saw-pit, which provided deposits of mixed periods, all classified as Phase A.

Indications are that under post-mediaeval conditions of climatic deterioration, erosion and depletion processes dominated and this pattern can be observed elsewhere on the machair. A terminus to this erosion period was recorded with rare precision on RUX6, as in 1898 a saw-pit was dug down from the contemporary land surface, through the degraded surface of the early Bronze Age horizons and into the late Neolithic levels.

Level V was the interface between Phases A and B. It was an erosion-enriched horizon comprising the redeposited 'concentrate' of the contents of all the missing degraded soil horizons down to early Bronze Age levels. It included the residues of horizons of which there was no surviving evidence, but which could have derived from any period between the middle to late Bronze Age and the nineteenth century AD. These vanished horizons may still survive in the unexcavated areas immediately adjacent inland, as RUX1 perhaps indicates.

The content of level V was very slight but without stone and was consistent with the view that human activity on *Rubha an Udail*, with the exception of ploughing, had moved inland some 200 m onto a prominent ridge 10 m high that had been building up during the early Bronze Age. The degradation and erosion of the land surface during the post-medieval period led to the (re)exposure of the latest Bronze Age monuments on RUX6, rendering them vulnerable to human interference.

Level V sealed earlier deposits and represented the final truncation horizon following the erosion of a now unascertainable depth of material, perhaps in a number of episodes. It is the final and only recordable horizon, being late nineteenth century AD in date (Table 2.1). Nevertheless, the first twenty years of the twentieth century was also a period of renewed sand deposition. This can clearly be observed on RUX6 as one to two metres of fine sand horizons accumulated interspersed with kelp ash horizons. This terminated in *c.* AD 1926 when an exceptionally well-dated kelp drying dyke was constructed on stable turf. This situation continued until 1974 when erosion recommenced and the salvage excavation became necessary.

Erosion events A, B, E and F were linked with stages in the subsequent build-up of new deposits and ground horizons from about 1900 to 1974, representing as they do very finely dated machair formation and degradation. These events are significant as they are rarely recorded and they are relevant to the recent geomorphological history of the area.

In addition to the processes described above, a number of important but random human factors occurred, including elements of crofting activity (cottage industries) during the late nineteenth century. Crucial to the build-up of the Phase A deposits was a kelping focus based on the plentiful supplies of tangle from the adjacent bay, *A' Croig Beag*. This example of a comprehensive kelp 'industry' and of a kiln, drying dykes and heavy ash concentrations, probably represents the entire township unit. The spread of ash clearly helped the build-up and stability of Phase A.

The major disturbance of the site, due to its length and its depth, was the saw-pit (feature AA) sited close to ship wrecks of wooden vessels for which evidence can still be seen after tidal scour. The wrecks provided the planking essential for the resettlement of the township of Grenetote at the turn of the nineteenth/twentieth century. The pit neatly bisected the excavation area in an NE/SW direction and provided a perfect stratigraphic section through the site. It seems most likely that a mason's hammer was used to smash through a standing stone (BD) that impeded the digging of the pit and hindered the saw working.

Some erosion of the turf-covered Bronze Age monument probably took place in the eighteenth-nineteenth centuries which revealed its stone construction. Due it its exposure a massive robber pit over 3 m in diameter was dug into the top of the Bronze Age cairn (feature BC) and down through its centre. The pit was backfilled with cairn debris but is otherwise an undateable event. Next, and probably contemporary with late nineteenth century exposure as well as the timber and kelp activity described above, another robber pit was dug into the cairn for stone. Finally, *sgeirean* or *bile* (kelp drying dykes) were constructed over the site, almost certainly using the highly suitable small slabs and boulders of the cairn as a quarry. A kelp dyke was built on top of the remains of the cairn by Mr D. MacInnes from Grenetote. The six kelp dykes on the site and in the immediate vicinity were calculated to have used some 600 robbed stones. In addition to the drying dykes and ashes, a stone-lined hollow on the shore by the low cliff face indicated the location of a kelp burning pit. The continued accumulation of anthropogenic deposits within those of geomorphological origin probably ended when the settlement focus moved to Grenetote. Crawford was armed with a significant amount of local history of recent events and the features described here cover the latest period at the site from about 1880-1910 to the present day.

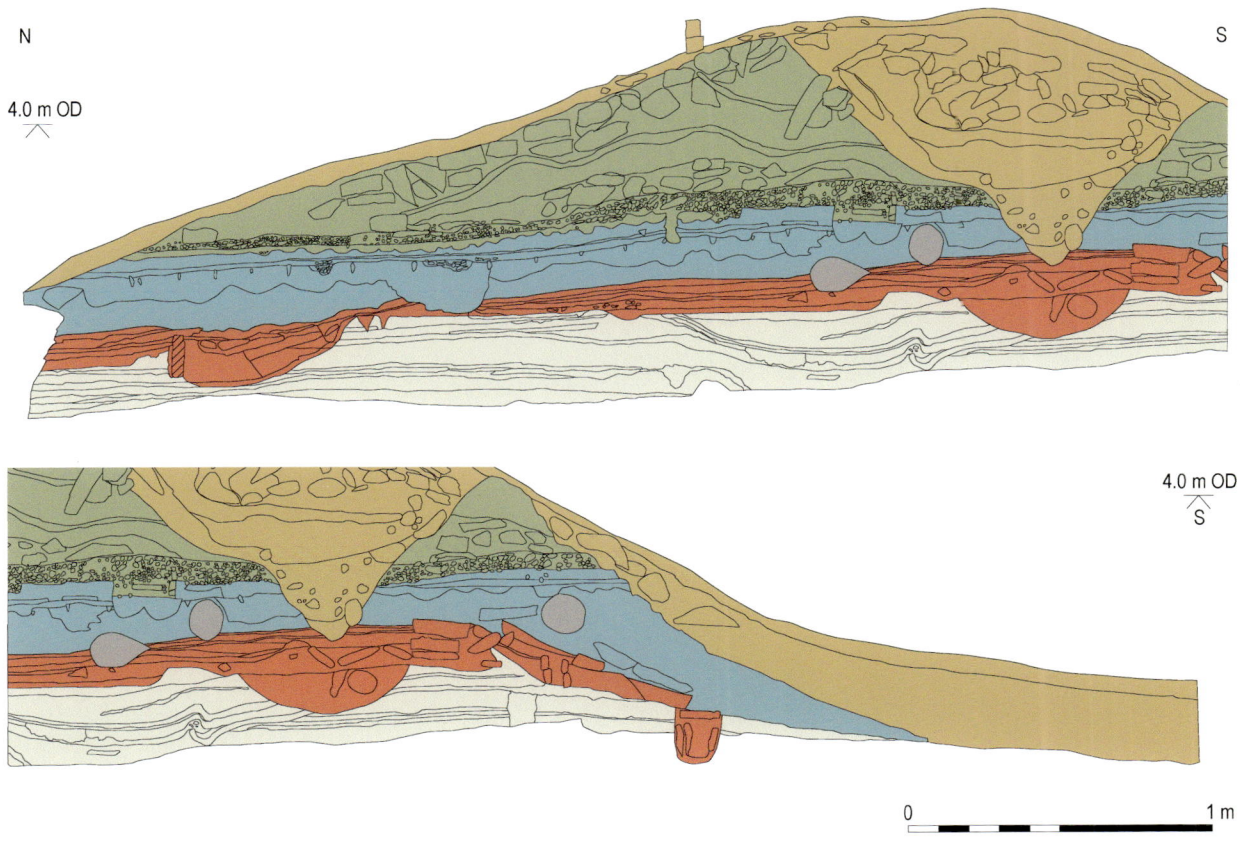

FIGURE 2.48:
Section 14 Phases A (brown), B (green), C (blue), D (red) and E (grey).

Levels (Table 2.1)

Levels I-V seen in the upper part of Section 14 (Figure 2.48), show the erosion interface between the uppermost early Bronze Age levels exposed most likely in c. 1900 AD by a saw-pit and the build-up during modern times of some four levels or horizons of blown sand with kelp burning ash. Across most areas of the site they had attained a depth of 2 m by 1974 except for the upper slopes of the cairns, where they were much thinner.

Features

Saw-pit (AA) (Figure 2.49)

This was a U-shaped trench oriented NE/SW that was a minimum of 0.8 m broad and c. 0.6 m deep dug down to bedrock and extended into the unexcavated south-west area of the site. It survived 13 m long and bisected the excavation into two unequal parts, a west (2/3) and east (1/3), and was identified as a saw-pit trench constructed in the late nineteenth century. Given the length of the feature, it is likely that cross supports were necessary for the timbers being sawn and it is suggested that the broken off shaft (AL) of the standing stone (BD) supported such a cross strut.

The saw-pit was dug after level VI was deposited (see Table 2.1), and spoil from it created a heap (AK), behind a low berm. On abandonment the feature was partly filled in by a collapse of Bronze Age levels from the west side, spoil that crept back into the pit from the east, and also by drifting sand (level IV.4). Finds located in its fill were from all the site phases that it cut through. The saw-pit provided two excellent sections right through the site but it was not realised when it was first encountered that it caused severe damage to the prehistoric structures.

Historical comment

Saw-pits probably had a limited life span and their disuse and subsequent infill by drifting sand followed rapidly, as both a timber supply (driftwood) and its need were transitory. According to oral tradition (Donald MacInnes from Grenetote pers. comm, in the autumn of 1980 to Iain Crawford), a series of

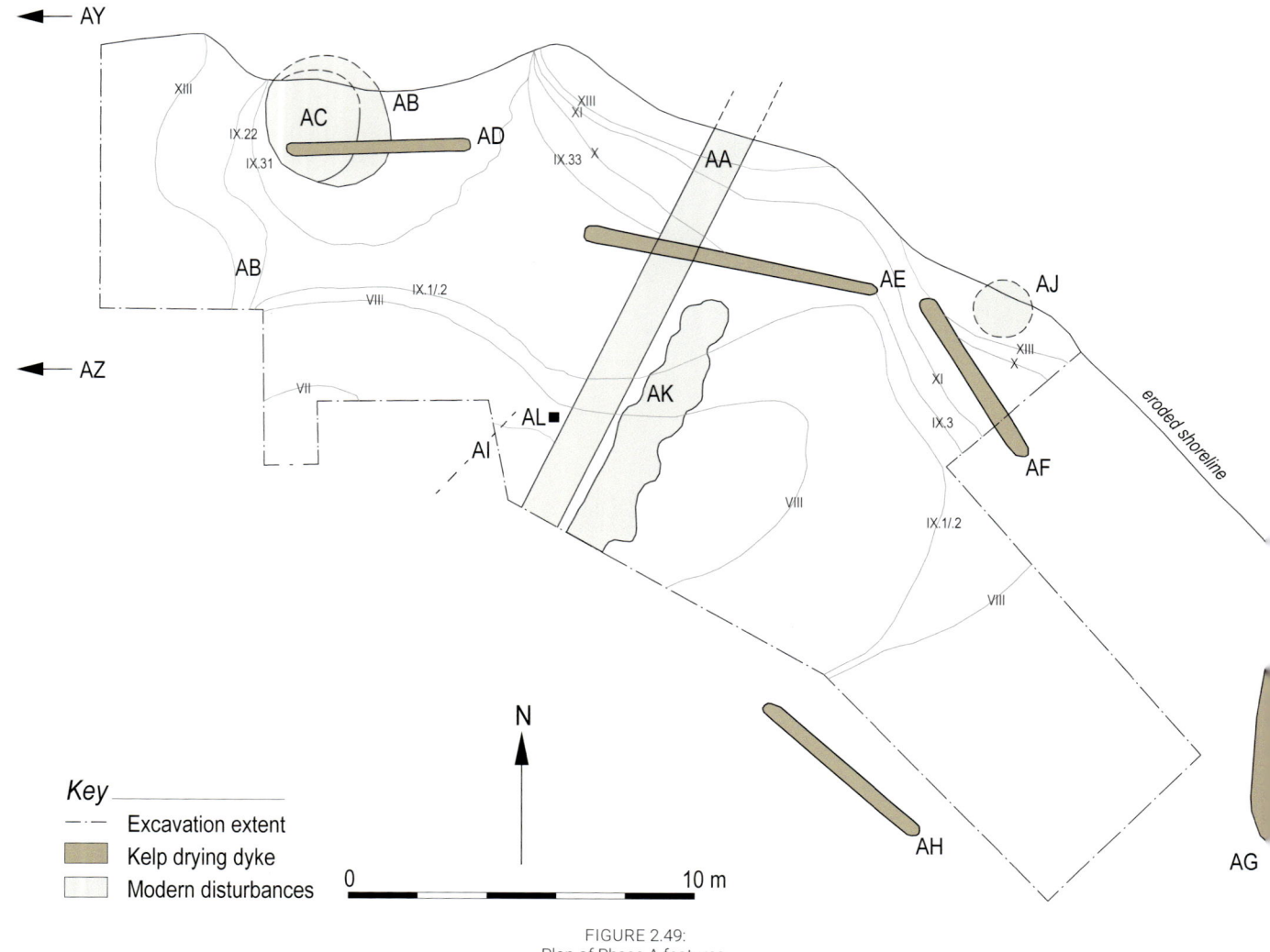

FIGURE 2.49:
Plan of Phase A features.

saw-pits were opened up in 1898. Their purpose was to utilise the resources of the substantial sailing ship wrecks lying along the west coastline of the *Ard a' Mhorrain* for building timber. The initiative was developed by the urgent need for planking to roof the proposed croft houses to be built as Grenetote was currently being resettled, after mid-nineteenth century clearance by the Congested Districts Board. The settlers were from Hougarry on the west side of North Uist.

Elements of these massive ships timbers, with large bronze bolts still in place, were seen by Crawford lying on the adjacent turf horizons in the mid-1960s. One wreck still lies embedded in the inter-tidal sand near the now tidal island of *Huilis* (NGR: NF 8265 7870) and is occasionally scoured and exposed at low tide.

Robber pit (AB) (Figure 2.49)

A splayed bowl-shaped pit or crater, c. 3 to 4 m in diameter and 1.5 to 2 m depth was dug through the centre of the stonework of the cairn (BC) complex and tapered to a butt-end suggestive of a trench. It created a deep bottomed crater in the estimated centre of the cairn and resembled the classic form of robber intrusion looking for grave goods in the centre of an artificial mound.

The lower portion of the feature reached the centre of the cairn complex, where a pit with a burial (Square G) resulted in the removal of all elements except for a scatter of human bone fragments left in the backfill of the pit. The indications are that the pit was either filled back in deliberately or that its spoil may have slipped or slumped back in.

The digging of the pit is not dated but would have been well known locally and it is supposed that the activity took place sufficiently prior to the next and unsuccessful attempt at digging a robber pit (AC) that it was either undetectable on the surface or forgotten about. It dates from any period following completion of the cairn (BC) until the crofting industry of recent times. However, it is most likely that this and the other robber pit (AC) belong to the industrial period of timber cutting and kelp drying.

Stone robbing pit (AC) (Figure 2.49)

This pit was c. 3 m in diameter but shallower than the earlier robber pit (AB) into which it was dug. It only attained a depth of c. 0.5 m and was probably naturally filled in with drifted sand. There is no further information on this feature except that it is later in date than AB and may have been another attempt at taking stone from the cairn.

Kelp-drying dyke (AD) (Figure 2.49)

A stone dyke aligned E/W was built on top of the crest and down side of the mound of the cairn complex (BA, BB and BC). It measured 5.4 m long and 0.5-0.6 m wide and was built of three courses with the second, middle course leaning. It was built by Donald MacInnes and his father Norman in 1924 (Donald MacInnnes of Grenetote, pers. comm). It is likely that this and other dykes were constructed in part by material produced from the stone robbing pits in the cairn complex and from stones along the shore. In this instance the construction of the dyke required the deliberate infilling of the robber pits (AB and AC) on the cairn with smaller stone and sand before its foundation could be laid. Unlike other examples (see below) this was of one build. It was also one of the latest built on the site and one of the last to be used in North Uist. It belonged to croft no. 20.

Kelp drying dyke (AE) (Figure 2.49)

This disused dyke was 8.5 m long and 0.5-0.6 m wide. It comprised three courses and lay partly under the level I turf and topsoil and was situated almost parallel to the coast between the dyke on the cairn (AD) and one to its immediate east (AF), which may have formed part of a drying system with it. It was noted when it was excavated that it had acquired another course of stone during its history to enable better drying conditions for the kelp.

Kelp drying dyke (AF) (Figure 2.49)

This dyke was built to the immediate east of AE, and was slightly shorter at 6 m in length, but the width of 0.5-0.6 m was maintained. Three courses of it survived, with the topmost added later. It too was aligned more or less parallel to the coast, and was the closest of the drying dykes to the kelp kiln or pit situated immediately north on the coastal edge. Its location to the latter may suggest it was one of the earliest of the dykes to be built, and demonstrated the necessity and close relationship of both features in the kelp industry.

Kelp drying dyke (AG) (Figure 2.49)

This dyke was situated just beyond the eastern extent of the excavated areas. It was slightly curved, oriented N/S in contrast to the other dykes and ran up the sand hill to its crest. Although it was not excavated it was measured to be 7.2 m long by 0.5-0.6 m wide, and was considered by Crawford to be one of the earliest dykes in this area and most likely for kelp drying.

Kelp drying dyke (AH) (Figure 2.49)

A final dyke was recorded south of the south-eastern most extent of the excavated area. It was not excavated but it was recorded as oriented NW/SE with a length of 6 m and width of 0.5-0.6 m.

Possible kelp drying dyke trench (AI) (Figure 2.49)

The robbed out foundation trench to a possible kelp drying dyke survived in the southernmost part of the middle of the excavated area. Its width was 0.5-0.6 m but no stone was present in the trench, only white sand, and its length could not be ascertained. If it had been a dyke it may have been robbed for the construction of another in a better location.

Historical comment

It is clear that as the dykes became overgrown their function was revived by the addition of another course. This enabled them to maintain their capacity to dry kelp above the turf and in full exposure to the wind and the sun. This is certainly true of dykes AE and AF, which were sectioned during the process of excavation. The remainder, AG, AH AI, and AZ were still visible in level I but probably out of use.

Kelp drying dykes were constructed through much of the latter part of this phase but it is not certain that all were constructed for the drying of kelp. All the dykes were visible and freestanding in or within level I (although, there is no additional photographic evidence surviving in the archive to confirm this). Kelp working was an ancillary crofting industry closely associated with seaweed collection and the factory processing of the alginate. On the Hebrides this industry persisted at a low level into the post-war years (1939-46) (Crawford 1962:106). In Harris and Barra chopped tangle was still to be seen on dykes as late as the 1960s, but this may have been intended as fertiliser for the machair. In total there were 24 kelp drying dykes on *Rubha an Udail* and the association with the 23 crofts In Grenetote, the home township, is that each croft may have had its own drying dyke.

Kelp kiln (AJ) (Figure 2.49)

This was an ill-defined but irregular sub-circular feature c. 2 m in diameter comprising orange and grey ash and fire cracked stones. It was an exploitation of an exposure of bedrock at the top of the shore, which was formalised by some slight flagstone enhancement to make it into a workable unit. Although unstratified in relationship to the rest of the site, the kiln is assumed to be the source of ash noted in the sand and organic horizons close by. Its purpose was for the communal firing of the dried kelp from the drying dykes to produce an alkaline ash for sale to the glass industry.

Saw-pit spoil (AK) (Figure 2.49)

The spoil from the digging out of the saw-pit (AA), was thrown parallel to the east side of its trench where it survived as a long low mound c. 2 m in width by c. 0.20 m in height and at least 9 m in length. Its north (seaward) end had been eroded away but its landward end disappeared into the deepening sands of unexcavated deposits. A narrow berm prevented it from drifting or slipping back into the saw-pit but that failed as the sides of the pit collapsed. It comprised mixed sand and soils from all levels but especially levels V-XV. A number of finds from the nineteenth century back to the Neolithic were noted in this feature.

Shaft of standing stone (AL)

This was a slab of light grey, gneiss that survived 0.4 m in length and measured 0.6 m in width by 100-130 mm in depth. It had been fractured at both ends and was noted to match the surviving in situ stub of an early Bronze Age standing stone (BD). The slab was probably a piece prised from a nearby rock outcrop and it may have been a recognisable monument up to the time of the industrial activity on the site at the turn of the nineteenth/twentieth century.

The remains of the standing stone (BD) were encountered during the digging of the saw-pit, where it was broken to remove an obstacle and facilitate the use of the pit. The broken fraction was either used to shore up the collapsing side of the saw-pit, eventually slipping into it, or used to support a cross piece.

Fresh water spring (AY)

A cleft in, and against, the bedrock formations on the shore front was identified as a possible spring. Its location was 20 m beyond the western edge of the site, and Crawford noted it as situated barely above HWMOST. Water flowed into a feature that measured 3 m in length by 0.5 m in width and whose depth of 0.2 to 0.4 m was variable. The feature was naturally dammed by sediments to form a fresh water pool fed by run-off, and had long been regarded by the local community as a well or spring and was much in use in modern times to water cattle and horses.

There were obvious grounds in thinking that a well or spring was a factor in the location of early settlement. However, the Machair Research Group working party who visited the site in September 1993, established the source of this water as run-off from *Rubha an Udail* that accumulated at this point due to the configuration of the underlying bedrock. Its relation to the early site location is uncertain as the availability of fresh water would then have been dependent on sand and rock configurations to an earlier, lower shoreline, and also to the height of ground water. These factors would have dictated whether fresh water run off became a seepage line onto the beach, or a small basin for water accumulation.

Unfortunately, the improvement of water flow was considered possible by a Grenetote resident who used a stick of dynamite to widen the supposed orifice. The result was a reduction in flow as the accumulation point was shattered.

AZ Kelp drying dyke

A typical dyke 9 m long and 0.5-0.6 m wide was noted to the west of the site. Its orientation was not recorded but it was considered a good example of the type, and was free of excavation disturbance (Figure 1.6).

PART 3 Dating and human remains

Early dating of the site

By Beverley Ballin Smith

From 1981 to 1994 Iain Crawford sent 17 samples for radiocarbon dating to the Quaternary Research Laboratory in Cambridge. The samples were mainly from the floor and hearth of the two late Neolithic buildings (DH and DJ), from the human remains in both cists (BB3 and BM), and the temporary structure on the east side of the site (BG24). Only some of the material, which included charcoal, marine shell, human and animal bone was dated. Most of the samples were probably never processed, and those that were, appear to have been lost with the closure of the laboratory.

During the course of the RUX6 post-excavation it was possible to re-date the same structures and burials using the best organic samples that were available. The dates produced are discussed below, and later in PART 6.

Radiocarbon dating and Bayesian modelling

By Derek Hamilton

A series of 16 radiocarbon dates are available from features associated with Neolithic and Bronze Age occupation at RUX6. The samples were pretreated, combusted, graphitised and measured by accelerator mass spectrometry at the Scottish Universities Environmental Research Centre, as described in Dunbar et al. (2016). The results have been calibrated from the conventional radiocarbon ages (Stuiver and Polach 1977) and are quoted in accordance with the international standard known as the Trondheim convention (Stuiver and Kra 1986). All results are given in Figure 3.1 with calibrations calculated using the probability method of Stuiver and Reimer (1993).

Stable isotopes and marine correction

The radiocarbon measurements on the whale vertebra and two bones from human skeletons were not calibrated using a terrestrial calibration curve (IntCal13). The whale was calibrated using the Marine13 curve (Reimer et al. 2013), while the human bones were calibrated using a linear mixed-modelling approach to calibration, detailed in Cook et al. (2015). The local reservoir correction (ΔR) used was derived from the work of Russell et al. (2015) for the Scottish coast, and is −47 ±52 ^{14}C years. Using dietary δ^{13}C end members of −21.0‰ for a wholly terrestrial diet and −12.5‰ for marine diet (Arneborg et al. 1999), the linear interpolation method produced an estimate of 21.2% marine diet for SUERC-69666 and an estimate of 24.7% for SUERC-69665. These numbers were used, in OxCal, to combine this percentage of the Marine13 curve to the IntCal13 curve, with a 1σ error of ±10 years, using the Mix_Curves function.

The samples and model

The samples consisted of human and animal remains, as well as charred cereal grains, charcoal, and charred food crusts on pottery sherds. Most of the samples were recorded as having come from pits or similar features, or occupation deposits. The two phases of activity were separated by sterile blown sand layers, so the dating for the Neolithic and Bronze Age is considered separately as well.

Sample Nr	Lab Code	δ¹³C	Context	Radiocarbon Age BP	95.4% probability
17642	SUERC-69665	-18.9‰	Human tooth from large cist below cairn (BB3)	3618±31	2119–1892 cal BC
26319	SUERC-69666	-19.2‰	Human remains in cist (BG24 structure) level IX.1	3432±30	1877–1658 cal BC
23557	SUERC-70211	-20.9‰	Sheep bone from floor of DJ structure level XI.01	3909±32	2474–2297 cal BC
23659/ 23667/ 23686	SUERC-70212	-21.0‰	Sheep bone from pit in DJ structure level XI.02	3432±32	2566–2307 cal BC
23829	SUERC-70216	-28.0‰	Carbonised food residues from pottery from floor of DJ structure level XI.02	4181±32	2888–2640 cal BC
25442	SUERC-70217	-25.8‰	Carbonised food residues from pottery from floor of DH structure level XI.01	4008±32	2618–2467 cal BC
26024	SUERC-70218	-21.7‰	Cattle bones from floor 1 and hearth of DH structure level XI.031	3859±32	2462–2209 cal BC
26298	SUERC-70219	-23.7‰	Carbonised barley seeds from BG4 structure, level floors IX, 1/2.2	3378±32	1751–1566 cal BC
26331	SUERC-70220	-29.4‰	Hazel charcoal from BG24 structure level IX.22, SE Quadrant	3457±32	1880–1691 cal BC
26364	SUERC-70221	-27.5‰	Hazel charcoal from BG24 structure level IX.2, NE Quadrant floors 1/2	3391±32	1759–1651 cal BC
26364	SUERC-70222	-25.1‰	Birch charcoal from BG24 structure level IX.2, NE Quadrant floors 1/2	3266±32	1622–1456 cal BC
26321	SUERC-71140	-20.7‰	Cattle bones from cist (BG 24 structure)	3188±32	1518–1409 cal BC
26831	SUERC-71141	-20.8‰	Sheep/goat molar from beneath wall of DJ structure level XI.2	3981±35	2580–2350 cal BC
26855	SUERC-71142	-20.9‰	Sheep/goat bone from beneath hearth of DH structure, level XI.2	4000±33	2617–2464 cal BC
26888	SUERC-71143	-13.3‰	Whale bone on plinth (DC) level XI.22	4176±33	2432–2195 cal BC
26893	SUERC-71147	-21.4‰	Cattle/deer bone within plinth shaft (DC) level XI.22/ib	3651±32	2136–1939 cal BC

TABLE 3.1:
Radiocarbon dates

Neolithic

There are four radiocarbon dates from Building 1 (DJ). A sheep/goat molar was recovered from under the stone wall of the building and the date (SUERC-71141) provides a *terminus post quem* for wall construction and house occupation. Two samples come from deposits in the lower occupation levels. There is a date (SUERC-70212) on a bone (SF 23686) from a sheep buried in a pit in the floor; and a charred food residue on a pot that was recovered from in or near the hearth produced SUERC-70216. The residue is significantly older than the other dates from the structure. While it could be a residual sherd, further possibilities are that either the residue was from the cooked remains of marine fish or mammal or that the sherd body contained old carbon that was incorporated into the sample when it was scraped from the surface. It is not possible to be sure the mechanism by which the date came to be too old, and so it has been excluded from the modelling. While the first two samples were in the lower fills of the building, SUERC-70211, from an articulated ruminant, dates the uppermost floor level in the structure.

There are three radiocarbon results from Building 2 (DH). The first date (SUERC-70218) is from the articulated remains of a cow, while the second date (SUERC-70217) is from a charred food residue on a pottery sherd. A sheep/goat astragalus from beneath the slab of the final hearth provides a further date (SUERC-71142) for its occupation.

Bronze Age

There are four dates from charred plant remains recovered from the floor levels of BG24, a temporary structure in the eastern part of the site. A fragment of hazel roundwood charcoal was dated (SUERC-70220) from the lowest floor. Three samples came from deposits at the interface of the middle and upper floor levels: a single barley grain (SUERC-70219), a fragment of hazel roundwood charcoal (SUERC-70221), and a fragment of birch roundwood charcoal (SUERC-70222).

There are two dates from the nearby cist constructed into the filled in temporary structure. SUERC-69666 is from the tooth of a poorly preserved human, while a cow long bone that was recovered next to the inhumed individual produced SUERC-71140. The tooth from an individual interred in a cist under a kerbed cairn in the west of the site was also dated (SUERC-69665).

Finally, there are two dates (SUERC-71147 and SUERC-71143) from Structure DC, a shaft of natural bedrock with a structured deposit containing a cow/deer ulna fragment placed below a whale vertebra. Even after correcting the date from the whale bone for a marine reservoir effect (MRE), the result is considerably older than the deer bone. This would suggest the vertebra was found by the inhabitants at RUX6 and used in this closing deposit, or it was intentionally curated for approximately two centuries. The result has been excluded from the modelling, since the death of the whale is not chronologically related to the deposition of its vertebra.

The results

A Bayesian approach has been adopted for the interpretation of the chronology for activity dated at RUX6 (Buck *et al.* 1996). Bayesian statistics provides a method of allowing different types of information (e.g. radiocarbon dates, phasing, and stratigraphy) to be combined to produce realistic estimates of calendar dates. The technique used is a form of Markov Chain Monte Carlo sampling, and has been applied using the program OxCal v4.2. Details of the algorithms employed by this program are available from the on-line manual or in Bronk Ramsey (1995, 1998, 2001 and 2009).

The algorithm used in the model described below can be derived directly from the model structure shown in Figure 3.1. The calibrated radiocarbon dates are shown in outline and the posterior density estimates produced by the chronological modelling are shown in solid black. Highest posterior density intervals used to summarise these distributions in the text are given in italics to reflect the fact that they are modelled, and emphasise that they are not absolute and would change given a different set of parameters or 'prior' beliefs.

The dated activities from the Neolithic and Bronze Age have been separated in the model as sequential but not contiguous. The ordering of the samples, based on observed and/or inferred stratigraphic relationships, has been included as described above.

The primary model for RUX6 has good agreement between the radiocarbon dates and archaeology that has been described (Amodel=111).

It estimates that the Neolithic activity began in *2700–2470 cal BC* (*95% probability*; Figure 3.1: *start: Neolithic RUX6 (main)*), and probably in *2585–2485 cal BC* (*68% probability*). It lasted for *20–475 years* (*95% probability*; Figure 3.2: *span: Neolithic RUX6 (main)*), and probably for *60–305 years* (*68% probability*). The Neolithic activity ended in *2465–2165 cal BC* (*95% probability*; Figure 3.1: *end: Neolithic RUX6 (main)*), and probably in *2450–2295 cal BC* (*68% probability*).

The Bronze Age activity began in *2270–1925 cal BC* (*95% probability*; Figure 3.1: *start: Bronze Age RUX6 (main)*), and probably in *2125–1965 cal BC* (*68% probability*). This activity ended in *1550–1215 cal BC* (*95% probability*; Figure 3.1: *end: Bronze Age RUX6 (main)*), and probably in *1495–1380 cal BC* (*68% probability*). The overall period of Bronze Age activity at RUX6 was *445–930 years* (*95% probability*; Figure 3.2: *span: Bronze Age RUX6 (main)*), and probably *510–740 years* (*68% probability*).

One concern was the seemingly early dates for the two carbonised residues on the pottery sherds. An alternative model was constructed that applied a marine reservoir correction to these two dates, assuming that perhaps the food stuffs cooked in the pots was fish or some other form of marine protein. This model has good agreement (Amodel=85) between the assumptions and the dates.

It estimates that the Neolithic activity began in *2780–2420 cal BC* (*95% probability*; Figure 3.3: *start: Neolithic RUX6 (sensitivity)*), and probably in *2605–2480 cal BC* (*68% probability*). This activity ended in *2220–1975 cal BC* (*95% probability*; Figure 3.3: *end: Neolithic RUX6 (sensitivity)*), and probably in *2155–2030 cal BC* (*68% probability*). The overall period of Neolithic activity at RUX6 was *255–710 years* (*95% probability*; Figure 3.4: *span: Neolithic RUX6 (sensitivity)*), and probably *360–560 years* (*68% probability*).

The Bronze Age activity began in *2125–1920 cal BC* (*95% probability*; Figure 3.3: *start: Bronze Age RUX6 (sensitivity)*), and probably in *2055–1960 cal BC* (*68% probability*). It lasted for *430–810 years* (*95% probability*; Figure 3.4: *span: Bronze Age RUX6 (sensitivity)*), and probably for *500–665 years* (*68% probability*). The Bronze Age activity ended in *1560–1240 cal BC* (*95% probability*; Figure 3.3: *end: Bronze Age RUX6 (sensitivity)*), and probably in *1495–1380 cal BC* (*68% probability*).

Discussion

The dating of the charred residues has reinforced the difficulty in interpreting the results from this sample type. A model that accepts both dates as accurately reflecting when people lived in the two Neolithic buildings at RUX6 has a low agreement (Amodel=25), which is largely the result of the obviously 'early' result on the sherd in Building 1 (SUERC-70216). As noted above, when that result is excluded the model has good agreement. Furthermore, the sensitivity analysis that calibrates these two dates using the Marine13 curve also has good agreement, suggesting the possibility that the charred food residues were possibly from cooking marine fish or mammals. Although the use of the Marine13 curve brings the dates from the sherd in Building 1 (SUERC-70216) more accurately in line with the other dates from this feature, the date from the sherd in Building 2 (SUERC-70717) becomes rather late compared to the other dates in that house. It is possible that the charred residue on the sherd in Building 1 reflects the cooking of marine animals, while the residue on the sherd in Building 2 reflects the cooking of terrestrial plant of animal products. Without undertaking residue analysis, this question remains open to debate, and so the main model (Figures 3.1 and 3.2) for the RUX6 site is the preferred chronology.

The burials on the site

by Beverley Ballin Smith

The human remains from the site have not been an easy subject to study but the authors of the following sections, Julia Beaumont, Solange Bohling, Jo Buckberry and Cassandra Hall have investigated them using detailed osteological and isotopic analyses.

Most of the human remains from the whole of the Udal project were sent to the Duckworth Laboratory in Cambridge by Iain Crawford as he had links with the university and the area. A colleague of his at the then Department of Physical Anthropology, C. Bernard Denstone, undertook preliminary analysis of the material but his work was never published. Dr Jo Buckberry at the Biological Anthropology Research Centre, Archaeological Sciences, University of Bradford was asked to identify and analyse the RUX6 remains, together with additional material found during the assessment between 2010 and 2012. Re-dating of the human remains was considered and undertaken (see previous section), and the possibilities opened up for the isotopic analysis of the skeletons.

The problem was that it was uncertain where all the skeletons had been originally located on the site. No written narrative or discussion of them had been found in the site archive except in the annual reports issued by Crawford after his excavation seasons. The best preserved, SF 17642, had been removed from its burial place within a large cist in 1974 when the stonework forming the cairn above it was removed during the first archaeological intervention on the site. Its location was straightforward, as a good photographic record of the cist, its lid and the skeleton survived along, with the site section drawings and plans.

The location of the second and less well-preserved skeleton (SF 26319) was less clear-cut. There had been continuing natural depletion of the site since 1974 and a cist with its contents had been found at the eroding coastal edge but it was not in good condition. It was discovered and excavated in 1983, but its location on the east side of the site was not immediately identifiable from the site plans or photographs. Crawford had fortunately written

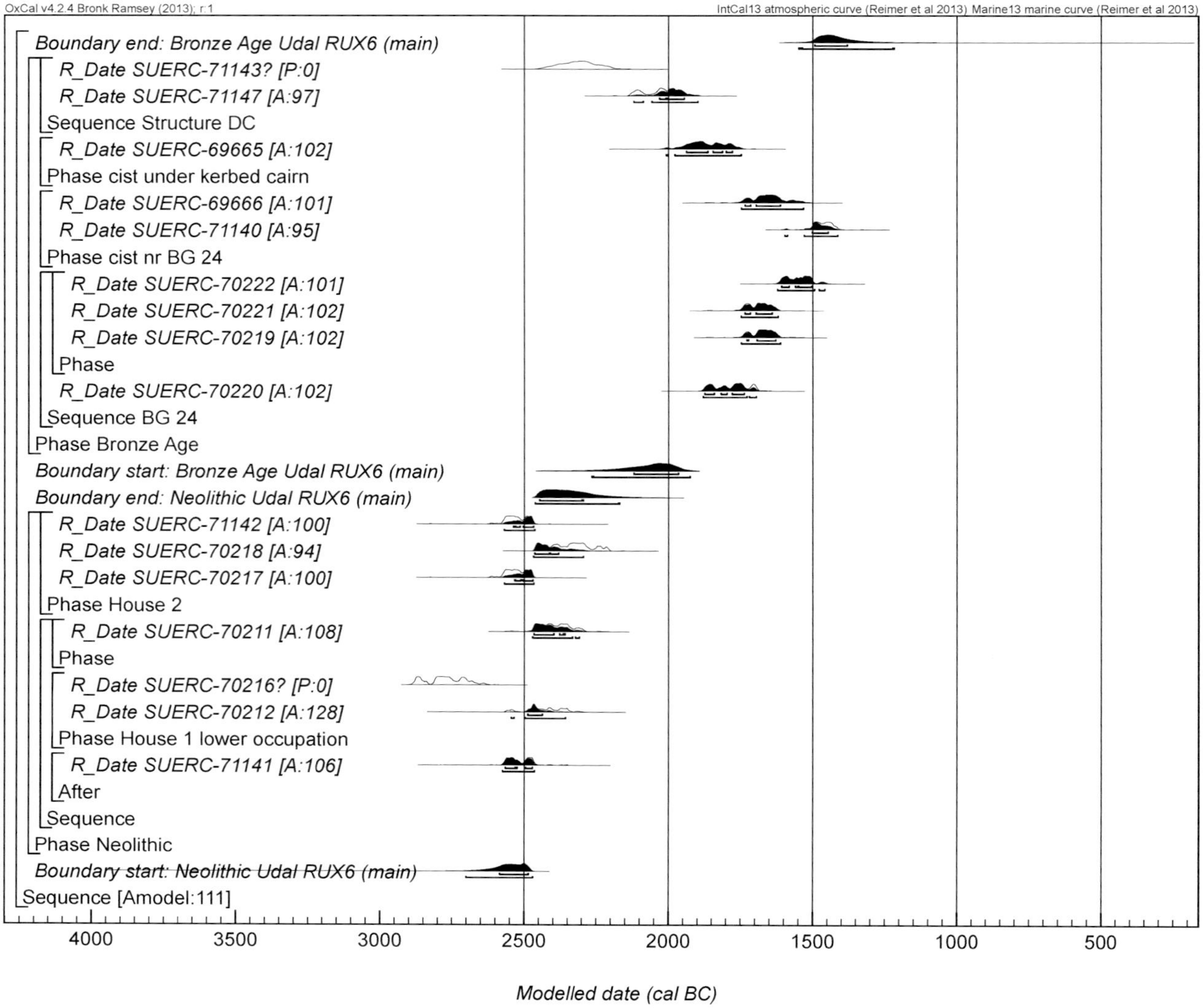

Part 3

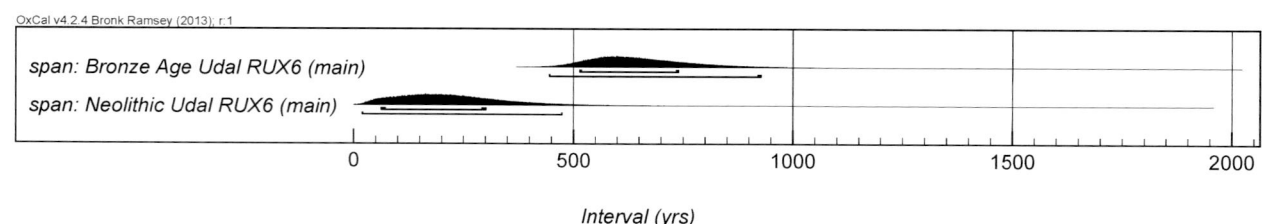

FIGURE 3.1:
Chronological model for RUX6. Each distribution represents the relative probability that an event occurs at a particular time. The distributions in outline show the calibration of each result by the probability method (Stuiver and Reimer 1993). The solid distributions are *posterior density estimates* derived from the chronological model. This model is exactly defined by the square brackets and OxCal keywords at the left of the diagram.

FIGURE 3.2:
Spans for the modelled activity at RUX6. The spans are based on the model shown in Figure 3.1.

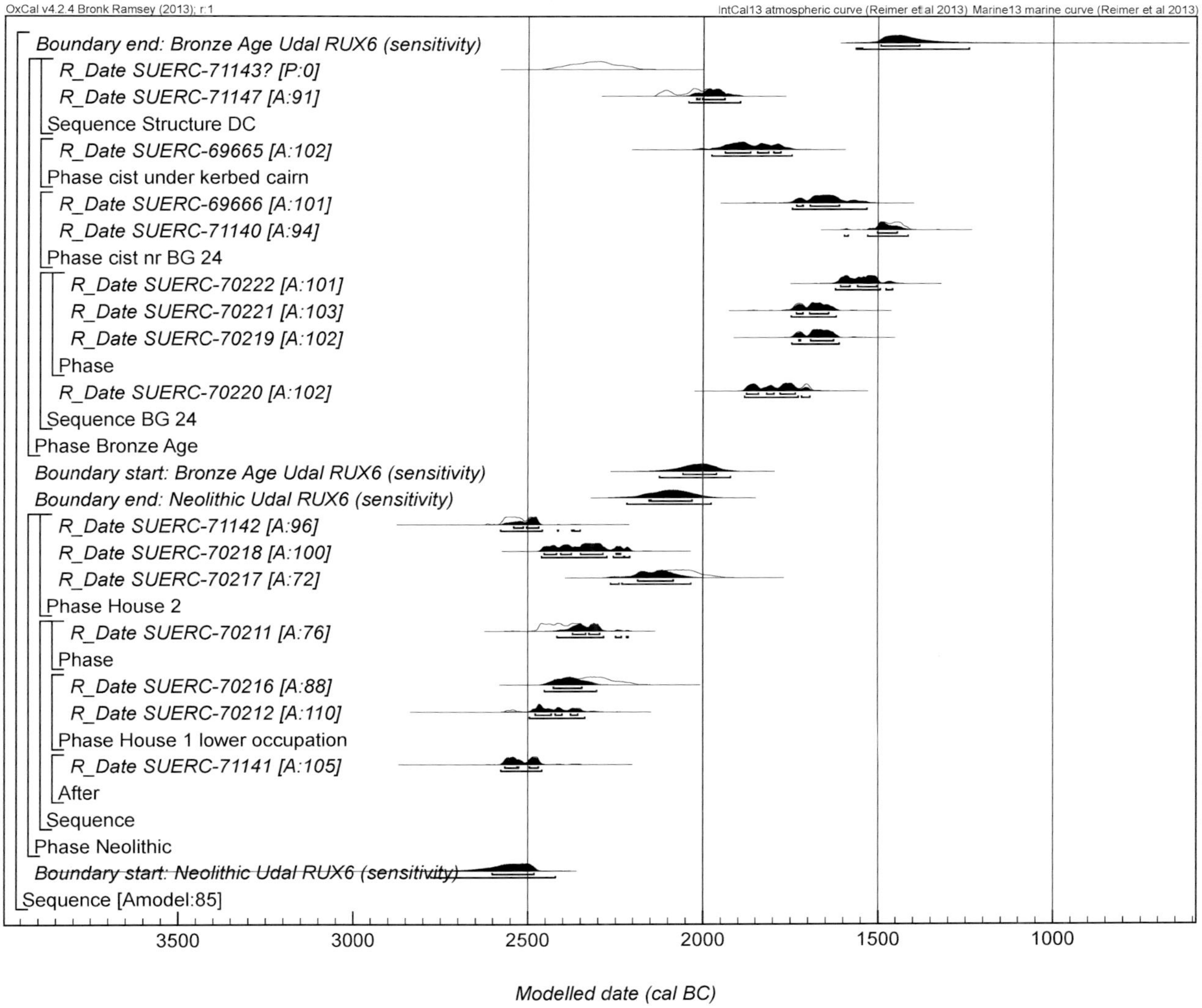

FIGURE 3.3:
Alternative chronological for the dates from RUX6, with the two charred food residue dates included as deriving from marine-based protein. The model is as described in Figure 3.1.

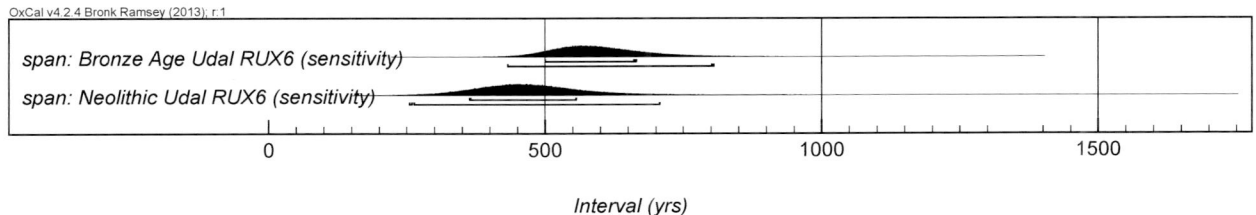

FIGURE 3.4:
Spans for the modelled activity from RUX6, based on the sensitivity analysis. The spans are based on the model shown in Figure 3.3.

about its discovery in the 1983 annual report and this was the fullest account there was. A summary plan, drawn in ink was used to illustrate the annual report. It showed the location and outlines of the Phase B features and stonework across the site and also included a second cist and an outline of how the skeleton was positioned. The photographic record was interrogated again, and the cist and its contents were eventually identified and confirmed as being that of SF 26319.

During the analysis of the human remains by Bohling and Buckberry the remains of a possible third burial were discovered amongst the disarticulated bone from the site. The question posed by them and Hall was, were did these remains come from? More precisely, where was Square G, as all the material identified as possibly belonging together had been excavated from this grid square in 1974. It was clear from this specific investigation that Crawford was still using imperial measurements at that date and worked in 12 by 12 foot grid squares (see PART 1), but there was no record of the positioning of the grid or its labelling. Further interrogation of the 1974 annual report indicated a 'central inhumation was in a pit disarticulated and partly removed by the tide' and that it was located below the west cairn and presumably right on the coast. Was this SF 17436? The site plans and sections produced nothing, but a series of four photographs of the main cist for SF 17642 revealed an excavated pit directly to its north (Figure 2.29), lying in the cliff edge, with only a few centimetres between it and the next eroding tide. A summary plan in the 1974 annual report confirmed the location of 'PIT INHUM CAIRN3', which has been identified as the only possible location of this third burial. Interestingly, this feature on the plan had been highlighted by Crawford, as if he too had problems locating it. There was no written or drawn record of the position of the bones, and some could have been lost to erosion prior to excavation. Crawford also suggests (PART 2 Robber pit AB) that the burial could have been disturbed by the deep digging of a stone-robbing hole. The position and shape of the pit was subsequently confirmed at the north end of Section 3 (Figure 2.28) after comparison with the site photographs. The position of this informal burial in a pit within the cairn is described in PART 2. The informality of this burial contrasts with the large cist and inhumation SF 17642, and although a sample of bone was taken for radiocarbon dating, its condition was so poor that the process failed. It could be argued that this was the earliest burial recorded on the site but this and other matters concerning the people who inhabited the site is discussed further in PART 6.

Skeletal analysis

By Solange Bohling and Jo Buckberry

Introduction

Two early Bronze Age individuals (SF 26319 and SF 17642) and 19 samples of disarticulated remains from RUX6 were analysed and inventoried following procedures and standards set out by the British Association for Biological Anthropology and Osteoarchaeology (BABAO) and the Chartered Institute for Archaeologists (CIfA) (Brickley and McKinley 2004).

Preservation

Individual SF 26319 was poorly preserved with 50-75% of the skeletal material remaining. The bone was extremely dry, very brittle, and friable. The skull was mostly present but in fragments. The thoracic and lumbar regions of the spine were completely absent and the cervical region was highly fragmented. The ribs, os coxae, hands, and feet were 50-75% complete but were highly fragmented and poorly preserved. All long bones were present with almost no preservation of the proximal or distal joint surfaces

Individual SF 17642 had good preservation with more than 75% of the skeletal remains present. Most of the cranium was intact and many of the long bones complete. However, the torso region was poorly preserved and no ribs and few vertebrae were recovered.

Sex assessment

The sex of each individual was assessed utilising the morphology of the pelvis and skull (Buikstra and Ubelaker 1994; Walker 2008; Klales et al. 2012), relying more on the pelvis as shape differences are population independent (Mays and Cox 2000; Steyn and Patriquin 2009; Moore 2013). For individual SF 26319, both ossa coxae were too badly damaged to be utilised for sex assessment and the skull was also fragmented. While cranial sex assessment is not as accurate as pelvic sex assessment, the fragments of the cranium were overall extremely robust thus individual SF 26319 was probably male. For individual SF 17642 both the pelvis and cranium were well preserved and both the assessments using the skull and pelvis agreed, indicating individual SF 17642 was male.

Age estimation

The age of each individual was estimated based on macroscopic methods involving the pubic symphysis (Brooks and Suchey 1990), auricular surface (Buckberry and Chamberlain 2002), cranial sutures (Meindl and Lovejoy 1985), and dentition (Brothwell 1981). The method outlined by Brothwell (1981) is thought to be appropriate for all British archaeological populations from the Neolithic to the medieval period, so this method was applicable to the individuals from Udal. Transition analysis (Boldsen et al. 2002), which is especially useful for estimating the age of fragmented remains, was also applied to both individuals.

As individual SF 26319 was so badly preserved, age estimation techniques involving the pubic symphysis, auricular surface, and cranial sutures were not possible. The dentition was intact so Brothwell (1981) could be applied and provided an age range of c. 33–45 years. However due to the severe wear on all non-molar teeth, the age of individual SF 26319 may have been over 45 years. Transition analysis was also applied providing an age range of 28.6 to 110 years, with a maximum likelihood age of 77.5 years. However, it must be noted that due to the fragmentary state of individual SF 26319, only two traits could be scored. Overall, individual SF 26319 was probably over 45 years old.

For individual SF 17642, the sternal ends of the claviculae and the sacral vertebrae were not fully fused giving an age range of <29 years and <32 years respectively (Scheuer and Black 2000). The age of individual SF 17642 was also estimated utilising the pubic symphysis (Brooks and Suchey 1990), which provided an age range of 19–34 years. The skull was still mostly articulated thus age estimation using cranial sutures was applied providing an age range of 18–45 years for the vault sutures, and 19–48 years for the lateral-anterior sutures (Meindl and Lovejoy 1985). The dentition was very well preserved, and provided an age range of c. 17–25 years (Brothwell 1981). Transition analysis provided an age range of 19.6 to 25.9 years with a maximum likelihood age of 19.6 years (Boldsen et al. 2002).

With these methods combined individual SF 17642 was probably a young adult, aged between c. 19 and 25 years.

Metric and non-metric analysis

Metric and non-metric traits were recorded (following Buikstra and Ubelaker 1994; Berry and Berry 1967; Finnegan 1978), to allow for comparison with other populations, however little significance can be placed on these traits for such a small population. Individual SF 26319 was too badly preserved to take any measurements except for two from the tibiae. The cranial and post-cranial measurements and non-metric traits for both individuals are included in the archived report.

A series of cranial and post-cranial indices were calculated for individual SF 17642 following Bass (2005), which describe the shape of the cranium, nasal aperture, orbits, palate, femora, and tibiae (Table 3.2). Some of the indices mentioned in Bass (2005) could not be calculated due to damage.

Index	Value	Range	Description
Cranial Index	77.9	Mesocrany	Average or medium
Fronto-Parietal Index	69.5	Metriometopic	Average or medium
Nasal Index	51	Mesorrhiny	Average or medium
Orbital Index	78.6	Chamaeconchy	Wide orbits
Maxillo-Alveolar Index	121.2	Brachycrany	Broad palate
Cnemic Index	80.1	Eurycnemic	Dorsoventrally flattened

TABLE 3.2:
Cranial and post cranial indices for SF 17642

Stature was calculated for individual SF 17642 following Trotter (1970). This method for stature estimation was developed on modern American individuals, and therefore may produce inaccurate results when applied to skeletal material from Bronze Age Scotland. However, a recent study indicated that this method was appropriate for medieval English archaeological remains (Mays 2016). Utilising the equation for the right femur and tibia, stature was calculated to be approximately 160 cm ± 2.99 cm (c. 5 feet 3 inches).

Skeletal remains from Bronze Age Scotland are rare, and in many cases are too poorly preserved to allow for calculation of stature. Table 3.3 provides a summary of metrical data collected from two comparable sites.

Site	Period	Male mean (cm)	Range (cm)	Number	Reference
Northton, Isle of Harris	Mid-Late Bronze Age	170	165–173	3	(Simpson et al. 2006)
Northton, Isle of Harris	Late Iron Age	171.5	168–173.5	3	(Simpson et al. 2006)
Barns Farm, Dalgety	Early Bronze Age	168	168	1	(Watkins 1982)

TABLE 3.3:
Inter-site comparison of stature

Individual SF 17642 is shorter than the mean height from these comparable populations, however, because these means are based on such small sample sizes it is difficult to place significance on this result.

Palaeopathology

Individual SF 26319

Although the skull was fragmented, a portion of the maxillary sinus was identified. The internal surface of the sinus was covered with diffuse, smooth but highly porous compact bone (Figure 3.5), indicative of chronic maxillary sinusitis (Boocock et al. 1995).

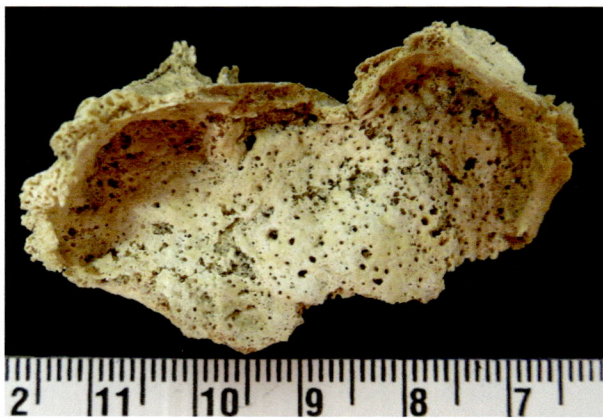

FIGURE 3.5:
Diffuse smooth but porous compact bone on internal surface of maxillary sinus.

Due to the severe wear to all teeth, only 14 teeth could be confidently identified. Examples of dental attrition are provided in Figures 3.6 and 3.7. There were seven probable pre-molar and canine teeth/roots present as well as nine maxillary molar roots, all of which could not be confidently identified due to the level of attrition. Of the teeth present, six teeth were worn to expose the pulp cavity, and four were just roots (assuming the nine maxillary molar roots represent three maxillary molars).

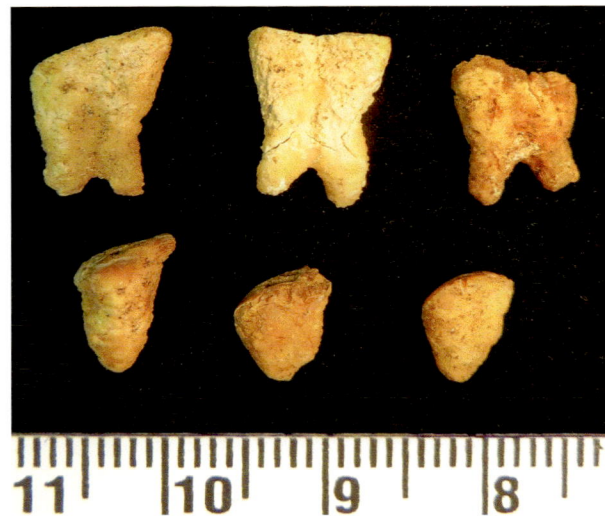

FIGURE 3.7:
Maxillary molar roots.

There was a circular lytic lesion with sharp edges (5 mm in diameter) immediately inferior to the left mandibular first pre-molar (Figure 3.8) which is characteristic of an apical granuloma (Dias and Tayles 1997).

FIGURE 3.8:
Apical granuloma inferior to left mandibular first pre-molar.

FIGURE 3.6
Seven probable canine and pre-molar teeth, four with pulp cavity exposure.

On the right side of the mandible, there was another possible oval-shaped lytic lesion immediately inferior to the canine and first pre-molar (Figure 3.9) with rough irregular edges approximately 20 mm in length. Inferior to the possible lesion, there was an area of irregular looking compact bone, however, due to the badly preserved nature of the mandible and the high amounts of taphonomic damage, it was impossible to say for certain whether or not this damage was indicative of another apical granuloma or if it was simply post-mortem damage.

FIGURE 3.9:
Possible apical granuloma inferior to right mandibular canine and first pre-molar.

The exposure of the pulp cavity increases the likelihood of pulp infection which can lead to infection of the alveolar bone (Rose and Ungar 1998). Thus, given the severe attrition of a majority of the individual's teeth, including the pulp exposure in six of the remaining teeth, it was not unlikely that the individual had more than one apical granuloma. Studies have shown that in some cases, dental infection can lead to chronic maxillary sinusitis (Boocock et al. 1995) as was observed in this individual. However, as the maxillary alveolar bone was badly preserved on the right side and absent on the left, it was impossible to determine whether there was a fistula or sinus present between the maxillary teeth and the maxillary sinus that would have allowed for the spread of the dental infection into the sinus. Thus it cannot be said for certain whether the chronic maxillary sinusitis was due to the individual's maxillary dental infection (Roberts 2007), but it remains a possibility.

There was osteophyte formation around the articular facet for the C1-C2 joint of the atlas and on the superior aspect of the odontoid process of the axis. While there was osteophyte formation on these fragments of the first two cervical vertebrae, there was no corresponding eburnation, porosity, or new bone formation so osteoarthritis could not be diagnosed (Waldron 2009). Most of the other spinal joints were not available for analysis.

Individual SF 17642

Linear enamel hypoplasia (LEH) is caused by non-specific stress and poor nutrition in childhood which results in the interruption of normal enamel development (Goodman and Rose 1991). After the period of non-specific stress has ended, normal enamel development resumes resulting in visible horizontal lines or grooves in the enamel, or less frequently plane or cuspal patterns (Ogden et al. 2007). LEHs were found on the lingual surface of all four mandibular incisors (Figure 3.10).

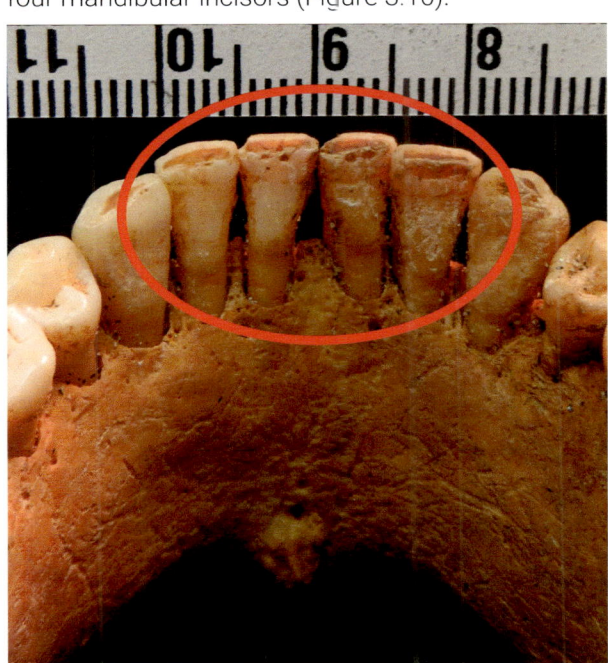

FIGURE 3.10:
Linear enamel hypoplasia on lingual surfaces of all four mandibular incisors.

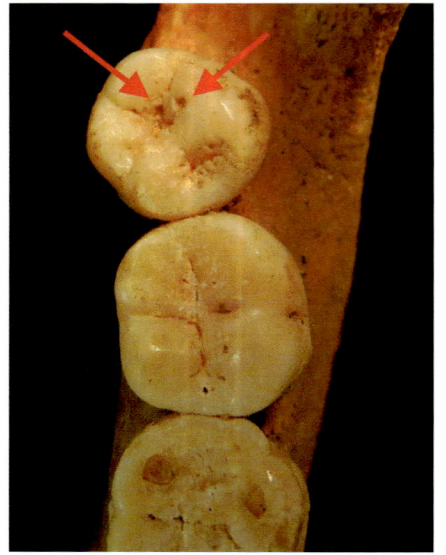

FIGURE 3.11:
Possible carious lesions on distal occlusal surface of left mandibular third molar.

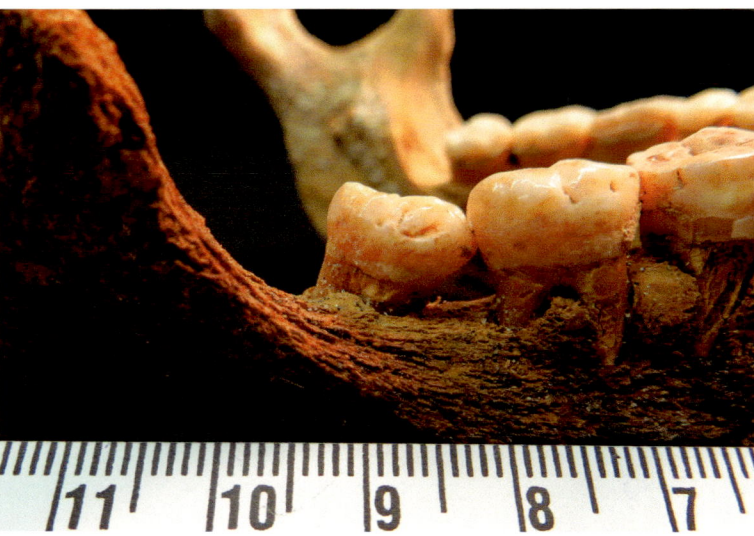

FIGURE 3.12:
Impacted right mandibular third molar.

There were two possible carious lesions on the distal occlusal surface of the left mandibular third molar (Figure 3.11).

The right third mandibular molar was tilted anteriorly and was not in full occlusion with the corresponding maxillary third molar (Figure 3.12). As the tooth was nearly completely erupted with no evidence of damage to the adjacent tooth, it is likely that this would have been symptomless, although food packing may have occurred in the large pocket created between the third and second molars (pers comm Beaumont 2016).

The heads of both femora were anteriorly displaced resulting in increased anteversion of both femora (Figure 3.13). The femoral neck angle (FNA), which is the angle between the longitudinal axis of the femoral neck and a horizontal line through the distal condyles, is usually between 15–20° in non-pathological adults (Cibulka 2004). The FNAs of the right and left femora were approximately 35° and 39° respectively. These measurements were taken manually from photographs.

Previous studies have found that existing methods to measure the degree of FNA are very difficult to reproduce accurately (Stirland 1994), and in many cases the various methods lead to different results (Rokade and Mane 2008). Despite this, it is evident that individual SF 17642 had a larger than average FNA. This increased FNA has been

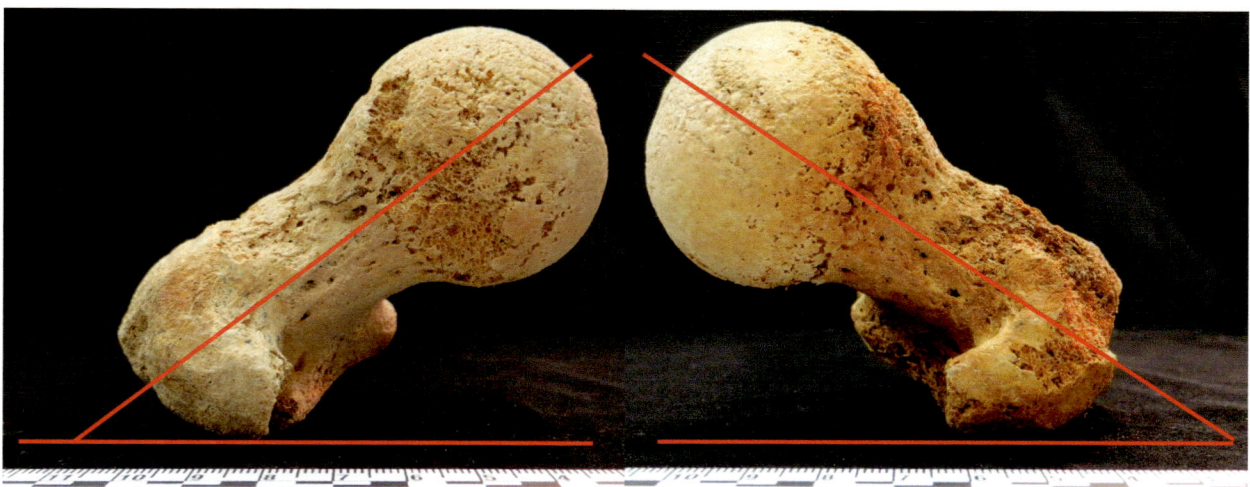

FIGURE 3.13:
Increased anteversion of right and left femoral heads.

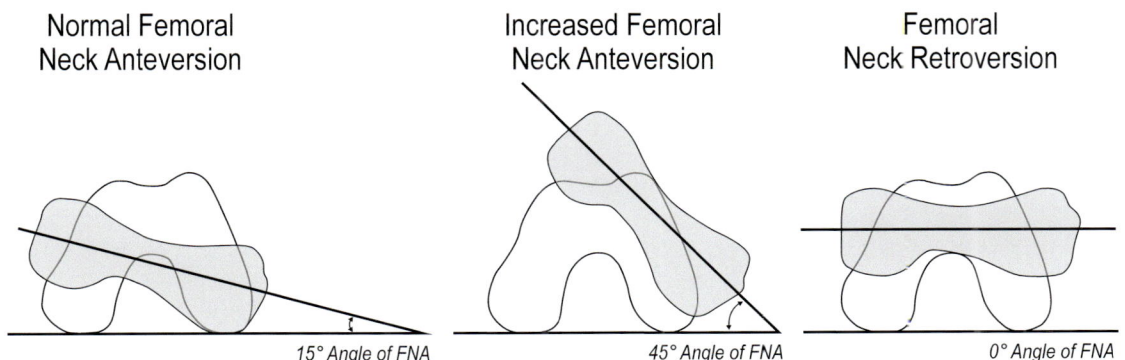

FIGURE 3.14:
Varying levels of femoral neck anteversion. Redrawn after Cibulka (2004)

clinically associated with orthopaedic conditions, including but not limited to, in-toeing (pigeon-toed), genu valgum (knock-kneed), pes planus (flat foot), and tibial torsion (Cibulka 2004; Djukic et al. 2014) (Figure 3.14), however these were unlikely to have had clinical significance in the life of the individual.

Both acetabula were more ovoid in shape than normal with supero-lateral elongation. It is unclear if this is related to the increased anteversion of the femora.

Enthesophytes were found on the proximal end of both the right and left tibiae (Figure 3.15). There was one enthesophyte present on the proximal end of the left tibia. It was smooth, approximately 15 mm in length, and was located immediately inferior to the fibular facet almost as an extension of the facet. Given its location, it was possible that this enthesophyte constituted ossified joint capsule of the proximal tibio-fibular joint.

There were two enthesophytes found on the right tibia. The first was adjacent to the medial aspect of the fibular facet and was approximately 15 mm in length. The second was located infero-laterally from the first enthesophyte and was approximately 5 mm in length. It is unclear whether the enthesophytes found on both tibiae would have had any clinical significance in life.

Miscellaneous

Large, rugose muscle attachments were found on many of the bones. Both claviculae had large attachment sites for *M. pectoralis major*. Both femora had large gluteal tuberosities for the attachment of *M. gluteus maximus* (Figure 3.16), and both humeri had large attachments for the *M. biceps bracchi* and *M. latissmus dorsi* muscles. The region where a tendon or ligament attaches to bone is known as an enthesis (Villotte et al. 2010). There are two types of entheses, and they are distinguished based

FIGURE 3.15:
Enthesophytes on proximal end of right and left tibiae.

on the type of tissue that attaches to the bone: fibrocartilaginous and fibrous entheses (Villotte and Knüsel 2013). Changes to the appearance of fibrocartilaginous entheses are associated with activity, and have been shown to increase in frequency with increasing age (Villotte *et al.* 2010), however Villotte and Knüsel (2013) found no association between activity and fibrous entheseal change. Many researchers have cautioned against linking enthesopathies with specific activities, especially with a lack of comparable clinical data from similar activities (Cardoso and Henderson 2010; Villotte *et al.* 2010; Meyer *et al.* 2011), and because individual SF 17642 is part of such a small population, nothing conclusive can be drawn about his rugose muscle attachment sites.

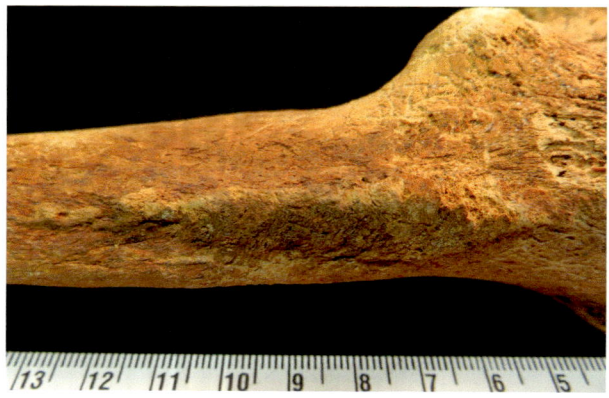

FIGURE 3.16:
Rugose gluteal tuberosity on posterior surface of right femur.

Disarticulated material

An analysis was performed on the disarticulated human remains from 19 small finds which were either single bones or groups of bones. The objective of this work was to determine the minimum number of individuals (MNI) present at the site as a whole as well as in the disarticulated samples, to examine associated groups, and to perform standard age and sex assessment on the human remains found.

Each sample was analysed and a detailed written inventory was recorded using Microsoft Excel (Table 3.4). Based on the limited contextual information, five groups of samples containing human bone were identified: individual SF 26319 and one associated sample, individual SF 17642 and three associated samples; seven samples from Square G; one tooth from Square H; and eight samples containing human bone which could not be linked to a specific area (these samples were probably from across the site, three of them are related to a cist but it is not clear if they were from individual SF 17642 or SF 26319; sample SF 23827 is noted as being from 'Neolithic building 1, hearth').

After an inventory was complete, the only elements found to be repeated in the disarticulated remains were the left intermediate cuneiform in samples SF 17638 and SF 17106 and the right third metacarpal in samples SF 17638 and SF 17436. This gives an MNI of two for the disarticulated samples. When the disarticulated remains and the two articulated skeletons were considered together, a MNI of three was obtained for the RUX6 site as there were three of the following elements present: right zygomatic, right clavicle, right humerus, left ulna, left radius, right tibia, and left talus.

Square E: samples SF 17638, SF 17624, SF 23861, and associated articulated individual SF 17642

The left navicular and intermediate cuneiform from sample SF 17638 articulated with the medial and lateral cuneiforms from the articulated individual SF 17642 (Figure 3.17). In addition, the third metacarpal from sample SF 17638 articulated with the second metacarpal from the articulated individual SF 17642. Sample SF 17624 (a single proximal hand phalanx) was found in in the cist fill of Square E. While no direct articulation was made, this bone probably came from articulated individual SF 17642.

FIGURE 3.17:
Articulation of left navicular and intermediate cuneiform from SF 17638 with left lateral cuneiform from individual SF 17642.

Group	SF No.	Context information	Inventory
1	26319	1983: 24 cist; Fl 1	- articulated skeleton- older adult male
	26243	From soil sample under 24 cist skeleton	- unidentifiable fragments
2	17642	Square E; cist	- articulated skeleton- young adult male
	17624	Square E; cist fill I/III	- proximal hand phalanx
	17638	Square E; cist fill VII	- R capitate, R lunate, R hamate - 2 proximal hand phalanges - R MC3 and 1 proximal fragment of a MC - L navicular, L intermediate cuneiform - 2 proximal foot phalanges, 1 intermediate foot phalanx, and 3 distal foot phalanges (one is for first toe) - 2 rib frags - ~70 unidentifiable frags + 3 unidentifiable phalanx fragments
	23861	Disturbed area under skeleton in NE cist	- 1 probable hand phalanx (proximal or intermediate)
3	16201	Tumulus collapse	- 8 cranial fragments (none diagnostic) - fragment of C1 - R upper premolar - R proximal humerus and part of shaft - fragment of L proximal radial shaft (fits with SF 17436) - 2 proximal hand phalanges - 1 ?MC shaft - 3 os coxae fragments - L talus - 3 unidentified fragments (2 with articular surfaces)
	17106	Square G; insert	- R tibia (in 2 fragments) - 1 distal fragment of a proximal hand phalanx - L intermediate cuneiform - 1 fragment of a pedal phalanx - 6 unidentified fragments
	17372	Square G; insert	- R zygomatic fused to part of maxilla - ?MT shaft - 2 molars (probably R and L maxillary second molars)
	17436	Square G	- lower premolar - L radial shaft fragment (fits with SF 16201) - unsided trapezoid - L MC1, L MC2; R MC3, 1 frag of MC head - 2 proximal hand phalanges, 2 intermediate hand phalanges, and 1 distal hand phalanx (for MC1) - R cuboid - 3 unidentified frags
	17437	Square G	- L ulna (in 2 fragments) - R ulna - ulna shaft fragment (probably part of R proximal ulna frag but cannot fit together) - L clavicle (missing lateral end) - L glenoid fossa - 4 probable skull frags (non-diagnostic) - ?L scaphoid - 1 proximal head of proximal hand phalanx - 3 phalanx fragments - ~67 unidentifiable frags
	17443	Square G	- MT5 shaft (unsided) - 2 unidentifiable bone fragments

TABLE 3.4:
Inventory of all RUX6 samples; NB: R= right, L= left, MC= metacarpal, MT= metatarsal, C= cervical

Group	SF No.	Context information	Inventory
3	24051	West half	- surface of R glenoid cavity of scapula
4	16778	Square H	- upper canine (?R)
5	15715	Shore tumulus; surface scatter	- lower molar (possibly third)
5	16195	N/A	- intermediate hand phalanx
5	23105	Cist insert; LBA – Neo	- 1 intermediate pedal phalanx and 1 ?distal pedal phalanx
5	23827	Neo Building 1; hearth + environment	- 3 unidentifiable tiny fragments
5	23859	346/352	- L proximal humerus (in 2 fragments)
5	24050	West half; surface clearance	- 1 probable hand phalanx (proximal or intermediate)
5	24052	Cist insert	- 2 scapula fragments - 1 MT fragment (unidentifiable)

TABLE 3.4 continued:
Inventory of all RUX6 samples; NB: R= right, L= left, MC= metacarpal, MT= metatarsal, C= cervical

Disarticulated individual from Square G

From the remaining disarticulated samples (excluding samples SF 17638, SF 17624 and SF 23861), there were no repeated elements, providing an MNI of one. The samples recorded as Square G or Square G insert (SFs 17106, 17372, 17436, 17437 and 17443) can be confidently assumed to represent a single disarticulated individual; there was no duplication of elements and the size, age and sex of all elements were consistent. In addition, the proximal fragment of the left radius from sample SF 16201, which was described as tumulus collapse, fit with a fragment of the radial shaft from sample SF 17436 (Figure 3.18).

The fragment of the right glenoid cavity from sample SF 24051 was compared with the left scapula fragment found in sample SF 17437 (Figure 3.19). The maximum width of the glenoid cavity of the right scapula was 32 mm and 31 mm in the left. It is likely that sample SF 24051 belonged to the individual from Square G.

Sex assessment was extremely difficult due to the fragmentary nature of the bones and the lack of a well-preserved pelvis and skull. In sample SF 16201, the external occipital protuberance was located on a small cranial fragment which was badly eroded. The area was protruding which very tentatively suggests that this individual was male. In the same sample, the maximum width of the right humeral head was measured with a digital caliper to 49 mm. This again tentatively suggests a male (Bass 2005). The only other fragment that could be used for sex assessment was the left glenoid cavity found in sample SF 17437 (Figure 3.19). The maximum width of the glenoid cavity was measured to be 31 mm, which also tentatively suggests male (Bass 2005). With the above evidence, it can only be said that the individual represented by the disarticulated samples from Square G was possibly a male.

Age estimation was similarly difficult due to the lack of pubic symphyses, auricular surfaces, and cranial sutures. There was a left clavicle found in sample SF 17437 with a fused medial epiphysis suggesting an age of 21+ years (Scheuer and Black 2000). The occlusal surfaces of the right and left upper second molars found in sample SF 17372, while taphonomically damaged, were not very worn with little enamel showing, tentatively suggesting this was not an old adult (Brothwell 1981). Overall this individual was probably a middle adult.

The right tibia in sample SF 17106 survived although it was fragmented towards the distal end of the shaft. The two pieces could be fitted back together and the maximum length of the tibia was found to be 39.4 cm giving a stature of 173 cm ±3.37 cm (c. 5'7" – 5'9"), assuming the individual was male (Trotter 1970).

FIGURE 3.18:
Associated radial fragments from two different samples (SF 16201 and SF 17436).

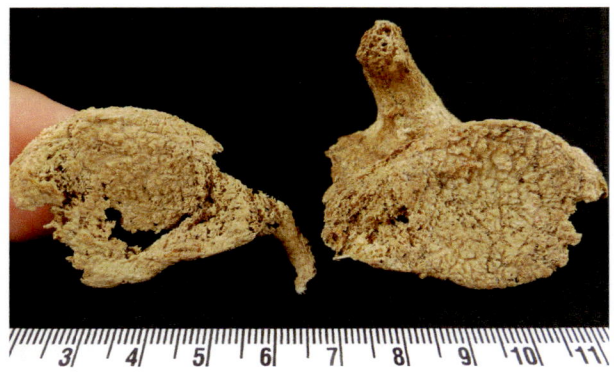

FIGURE 3.19:
Comparison between right glenoid fossa from sample SF 24051 and left glenoid fossa from sample SF 17437.

Remaining disarticulated samples

The remaining seven samples (SFs 15715, 16195, 23105, 23827, 23859, 24050, and 24052) could not be confidently associated with either of the articulated individuals (SF 17642 or SF 26319) or the disarticulated individual from Square G, but this remains a possibility. Samples SFs 23105 and 24052 were noted as associated with a cist, but it is not clear if these relate to individual SF 17642 or SF 26319.

The left humeral shaft and fragment of the humeral head from sample SF 23859 were compared to the right humeral head and proximal shaft from sample SF 16201 which belonged to the disarticulated individual in Square G (Figure 3.20). However, due to the severe taphonomic damage to the left humeral shaft, a visual comparison, while possible, was not particularly helpful. A comparison of the maximum width at the surgical neck of both humeri gave measurements of 20 mm for the left, and 25 mm for right. These may have come from different individuals, but size differences of this magnitude are possible within a single individual with strong directional asymmetry. If the humerus from sample SF 23859 was from a different person, it would increase the number of individuals present at RUX6 to four as both articulated individuals had both proximal humeri present.

Conclusion

In conclusion, individual SF 26319 was a poorly preserved, probable older male adult with severe wear to all remaining teeth, two apical granulomas, and chronic maxillary sinusitis. Individual SF 17642 was a well preserved young adult male with large muscle attachment sites, linear enamel hypoplasia, two possible carious lesions, an impacted third molar, and increased anteversion of both femora. Parts of the left foot and right hand of this individual were found among the disarticulated samples associated with RUX6.

As there were no repeating elements within the disarticulated samples found in Square G, it can be confidently said that these samples represent a third, disarticulated individual. Various skeletal elements throughout the samples suggest that this adult individual was possibly male. The lack of duplication of elements across the remaining samples indicates these are likely to be from these three individuals, however the size difference observed between the humeri from samples SF 16201 and SF 23859 tentatively indicate a fourth person may have been present.

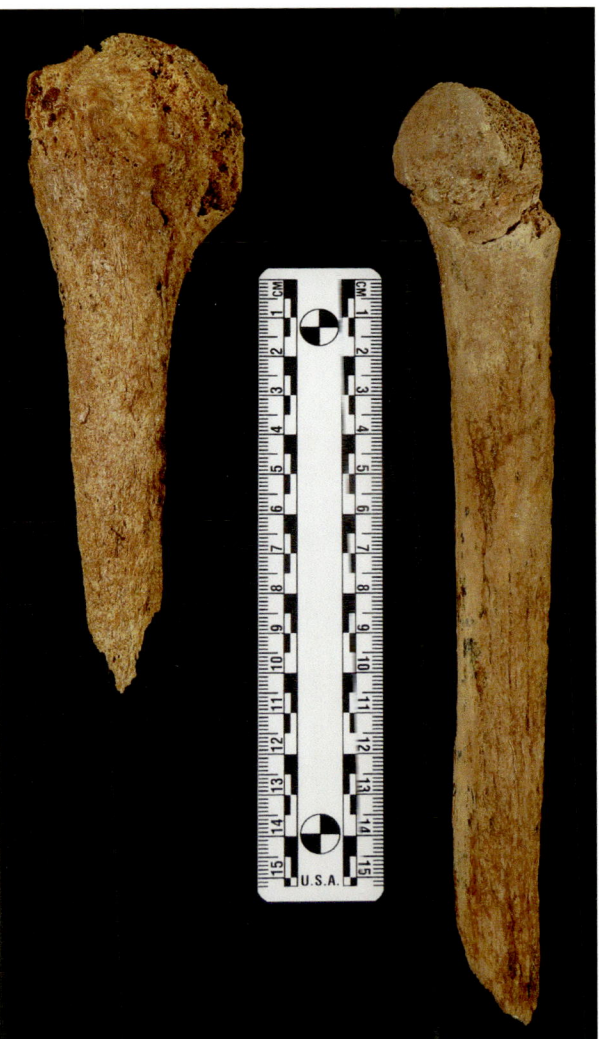

FIGURE 3.20:
Comparison of right humerus from sample SF 16201 and left humerus from sample SF 23859.

Isotopic analysis of the human skeletal remains

By Julia Beaumont, Cassie Hall, Derek Hamilton and Jo Buckberry

This chapter summarises the isotope analysis undertaken on the human remains from RUX6, in order to estimate their diet and physiological changes during life. Carbon ($\delta^{13}C$) and nitrogen ($\delta^{15}N$) stable isotope ratio analysis was carried out using incremental dentine sectioning for collagen for individuals SF 26319 and SF17642 (giving values for childhood and early adolescence) and bulk sampling of bone collagen for these two individuals plus an ulna from a third, disturbed burial from Square G, SF 17437 (providing an average for each of the diet during last 10 years of life). Contemporary faunal bone collage samples were also measured in order to provide an isotopic baseline. In addition, enamel from the teeth of SF 26319 and SF 17642 was sampled for strontium isotope ratios to assess individual migration.

Stable isotope measurements for diet, physiology and migration

The interpretation of variation in the ratios of stable light isotopes of carbon ($\delta^{13}C$) and nitrogen ($\delta^{15}N$) in skeletal and dental tissues has been well-established as a robust method of reconstructing the dietary habits of individuals, and has parallels in modern clinical studies. Fractionation is the key to understanding the dietary pathway of carbon and nitrogen within the biosphere through the food chain, and allows the reconstruction of the diets of fauna and humans.

When an element is incorporated in the tissues of a plant or animal, fractionation occurs and the newly-formed tissue will contain more of the lighter isotope. This means that ratio of the heavier isotope of nitrogen to the lighter $\delta^{15}N$ increases as a result of metabolic fractionation by 2–5‰ and the ratio of the heavier isotope of carbon $\delta^{13}C$ by about 1–2‰ at each trophic level of a food chain and can therefore provide information about the relative consumption of plant, animal and marine protein (Schoeninger and DeNiro 1984). For that reason, faunal bone and tooth from the site have also been analysed.

Isotopic values can also be affected by physiological changes. Raised $\delta^{15}N$ values of body tissues may be the result of extreme nutritional stress (Hobson et al. 1993; Guthrie and Picciano 1995). Mekota et al. (2006) showed how body mass index (BMI) was related to changes in the $\delta^{15}N$ values of individuals deliberately depriving themselves of food, while Duška et al. (2007) demonstrated a negative nitrogen balance (catabolism) in acute starvation in a clinical setting. Hatch et al. (2006) advised the use of isotopic values from hair as a diagnostic tool for anorexia and bulimia. Powanda and Beisel (2003) noted that the metabolic effects of infection on nitrogen balance were masked or reduced when protein and calorie intake was increased, suggesting that the effects of any illness on nitrogen balance would be greater when nutrition was also insufficient. It has also been suggested that chronic illness may cause changes in the $\delta^{15}N$ values of pathological bone (Katzenberg and Lovell 1999). Finally, dietary changes revealed by temporal changes in the isotopic values may also be evidence for migration of an individual (e.g. Müldner et al. 2011; Beaumont et al. 2013).

Materials and methods

The human samples consisted of rib bone from two articulated individuals (SF 26319 and SF 17642), the disarticulated ulna of a third (SF 17437) the upper left second premolar (ULPM2) from SF 17642, the lower right first permanent molar (LRM1) from SF 26319. The faunal samples were bone from one cattle (SF 23102) and a sheep (SF 26514), and dentine from a juvenile cattle (SF 26321) and dentine from a red deer tooth (SF 26185).

All samples were cleaned with a rose-head bur in a slow-speed rotary hand drill to remove adherent surface contamination. The human teeth were sectioned to select a single root from the M1 and divide the root of the PM2 in half. Any secondary dentine and cementum were removed, and a sample of the enamel removed for strontium analysis. The animal teeth were cleaned and as much enamel removed as possible without damaging the dentine.

The bone and dentine samples were demineralised in 0.5M HCl at 4°C. The softened human tooth samples were then divided into 1mm sections using a scalpel (Beaumont et al. 2013: method 2). All samples were denatured at 70°C in a pH3 solution, frozen at −80°C and then freeze-dried to produce collagen (Brown et al. 1988).

The collagen was weighed into tin capsules (in duplicate where possible) and measured combustion

Sample Nr	Species	Tissue	Age	Sex	δ15N (‰)	δ13C (‰)
26319	human	bone (rib)	c. 45+ years	M	11.1	−17.7
26319	human	dentine ULPM2 (average)			10.3	−19.4
17642	human	bone (rib)	c. 19–25 years	M	9.6	−20.1
17642	human	Dentine LRM1 (average)			9.7	−19.4
17437	human	bone (ulna)			9.9	−20.5
23102	cattle	bone			5.0	−21.8
26514	sheep	bone			6.2	−21.4
26321	juvenile cattle	bone			7.4	−21.4
26321	juvenile cattle	dentine			7.4	−21.1
26185	deer	bone			5.4	−22.2

TABLE 3.5:
RUX6 human and faunal samples, age and sex (Bohling and Buckberry 2016) and bulk δ15N and δ13C.

in a Thermo Flash EA 1112 and introduction of separated N_2 and CO_2 to a Finnigan Delta plus XL via a Conflo III interface. The collagen samples were interspersed throughout the run with both internal standards and international standards. Calibrated against these standards, the analytical error at 1 standard deviation was ±0.2‰ or better.

Results

The results of the isotopic analysis of nitrogen ($δ^{15}N$) and carbon ($δ^{13}C$) in the skeletal and dental tissue samples can be found in Table 3.5.

Animal baseline

The isotopic animal baseline can indicate or confirm a trophic level shift within diet.

Specimen SF 26321, a calf (juvenile cattle), was located in close proximity to individual SF 26319. SF 26321 has two data points representing the bulk sampling for dentine collagen and bone collagen, with very similar $δ^{15}N$ and $δ^{13}C$ values reflecting that these tissues are forming at the same time. As seen in Figure 3.21, SF 26321 fits within the trophic level shift offset explained above.

In Figure 3.21 the faunal data is situated in the lower left portion of the plot, with relatively low $δ^{15}N$ values, in comparison to the human $δ^{15}N$. The offset is described by Szpak et al. (2012) as 3.6 ± 1.3‰ for $δ^{15}N$ and 3.7 ± 1.6‰ for $δ^{13}C$.

This supports the hypothesis that this small sample of individuals may have had a diet which included the consumption of sheep, and possibly deer. However, the bone collagen $δ^{15}N$ and $δ^{13}C$ values for SF 26319 are much higher and would suggest that this individual has had input from marine sources in his diet as an adult, reflected in the bone collagen in the rib, which reflects the last 3–5 years of life (Beaumont and Montgomery 2016).

Bone collagen data from contemporaneous British sites

Two of the individuals from Udal (SF 17437 and SF 17642) fall into the terrestrial $δ^{13}C$ range of the Neolithic/Bronze Age/Iron Age individuals from broadly contemporaneous British sites; Sumburgh (Montgomery et al. 2013), Quarterness (Schulting et al. 2010), Cladh Hallan (Pearson et al. 2005) and Yarnton (Lightfoot et al. 2009). The third, SF 26319, appears to be similar in his $δ^{13}C$ range to the Iron Age male adults from Newark Bay (Richards et al. 2006), suggesting this individual had a more marine-based diet. This could be interpreted as evidence that he lived at a different time period, or that he resorted to a marine diet during a time of nutritional distress: the marine value is from rib (turnover time last 3–5 years of life) in contrast to early life values from the dentine, and a further short term period of marine input age 6.5–8 years in the dentine profile. All three have lower $δ^{15}N$ values than the sites that they match for $δ^{13}C$ (Figure 3.22).

Dentine profile SF 26319

For individual SF 26319, the incremental dentine $δ^{15}N$ and $δ^{13}C$ data is presented in Figure 3.23 using a time line to show the data against approximate age in years. The $δ^{15}N$ and $δ^{13}C$ values for each incremental sample generally appear to co-vary suggesting that these are dietary values.

Part 3

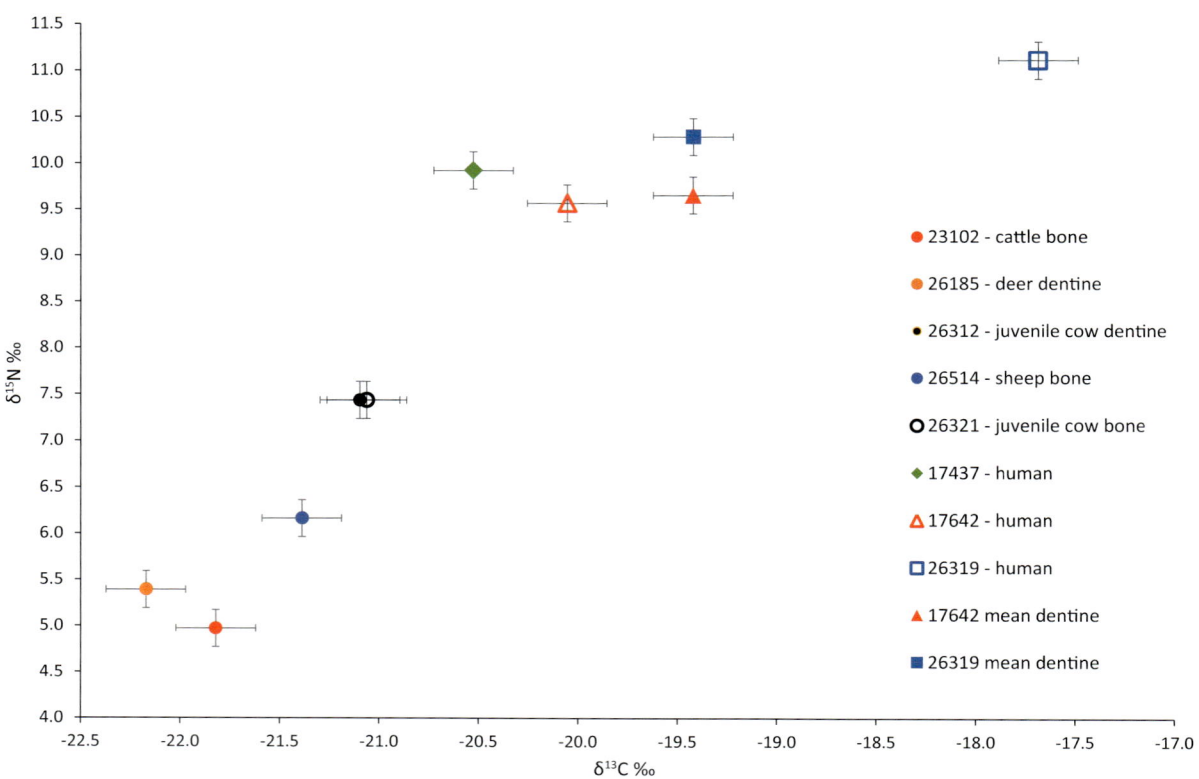

FIGURE 3.21:
A plot of animal dentine and bone collagen δ15N and δ13C alongside the human dentine collagen (averaged) and bone collagen. Error bars set ±0.2‰

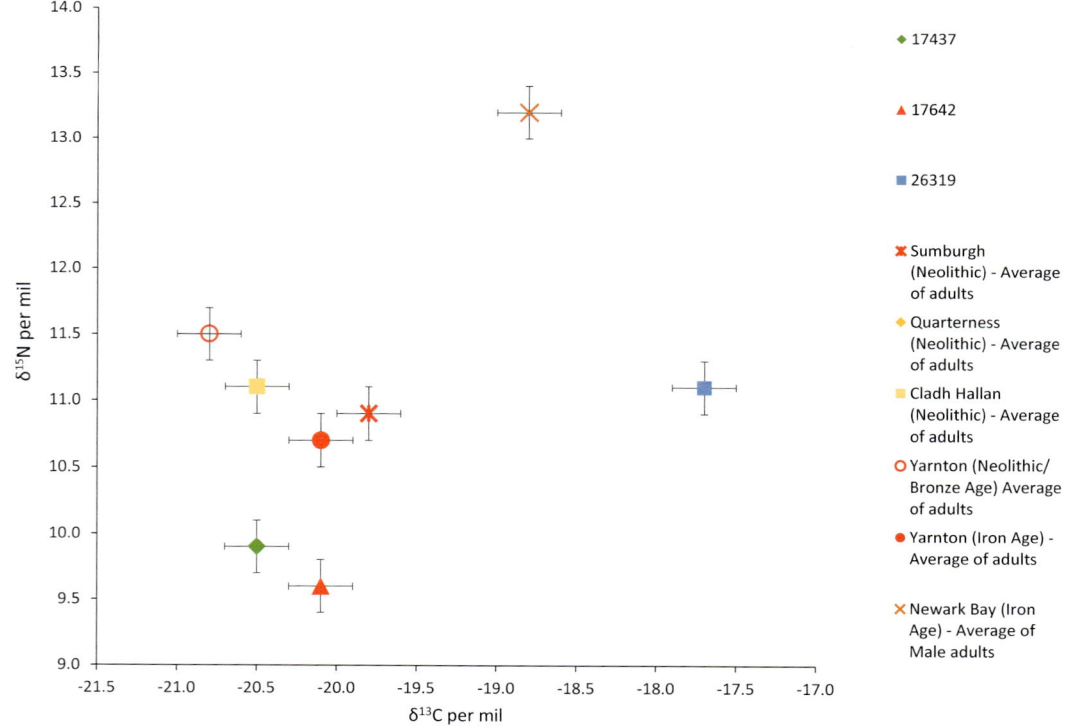

FIGURE 3.22:
Plot showing bone collagen δ15N and δ13C for SF 17437, SF 17642, and SF 26319 with bone collagen data from contemporaneous British sites. Sumburgh (Montgomery *et al.* 2013), Quarterness (Schulting *et al.* 2010), Cladh Hallan (Pearson *et al.* 2005), Yarnton (Lightfoot *et al.* 2009) and Newark Bay (Richards *et al.* 2006). Error bars set at ±0.2‰.

In more detail, the early increments suggest a short period of breastfeeding between birth and 6 months of age (a falling value for both $\delta^{15}N$ and $\delta^{13}C$), followed by a period from 0.5 to 1.5 years of opposing covariance which suggests a short period of nutritional distress (Beaumont et al. 2015; Beaumont and Montgomery 2016). Both isotope values then remain fairly stable until the age of 6 when both rise reaching values over −19‰ for $\delta^{13}C$ and over 11‰ for $\delta^{15}N$, suggesting a marine input to the diet (Montgomery et al. 2013). This is consistent with the bone collagen values for SF 26319, which also suggest that the marine input continued into adult life (and see Figure 3.21). There is a divergence of the values at the age of 9 years with rising $\delta^{15}N$ and falling $\delta^{13}C$, suggesting further nutritional distress.

Dentine profile SF 17642

Compared to SF 26319, the $\delta^{13}C$ values for both the bone (−20.1‰) and dentine (bulk −19.4‰) from SF 17642 remain within the expected terrestrial values from the faunal remains (Figure 3.24). There is evidence for a high marine input to the diet from 8.5–10 years, although the bulk isotope ratio values suggest that the individual was eating meat throughout life. The bone values are close to the bulk dentine values.

In detail, it appears that there is a period of opposing covariance from the age of 3.5 to 6.5 years, which could be a period of nutritional distress with marine input followed by recycling of body tissues shown by continued rise in $\delta^{15}N$ and fall in $\delta^{13}C$. A further period of opposing covariance at the end of the profile suggests some nutritional distress (13.5–14.5 years).

Strontium isotope ratios

Lab No	Sample No	Type	Species	$^{87}Sr/^{86}Sr$	std Error
H667	GU42089	Enamel	Human	0.7095 ± 0.0016	0.002
H668	GU42090	Enamel	Human	0.7095 ± 0.0015	0.002

TABLE 3.6:
Strontium isotope ratios for the two individuals, SF 26319 and SF 17642

Enamel from the teeth of SF 26319 and SF 17642 was sampled for strontium isotope ratios. Both individuals had a Sr value of 0.7095 (see Table 3.6). This is similar to the value of 0.7092 for individuals from Cnip (Montgomery et al. 2003) and for the plants living on the machair of South Uist (0.7100). Sr isotopic values for plants growing on the gneiss were reported as 0.7095, 0.7102 and 0.7103 (Montgomery et al. 2007). Overall the Sr results are consistent with individuals from the Uists.

Discussion

The individuals from RUX6 have a range of $\delta^{15}N$ and $\delta^{13}C$ values. Comparing these values with those from the contemporaneous faunal remains at the site, it appears that all three were consumers of these animals, with SF 23619 changing from a terrestrial diet from about the age of 6 years to a diet with a higher marine input and SF 17642 with a short-term episode of marine input between about 8.5–11.5 years. The matching variance in $\delta^{15}N$ and $\delta^{13}C$ confirms this to be a dietary change.

Both SF 23619 and SF 17642 show opposing covariance (when the $\delta^{15}N$ and $\delta^{13}C$ change but in opposing directions) within their dentine profiles, suggesting that each had a period of nutritional insufficiency during the period of tooth formation (Beaumont and Montgomery 2016). It is not possible to diagnose what caused this, but SF 23619 is recorded as having enamel hypoplasia present on his dentition at approximately the same age as the opposing covariance in the dentine profile, about the age of 1 year, another sign of non-specific stress during childhood (Bohling and Buckberry 2016). SF 17642 has the start of a similar pattern at the end of tooth formation, at about 13.5 years. This is similar to individuals reported at Sumburgh cist (Neolithic) where the short-term change was interpreted as a response to a failure of the terrestrial food sources (Montgomery et al. 2013).

Two of these individuals, SF 17642 and SF 17437, have bone collagen $\delta^{13}C$ values (reflecting their adult diet) which are consistent with the values from adults from contemporaneous British sites where the diet is mainly terrestrial. The third, SF 23619, has $\delta^{13}C$ values which suggest a higher marine input to the diet as an adult, similar to the isotopic values from the Iron Age adult males from Newark Bay.

It is possible that the two terrestrial consumers were near contemporaries, sharing the same diet, and that SF 23619 lived and died at a different time when the local diet was consistently more marine-based.

In all three cases, the $\delta^{15}N$ is lower than the contemporary sites (Figure 3.22). This could reflect the consumption of lower trophic level foods at

Part 3

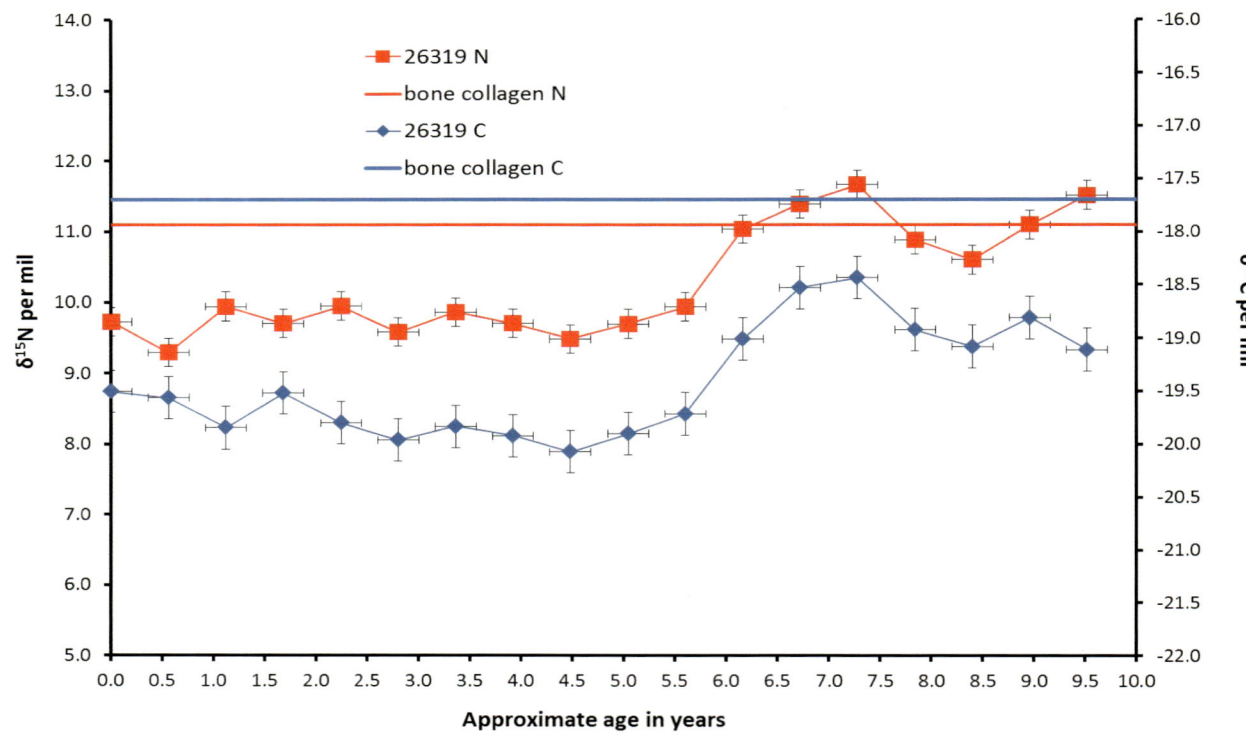

FIGURE 3.23:
Incremental dentine and bone collagen δ^{15}N and δ^{13}C profile for SF 26319 plotted against approximate age. Error bars set to 0.2‰.

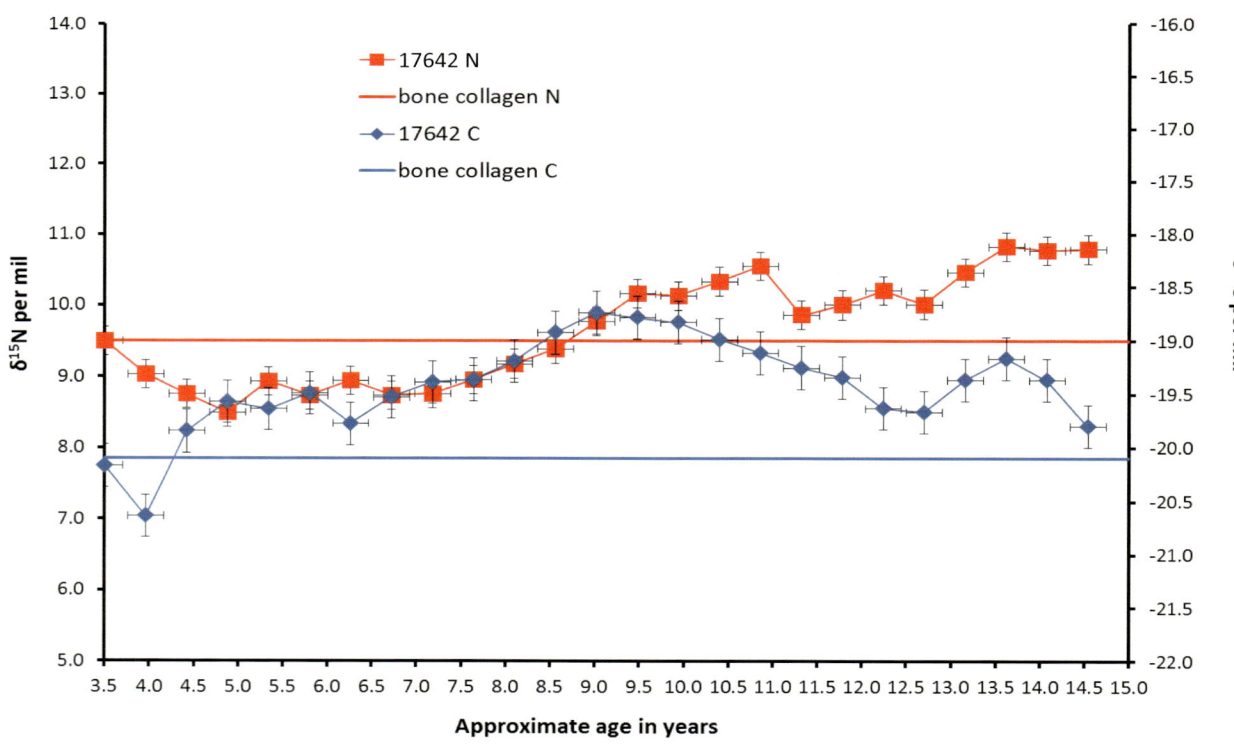

FIGURE 3.24:
Incremental dentine and bone collagen δ^{15}N and δ^{13}C profile for SF 17642 plotted against approximate age. Error bars set to 0.2‰.

RUX6, or a difference in the management of crops and animals: for example, manuring of crops or vegetation fed to the animals would result in higher $\delta^{15}N$ values in the faunal and human tissues, as would growing crops or feeding animals in areas of salt marsh (Bogaard *et al.* 2007; Britton *et al.* 2008).

While there is no evidence of migration between areas with different geology in the Sr results, he may have migrated from a location with the same Sr isotope ratio.

PART 4 The changing natural environment and subsistence farming

Introduction

By Beverley Ballin Smith

In this section, research has been undertaken by a number of specialists to understand changes in environmental conditions experienced at the site over time, and how these changes affected the subsistence farming of the inhabitants. Their results, and the implications they had for the survival of the settlement, are discussed further in PART 6.

One of Iain Crawford's main aims (see PART 1) in excavating RUX6 was to understand the environmental changes that had taken place, both in the past and more recently, across the Udal peninsula that affected prehistoric settlement established there, and also later activities. In 1974 when he began the investigation of the Bronze Age cairn, the complex stratigraphy of layers of sand on the top of bedrock and subsoil deposits were exposed in the low eroding coastal cliff. Further vertical stratigraphic exposure was revealed in the sides of a nineteenth century saw-pit that bisected the site. Crawford's work on other sites on the Udal peninsula had given him extensive experience in understanding both the accumulation and deflation of sand layers and in their interpretation. To some extent the natural machair environment also enabled him to pursue his interests in sampling techniques, and his methodologies are described in PART 2. At that time, he may have considered himself leading the way in the archaeological understanding of the natural (sand/machair) environment in Scotland and its effects on settlement patterns and land use (Crawford 1965; Crawford and Switsur 1977).

The assumption that Crawford's methods of sample recovery were consistent throughout the project and that samples were treated in the same way, may not be correct. It is likely that his methods changed over time, and occasionally after rare input from external specialists. In Crawford's *Udal Small Finds: notes and information* (Site Archive, dated 02/09/1997), intended for specialists he mentions 'a separate register concerned solely with the methodology and sampling for specific techniques: C14, TL, Soil Flotation'. Unfortunately, that register has not been located in the archive, and the investigation of environmental samples for this volume revealed some inconsistencies in methodologies and aberrations in results that might have been explained in the register.

In correspondence with Dr Richard Preece, Department of Zoology, Cambridge in 1996, Iain Crawford wrote that he was looking for 'environmental change patterns' by 'putting a chronological sounding through the anthropogenic deposits', to be able to date other sites and that he wanted a 'dating column, similar to that sought by pollen, tephra, varve and other specialists' (I A Crawford personal archive). Preece had been asked to identify the terrestrial snails from the remains of a wet sieved sample from the late Neolithic Building 2 (DH) from Phase D. His unpublished list was very similar to that produced by Ruby Ceron-Carrasco (Terrestrial Snails, below), but in his response, he advised Iain Crawford that the sample was not representative, as some species less than 1 mm in diameter were not present in the assemblage. During the excavation, all the samples from the late Neolithic levels had been wet sieved using a 5 mm mesh. By 1996 it was too late to resample using a smaller mesh size as Crawford had completed his excavation of the site ten years previously.

Similar problems had occurred with the botanical samples. Again, in 1996 he had admitted that a very coarse sieve had been used on-site during hand sieving of a Phase D sample (SF 27206 – not recorded in the database) from the upper floor of Building 2 (DH), and no seeds had been recovered (I A Crawford personal archive). However, he had expectations of seed survival in the wet residue 'dust' that was normally retained after sorting. Unfortunately, this 'dust' was not retained and the potential for seeds was lost. It seemed to be a problem during the processing of the Phase D (Neolithic) samples that some of the residues from the flotation machine were not retained or did not contain seeds. Crawford expected to find seeds from the floors of buildings, and when other Phase D samples produced none it left him wondering what the problem was. He considered that the methodology of samples processing had not been adhered to, or that there were issues with the flotation machine, or even the pH of the floor composition had caused problems in the non-survival of seeds. As the floors of the Phase D building overlay subsoil and bedrock, they were damp and likely to have been more acidic than the overlying calcareous layers of machair shell sand in Phase C. In the identification of botanical remains (below) by Susan Ramsay, seeds from the Neolithic settlement were noted as not present, but they did survive the on-site sorting of wet and coarse residues from Phase B and its machair deposits. The possibility of acidic conditions in the late Neolithic, combined with on-site mistakes and other factors, such as wind ablation and sand accumulation, have left an incomplete environmental record for Phase D (discussed further in PART 6). Unlike today's excavators who would normally sub-sample a bulk soil or sediment sample prior to processing, Iain Crawford did not do this. He might have considered it unnecessary, and certainly more samples would have added significant weight to the amount of equipment and finds he transported south after each excavation season. However, we are left in the position of not being able to revisit or interrogate problems such as this.

When Crawford began excavating RUX6 he altered the site collection policy so that all marine shell (and crustacea) from all levels of the excavation was kept. From a site that lay on the coast, this was not only a massive undertaking, but it meant that all shell from soil samples that were flotted was also kept, creating not only a huge volume of marine shell but also a collection that was of considerable weight and volume.. Some attempt on site was made to sort shell by species from Phase D (see the description by Catherine Smith in Non-worked Marine Shell, below). However, the analysis of the collection had to be prioritised and sampled in order to try to answer questions concerning its occurrence in features and levels associated with buildings and other areas of defined activities, compared to shell from general sand build-up and background scatter. Smith has attempted to discriminate shell deliberately brought into the site from naturally occurring shell but that has not been easy. Given the volume of this assemblage, with the vast majority of it coming from contexts not associated with structures or specific activities, its future information value will have to be evaluated (see PART 6).

In her faunal analysis (below), Judith Findlay Aird comments on some of the problems she encountered with Iain Crawford while working on the material as a post-graduate student. She was not the only one, as Dale Serjeantson (2013) who studied part of the Udal North faunal remains for her PhD also received restricted information and was not encouraged to publish. Judith was granted access to study the RUX6 material in the early 1980s but a lot had happened to the faunal remains between then and 2012 when the material was reassessed. Although faunal remains had been separated on site into categories of 'Bone-faunal, Bone-fish, Bone-worked and Bone-human, some of material had been removed and is now unexplainably missing from the collection, such as all the small mammal, and most fish and bird bones. In fact, it appears that most of this material had already been lost by 1995 (Site Archive). We are grateful that Judith, together with the project's other specialists, are hoarders of their own information, and without their records from over 30 years ago, important information from this site would have been irretrievably lost.

In spite of these inconsistencies and problems, Crawford's understanding of the natural processes affecting the site was considerable. His collection policies and methodologies for these somewhat fragile materials largely worked, even though we are left with an incomplete collection and in some cases minimal site information. Crawford's wife Imogen, an ecologist and environmentalist, also played a large role in providing scientific depth to his understanding of the natural and anthropogenic environments developed at the site, and in the

Botanical remains

By Susan Ramsay

Introduction

This archaeobotanical report details the analysis and interpretation of botanical remains recovered from samples taken during the excavations at RUX6, which were carried out during 1976 and the 1980s. However, due to the time interval since excavation and the lack of electronic records, it has not been possible to determine the exact location or context of all the samples that were excavated. The site was divided into five phases, with samples allocated to a phase, where possible.

Methodology

It is not clear how the original samples were processed for the recovery of botanical remains. Some may have been hand collected fragments of charcoal, but sieving appears to have been used for the collection of smaller remains, such as seeds.

The samples provided for botanical analysis were examined using a binocular microscope at variable magnifications of x4-x45. For each sample, estimation of the total volume of carbonised botanical material >2 mm and >4 mm was made. For each sample, all charcoal fragments were identified, together with any carbonised seeds or other plant macrofossils present within the samples.

The testa characteristics of small seeds and the internal anatomical features of all charcoal fragments were further identified at x200 magnification using the reflected light of a metallurgical microscope. Reference was made to Schweingruber (1990) and Cappers *et al.* (2006) to aid identifications. Vascular plant nomenclature follows Stace (1997).

Results and Discussion

The samples have been grouped by phase and the results are shown in Tables 4.1-5. Samples that contained no botanical remains or only burnt peat/turf/soil are not included in the tables but are listed separately.

Phase A – Modern (Table 4.1 and Figure 4.1)

The botanical remains from Phase A were all uncarbonised wood fragments and these were very well preserved, indicating their modern origins. Scots pine type was the commonest taxon present, with the fragments from SF 25051 also having the remnants of what appeared to be cream coloured paint on some edges. The sample of Scots pine wood from SF 23783 was in the form of a circular 'plug' or possibly a section cut horizontally across a circular wooden peg. Several of the other fragments also had the appearance of tapering wooden pegs. There is a possibility that some of these fragments are actually from the earlier phases of trial trenching at the site.

	SF No	23783	23811	23887	23915	25001	25051	23872
	Level	I/VIII						
	Context	Sills to top of VIII; 346/352	Ditch infill	Ditch infill	Ditch infill + cleaning saw-pit	Top BA - 19th C.	Saw-pit dig out; Neolithic - 19th C.	Ditch infill, near bottom of saw-pit
Uncarbonised wood								
Betula spp	birch	-	1 (0.36g)	-	-	-	-	-
cf. *Corylus* spp	cf. hazel	-	-	-	-	1 (3.25g)	-	-
Pinus sylvestris type	Scots pine type	1 (1.87g)	-	16 (1.65g)	1 (4.89g)	-	8 (6.10g) paint	-
cf. *Quercus* spp	cf. oak	-	-	-	-	-	1 (3.00g)	1 (21.71g)
Misc								
Bone		-	-	-	-	-	1 (1.19g)	-

TABLE 4.1:
Botanical remains from Phase A (modern)

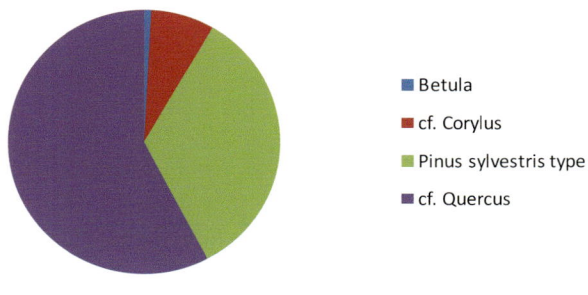

FIGURE 4.1:
Woody taxa from Phase A (modern)

Phase B – Bronze Age cist with kerbed cairn and other structures (Table 4.2 and Figure 4.2)

The contexts examined from Phase B showed significant amounts of burnt peat/turf/soil present. In addition, carbonised rhizomes are further evidence for the burning of grassland or peatland turves. Charcoal types present include birch, hazel, heather type and cf. spruce. Spruce is not native in the UK and, unless there is modern contamination, this taxon indicates the use of driftwood as fuel. Coniferous driftwood appears to have been commonly collected for fuel throughout the prehistoric period in the Western and Northern Isles, where other wood types may have been relatively scarce (Dickson 1992). Although spruce grows in mainland Europe, Dickson (ibid) considered that the finds of spruce charcoal, on coastal archaeological sites in Scotland, almost certainly originated from North America because of the direction of sea currents flowing up the west coast of Scotland.

It is impossible to be sure whether the birch and hazel charcoal originated from trees growing on the island or were also collected as driftwood. However, palynological evidence from a sub-peat stone bank near Loch Portain, North Uist, (Mills *et al.* 1994) provides evidence for birch and hazel growing on the island from the Neolithic period through until the Iron Age and possibly beyond. AMS dates were obtained on hazel charcoal from SF 26331, together with hazel and birch charcoal from SF 26364. Both were from the floors lower and upper burnt ashy floors respectively of the BG24 structure. The dates obtained from these charcoal fragments were 1880-1691 cal BC (SUERC-70220), 1759-1651 cal BC (SUERC-70221) and 1622-1456 cal BC (SUERC-70222) respectively (see Table 3.1). The heather type charcoal is almost certainly local but it is not clear whether it is evidence for the burning of above ground heather stems or a consequence of the burning of heathy turves that contain heather type woody roots. The fragments were too small to determine whether they were above or below ground stems but the lack of evidence for any burnt leaves or seed capsules might suggest that turves are a more likely source for these heather type remains. The grass, sedge and spike-rush seeds and grass stems also suggest that turves of some sort had been burned. The presence of a possible parsley-piert seed may be evidence for the onsite vegetation, since this is a species that prefers dry, nutrient-poor sandy soils.

Of particular interest are the two samples, SF 23397 and SF 26298, from the eastern half of the site that contain carbonised cereals. The majority of the cereal grains were too poorly preserved to be further identifiable, but a few could be identified as either hulled barley or just barley. This would be the expected cereal type for a site dating to the Bronze Age in Scotland and AMS dating of barley from SF 26298 gave a date range of 1751-1566 cal BC (SUERC-70219). This sample was found in the upper burnt and ashy floors associated with BG24 structure. There is no evidence for cereal processing in these samples but it is possible that any chaff or weed seeds may have been lost due to the method of sample processing.

Samples with only burnt peat/turf/soil included SFs 26318, 26474, 26480, 26489, 26513, 26528, 26576, 26169, 26243, whilst SF 26502 produced nothing identifiable.

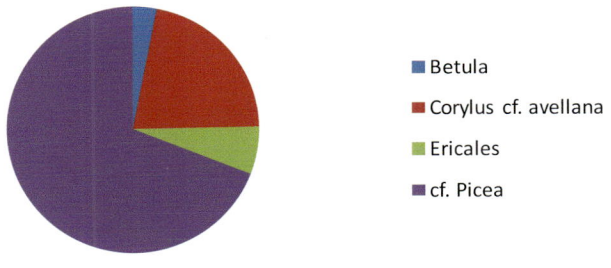

FIGURE 4.2:
woody taxa from Phase B (Bronze Age)

Part 4

	SF No	23397	25249	26331	26354	26364	26416	26501	26264	26298	26382
	Level	IX.1	IX.1	IX.22=32	IX.1/2	IX.2	IX	IX.II		IX.1/2	IX.22
	Context	East half		BG24 fl 3, SE quadrant	BG24 Cist	BG24 NE quadrant floors 1/2	BG24 west half	BG24 floor 1, west half	BG24 Cist	1983: BG24 SE Quadrant; Floors 1/2.2	1983: BG24 SE Quadrant; Floors 2/2.3
Volume of charcoal 2-4 mm		2.5ml	-	-	-	-	-	-	-	5ml	-
Volume of charcoal >4 mm		2.5ml	-	<2.5ml	<2.5ml	5ml	<2.5ml	5ml	<2.5ml	5ml	-
Charcoal	Common name										
Betula spp	birch	-	-	-	-	1 (0.03g)	-	-	-	-	-
Corylus cf. *avellana*	hazel	-	-	1 (0.05g)	3 (0.14g)	2 (0.02g)	-	-	-	-	-
Ericales	heather type	-	-	-	-	-	-	1 (0.06g)	-	-	-
cf. *Picea* spp	cf. spruce	-	-	-	-	7 (0.16g)	-	4 (0.32g)	-	1 (0.19g)	-
Indet rhizomes	indet rhizomes	9 (0.36g)	-	-	-	5ml (0.39g)	1 (0.14g)	1 (0.18g)	1 (0.03g)	30 (0.96g)	-
Burnt peat/turf/soil	burnt peat/turf/soil	-	-	10ml (3.34g)	2.5ml (1.19g)	20ml (7.66g)	2.5ml (0.97g)	5ml (1.78g)	<2.5ml (0.28g)	-	15ml (3.98g)
Uncarbonised wood											
cf. Coniferales bark	cf. conifer bark	-	1 (0.46g)	-	-	-	-	-	-	-	-
Carbonised cereals											
Hordeum vulgare var vulgare	hulled barley	-	-	-	-	-	-	-	-	2	-
Hordeum vulgare sl	barley	1	-	-	-	-	-	-	-	6	-
cf. *Hordeum vulgare sl*	cf. barley	-	-	-	-	-	-	-	-	11	-
Cereal indet	indet cereal	3	-	-	-	-	-	-	-	131	-
Carbonised seeds											
cf. *Aphanes arvensis*	parsley-piert	1	-	-	-	-	-	-	-	-	-
Carex spp (biconvex)	sedge (biconvex)	-	-	-	-	-	-	-	-	8	-
Carex spp (trigonous)	sedge (trigonous)	-	-	-	-	-	-	-	-	400	-
Chenopodium album	fathen	-	-	-	-	-	-	-	-	3	-

TABLE 4.2:
Botanical remains from Phase B (Bronze Age cist with kerbed cairn and other structures)

	SF No	23397	25249	26331	26354	26364	26416	26501	26264	26298	26382
cf. *Danthonia decumbens*	heath grass	1	-	-	-	-	-	-	-	-	-
Eleocharis spp	spike-rush	-	-	-	-	-	-	-	-	5	-
Poaceae	grass	-	-	-	-	-	-	-	-	1	-
Poaceae stems	grass stems	10	-	-	-	-	-	-	-	1	-
Ranunculus spp	buttercup	-	-	-	-	-	-	-	-	1	-
Stellaria / Cerastium	stitchwort/ mouse-ear	-	-	-	-	-	-	-	-	2	-
Misc											
Bone			-	-	-	-	4 (1.23g)	-	1 (0.15g)	-	-	1 (0.12g)

TABLE 4.2 continued:
Botanical remains from Phase B (Bronze Age cist with kerbed cairn and other structures)

Phase C – Bronze Age pits (Table 4.3)

Only two samples were analysed from Phase C. Unfortunately, little can be interpreted from these since SF 25751 produced no botanical remains and SF26250 contained only traces of carbonised rhizomes and a few uncarbonised seeds that may be relatively modern.

	SF No	25751	26250
	Level	IX.3	IX.3
	Context	East of saw-pit	1983: East of saw-pit; pits + level
Volume of charcoal 2-4 mm		-	-
Volume of charcoal >4 mm		-	<<2.5ml
Charcoal	Common name		
Indet rhizomes	indet rhizomes	-	6 (0.02g)
Uncarbonised seeds			
Brassicaceae	cabbage family	-	1
Cerastium / Stellaria	mouse-ear / stitchwort	-	1
Rumex acetosa	sheep's sorrel	-	1

TABLE 4.3:
Botanical remains from Phase C (Bronze Age pits)

Phase D – Neolithic structures and activity (Table 4.4 and Figure 4.3)

RUX6 - Phase D (Neolithic)

FIGURE 4.3
Woody taxa from Phase D (Neolithic)

The samples examined from Phase D contained relatively small quantities of carbonised remains, with heather type being the main charcoal type represented, although there was little evidence for the burnt peat/turf/soil that was common in the later phases. This might suggest that 'above ground' heather stems were being burnt and this would be supported by the findings of carbonised heather seed capsules in SF 23744. This sample also produced a few carbonised grass seeds and stems but also three small fragments of burnt Fucoid seaweed. It is not clear whether this seaweed has been burned deliberately as fuel or is simply an accidental incorporation as a result of a fire having been set on a seashore environment. The only tree taxa that were identified were cf. larch and cf. spruce, both non-native species and so indicating the collection and burning of driftwood. The lack of other tree taxa is difficult to understand since the palynological work from Loch Portain, North Uist, (Mills *et al.* 1994) and Loch Laing, South Uist (Bennett *et al.* 1990) produced a range of broad-leaved taxa such as hazel, birch, alder and willow that should have been present on the island at this time.

This might suggest that the fires set during this period were relatively ad hoc and so simply used whatever fuel was immediately to hand. There are no cereal grains from this phase of occupation, perhaps suggesting that either cereal processing was carried out elsewhere or that the site was not consistently occupied and may have had a non-domestic use. This lack of evidence for cereal grain was also noted at An Doirlinn (Garrow and Sturt 2016: 59), which was considered unusual, and

	SF No	26615	25147	25156	27214	26821	25496	25621	25739	25993	23663	23744	23694	27023
	Level	XI-XI.22	XI.01	VII/ top BA	XI.31	IX.2 early	IX.01	IX.1	IX.1		XI.02	XI.02	XI.02	XI.2
	Context	East of saw-pit, libation pit	Neo Building 2	Neo Building 2	Neo Building 2, floor 1 hearth & east of hearth	West of saw-pit, hearth in IMP of Neo 2	Neo Building 2 f1.1 - shell pit	Neo Building 2, top of f1.1	Neo Building 2, NW side, top f1.21	Neo Building 2 f1.1 - west of hearth	Neo Building 1; pit	Neo Building 1; floor	Hearth + circa (BB1)	1986: East external green b.; Neo
Volume of charcoal 2-4 mm		-	-	-	-	-	-	-	-	-	-	<2.5ml	-	-
Volume of charcoal >4 mm		2.5ml	-	-	<2.5ml	2.5ml	5ml	<2.5ml	<2.5ml	-	5ml	-	<2.5ml	-
Charcoal	Common name													
Ericales	heather type	6 (0.52g)	-	-	-	8 (0.22g)	27 (0.56g)	1 (0.08g)	4 (0.08g)	-	4 (0.48g)	4 (0.01g)	4 (0.03g)	-
cf. *Larix* spp	cf. larch	-	-	-	2 (0.04g)	-	-	-	-	-	-	-	-	-
cf *.Picea* spp	cf. spruce	8 (0.20g)	-	-	-	-	-	-	-	-	-	-	-	-
Indet cinder	vesicular cinder	-	-	-	-	-	-	-	-	-	-	14 (0.15g)	-	-
Burnt peat/ turf/soil	burnt peat/ turf/soil	-	-	2.5ml (0.66g)	-	-	-	-	-	-	-	-	-	-
Carbonised seeds														
Calluna vulgaris seed capsules	heather seed capsule	-	-	-	-	-	-	-	-	-	-	4	-	-
Fucoid seaweed	fucoid seaweed	-	-	-	-	-	-	-	-	-	-	3	-	-
Poaceae	grass	-	-	-	-	-	-	-	-	-	-	2	-	-
Poaceae stems	grass stems	-	-	-	-	-	-	-	-	-	-	7	-	-
Uncarbonised seeds														
Carex spp (biconvex)	sedge (biconvex)	-	-	-	-	-	-	-	-	-	-	-	-	1
Carex (trigonous)	sedge (trigonous)	-	-	-	-	-	-	-	-	-	-	-	-	1
Cerastium / Stellaria	mouse-ear /stitchwort	-	-	-	-	-	-	-	-	-	-	-	-	6

TABLE 4.4:
Botanical remains from Phase D (Neolithic structures and activity)

where the authors considered that different social practices or preferences may have been responsible as well as the soil conditions.

The uncarbonised seeds of sedge and mouse-ear / stitchwort from SF 27023 may be relatively modern since no other botanical remains were recorded from this sample.

Other samples (Tables 4.5a and 4.5b. and Figure 4.4)

Unfortunately, a large number of samples from this site could not be attributed to any particular phase or exact location. However, many of these samples produced only burnt peat/turf/soil (SFs 23122, 23132, 23253, 23532, 23544, 23581, 23996, 24993, 25081, 25097, 25118, 25162, 25196, 25258, 25275, 25297, 25323, 25338, 25359, 25364, 25395, 25426, 25469, 25482, 25502, 25514, 25518, 25557, 25600, 25647, 25777, 25812, 25936, 25958, 26198, 26230, 26914), or nothing identifiable (SFs 17605, 23272, 23318, 23367, 23368, 23888, 23948, 24020 - 24030, 25808 - 25810, 25820 - 25823, 25825, 26243, 26248, 26297, 26412, 26470, 26507, 26509, 26534, 26560, 26770, 26894 - 26902, 26908, 26909, 27206, 27207).

Overall, these samples produced mainly heather type charcoal, but there were concentrations of Scots pine type charcoal in SF 25018 and SF 25803. These samples also produced a type of vesicular cinder, with no obvious cellular structure, that may be the remains of a resinous substance from the original pine wood. It is impossible to say whether this Scots pine is the remains of driftwood or whether it was collected from local woodlands. Probable spruce charcoal was present in SF 25365 and SF 27218, with cf. larch also present in SF 27218, indicating the presence of driftwood in both of these samples. SF 24813 produced the only piece of willow charcoal from this site whilst SF 23085 contained a single large fragment of charred botanical material, which may be the remains of a holdfast from a relatively large species of seaweed.

These unstratified samples did not produce any carbonised cereal grains or other seeds and so add little to the evidence for domestic occupation on the site.

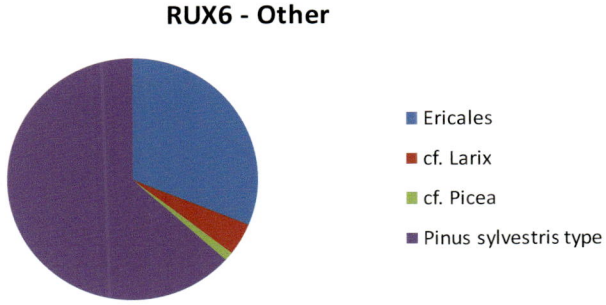

FIGURE 4.4:
Woody taxa from other samples

	SF No	23085	23119	23361	24813	25018	25198	25231	25232	25365	25409
	Level	I	II/III	IX.I	BA		VII/VIII	VI/IX.1/2	VI/IX.1/2	IX.1	IX.3/X
	Context	Under turf, west half	West half	East half	Andrew's trench, B Age level	Top BA/19th C.	East of saw-pit				
Volume of charcoal 2-4 mm		-	-	-	-	-	-	-	-	-	-
Volume of charcoal >4 mm			20ml	-	<2.5ml	<2.5ml	10ml	<2.5ml	-	<2.5ml	10ml	<2.5ml
Charcoal	Common name										
Coniferales	conifer	-	-	-	-	-	-	-	5 (0.12g)	-	-
Ericales	heather type	-	-	1 (0.33g)	-	-	1 (0.06g)	-	-	-	-

TABLE 4.5a:
Botanical remains from other samples

	SF No	23085	23119	23361	24813	25018	25198	25231	25232	25365	25409
cf. *Picea* spp	cf spruce	-	-	-	-	-	-	-	-	46 (1.22g)	-
Pinus sylvestris type	Scots pine type	-	-	-	-	43 (0.76g)	-	-	-	-	-
Salix spp	willow	-	-	-	1 (0.07g)	-	-	-	-	-	-
cf. seaweed holdfast	cf. seaweed holdfast	1 (3.22g)	-	-	-	-	-	-	-	-	-
Indet rhizomes	indet rhizomes	-	-	-	-	-	-	-	-	-	1 (0.16g)
Indet cinder	vesicular cinder	-	10ml (1.38g)	-	-	3 (0.22g)	-	-	-	-	-
Burnt peat/turf/soil	burnt peat/turf/soil	-	-	-	-	-	5ml (1.53g)	10ml (2.70g)	-	-	-

TABLE 4.5a continued:
Botanical remains from other samples

	SF No	25431	25491	25659	25720	25803	25982	26772	27069	27203	27218
	Level	IX.1/2		IX.2		IX.1/2	XI.3-IX.1	IX.22	IX.1/2	IX.2	
	Context	East of saw-pit		East of saw-pit	Chuck out end of saw pit/ Neo floor	East of saw-pit	Burnt black outside Neo Building 1&2	East of saw-pit	East of saw-pit	Surface collection	
Volume of charcoal 2-4 mm		-	-	-	-	10ml	-	-	-	-	-
Volume of charcoal >4 mm		<2.5ml	2.5ml	<2.5ml	-	30ml	10ml	<2.5ml	-	-	10ml
Charcoal	Common name										
Ericales	heather type	1 (0.02g)	8 (0.42g)	1 (0.05g)	5 (0.06g)	-	21 (1.88g)	-	-	-	20 (0.78g)
cf. *Larix* spp	cf. larch	-	-	-	-	-	-	-	-	-	16 (0.44g)
cf. *Picea* spp	cf. spruce	-	-	-	-	-	-	3 (0.12g)	-	-	-
Pinus sylvestris type	Scots pine type	-	-	-	-	83 (6.50g)	-	-	-	-	-
Indet rhizomes	indet rhizomes	1 (0.07g)	-	-	-	-	-	-	-	-	-
Indet cinder	vesicular cinder	-	-	-	-	12 (0.42g)	-	-	-	-	-
Burnt peat/turf/soil	burnt peat/turf/soil	10ml (2.63g)	-	20ml (11.39g)	-	-	-	-	<2.5ml (0.04g)	-	-
Misc											
Bone		-	-	-	-	-	-	-	-	4 (1.49g)	-
Burnt inorganic material		-	-	-	-	-	-	-	1.5ml (0.89g)	-	-

TABLE 4.5b:
Botanical remains from other samples

Pollen

By Beverley Ballin Smith and Keith Bennett

Iain Crawford was interested in the survival of pollen as an indicator of past vegetation and climate conditions. In 1995-6 he provided a colleague in Cambridge, Keith Bennett (now Professor of Environmental Change, University of St, Andrews) with nine soil samples spanning Phases B to E from the east side of the site. They were collected in 1984, with the aim of seeing whether pollen survived at the site. He had previously approached Professor Richard G West, Head of Department of Botany at Cambridge who recommended Dr Bennett.

The following information and summary pollen data has been kindly supplied by Professor Bennett during a short email exchange in 2012. The data (Table 4.6) indicates that grains of pollen were present in the soil samples, and that they were counted and identified (by Steve Borham, Cambridge). According to the data there was very little pollen, and what there is was surprisingly dominated by tree species. Their occurrence was interpreted as being the more resistant pollen grains that survived burial and taphonomic conditions. An alternative explanation was that they could have been modern contamination from elsewhere. Unfortunately, we have no information on how the samples were taken, and detailed site information was not recorded for all of them. We do not know how long the soils had been exposed on site or whether it was, for example, a windy day when the samples were taken. However, what is recorded is that all were contaminated with modern glass.

SF 26507 was taken from the NW quadrant of the second ash floor in the structure BG24. SF 26895 was from the lower level of shingle deposited during a period of Bronze Age erosion and is not clearly associated with Phase D, the late Neolithic. SF 26896 is associated with the ash floor of one of the two late Neolithic buildings. SF 26897 and 26898 are the anthropogenic upper portions of the natural subsoil (Phase E) with the remainder being weathered glacial till with traces of windblown shell sand (see Table 2.1).

Although the potential for pollen may have existed in the subsoil layers of Phase E, it is highly likely that the results show contamination from elsewhere. The low number of grains reinforces this suggestion. This data is presented here as it was an attempt by Crawford to gain further information about the past environment of the site. If he had used modern retrieval methods and recorded the conditions of sample retrieval, he might have had more success.

Small find No.	Level	Phase	Concentration (grains/cm3)	Types present
26507	IX.2	B	1242	Gram., (Corylus, Ericales, Plantago, Sphagnum)
26895	IX.3	B/D	474	Alnus, Corylus (Ulmus, Sphagnum)
26896	XI.2	D	1120	Alnus
26897	XI.3	E	49	Alnus
26898	XI.4	E	191	Alnus
26899	XII	E	719	Alnus
26900	XIII.1	E	595	Alnus
26901	XIII.2	E	139	Alnus
26902	XIV.I	E	1180	Alnus, (Pinus, Corylus, Ericales, Plantago

TABLE 4.6: Summary of pollen analysis from soil samples

The faunal remains

By Judith Finlay Aird

Introduction

The prehistoric faunal material from the Udal was studied between 1981 and 1984 and formed part of the source material for a PhD thesis (Finlay 1984). An interim report was never completed as Iain Crawford did not feel he was able to confirm final contexts for the material and could not communicate a timescale for publication within which a report would be framed. Material excavated from prehistoric contexts at the site after 1983 were analysed in 2016 and the data added to the 1984 listings. In 2016 a careful comparison was made between the material identified in the 1980s and that extant now and it is clear that the past 35 years have not been kind to the material as the collection has been picked through on at least one occasion. Some of the material is crushed or even missing – including almost all of the bird bone and fish bone - and there are absences of bones previously recorded against specific Small Finds numbers - often with a small cardboard label left behind in the bag of other bones from that SF number anonymously stating "removed for ref."

or similar. As a result, the identifications made in the 1980s have been used for the current analysis, augmented by the more recently excavated material not previously seen.

Methodology

Unless otherwise noted, all identification and analysis has been undertaken by the author, under the principles and methods outlined below. All the faunal remains were identified through the use of modern comparative material; textbooks of comparative faunal anatomy and the generous aid of a number of people (see Acknowledgements - Judith Finlay Aird).

Bones were identified with regard to species, anatomical element, position left or right of the axial skeleton, state of epiphyseal fusion and the stage reached in the sequence of tooth eruption and replacement at death. On the whole, ribs and vertebrae have not been identified as to species, since the value of the results so obtained does not generally justify the time involved.

Measurements of the mammal bones were taken where possible, following the standardised scheme of von den Driesch (1976) and care was taken to measure only bones of mature animals which showed no sign of erosion at the points of measurement. Burnt and chewed bones, along with those showing any abnormality, were not measured. Metrical analysis was taken for two purposes: to differentiate between species, type and sex, and to assess the size and quality of the main food animals and to note any variation over the time span under consideration. Measurements were statistically analysed but due to the small size of the sample available, the results were statistically insignificant.

Recovery methods can considerably affect the nature of the faunal assemblage available for study (Payne 1972), with un-sieved samples biased towards the larger elements as smaller bones are easily overlooked in hand excavation. At the Udal, dry-sieving was standard for all deposits and wet-sieving was carried out on a sample basis. From personal experience at the site in the early 1980s it was noted how often bird, fish and the smaller elements of larger mammals were recovered in the sieves rather than by hand collection. As mentioned earlier, this positive feature has been severely reduced by post-excavation factors of storage, loss and interpretation, which reduces the material available to those from contexts securely identified by 1983 or in 2016.

Various methods of estimating the sex ratios of faunal assemblages have been established widely and may consist of morphological differences between the sexes but in many cases the size and proportions of the bones may be the only distinguishing criteria. The RUX6 sample is too small to be statistically valid for measurement and the results are inconclusive. There are no also no horn cores present in the RUX6 material.

An estimation of the age of an animal at death can provide the basis for consideration of several aspects of the economy or even the function of a site, giving information about possible hunting/herding patterns, dietary preferences and ecological background. In the present study, ageing was investigated using a review of epiphyseal fusion (Silver 1969), the sequence of tooth eruption and replacement (ibid) and the degree of tooth wear (Grant 1982). The study of incremental structures in teeth was not applied to this material. Tooth wear analysis was considered using the schemes proposed by Payne (1973) and Grant (1975) but the results only served to augment the tooth eruption data and could not be analysed independently, owing to the small number of mandibles available. There are no known comparative studies of aging and sexing of "unimproved" breeds under the nutritional conditions naturally prevalent in the Outer Hebrides and so any suggestions of ageing and sexing of the RUX6 sheep and cattle must be seen as part of an overall scheme of relative values.

The quantifying of the faunal material from RUX6 has been estimated using both NISP and MNI methods. The Number of Individual Specimens (NISP) method records each bone or fragment which can be confidently attributed to a particular species, and counts loose teeth separately. The Minimum Number of Individuals (MNI) method is based on counting the most frequently represented skeletal element (taking into account the factors of age, size and left/right) for each context. The small size of the assemblage makes any estimation of numbers questionable, regardless of the method used, as we are unable to determine the relative importance of the many factors affecting its survival: human and animal transportation; chemical and physical forces while buried/exposed; site erosion and the vagaries of excavation and post-excavation work. All of these are known agents in the creation of the faunal collection reported here.

Phase	Context	O/C	Bos	Equus	Canis	Cervus	Antler	Rod	Lag	Aves	Marine mammal	Fish	NID
A	total	x	x	x	x	x	x	x	x	x		x	x
B	total	x	x		x	x	x	x	x	x	x		x
B	Bronze Age kerbed cairn and cist	x	x			x			x			x	x
B	Bronze Age structure (BG4 also known as the tadpole and B24)	x	x		x	x				x			x
C	total	x	x		x	x	x	x	x				
C	Phase C pits	x	x			x						x	x
D	total	x	x		x	x	x		x	x			
D	Neolithic Building 1	x	x			x	x			x		x	x
D	Neolithic Building 2	x	x			x	x			x		x	x
D	Neolithic pits (associated with Libation pit and stone settings)	x	x										
D	Neolithic ritual structures (Libation pit & stone settings??)	x	x				x						x
E	total	x	x			x							x

O/C	Sheep/goat	Antler	Antler
Bos	Cattle	Rod	Rodent
Equus	Horse	Lag	Rabbit/Hare
Canis	Dog	Aves	Bird
Cervus	Deer	NID	Not identified

TABLE 4.7:
Species by phase.

Interpretation of the evidence

Condition of preservation

The calcareous machair soils preserve bone to a high degree and generally the faunal material shows little signs of degradation. Some RUX6 bone shows dark staining and some appears extensively weathered, presumably due to exposure at the surface or perhaps from the action of wind/water uncovering and recovering deposits with blown sand. This is particularly true in Phase A but it did not hinder identification factors.

Species present at RUX6

Species excavated from RUX6 (by phase and context) are shown in Table 4.7.

Domesticated species represented

Throughout the phases at RUX6, ovi-caprids (sheep/goat) and cattle are the predominant species present. There is no evidence of goat in the form of horn cores or of the slight but significant differences in identification of certain anatomical elements (after Boessneck 1971), so all small ruminant elements have been considered as sheep. There is no evidence to suggest that either sheep or cattle were wild species and they are therefore presumed to be domesticated animals, husbanded for food.

Horse is represented only by 2 teeth from Phase A (up to the nineteenth century) and there is no evidence of pig among the assemblage from RUX6, although pig bones and teeth are found throughout the later prehistoric phases at the later Udal sites nearby (Finlay 1984).

A range of sizes of domesticated dog appear to be present in Phases A to D inclusive, although it is difficult to assess this from a few bones and teeth. Certainly there is evidence of an animal the size of a whippet, one like a modern collie and a smaller terrier-sized type. None of the dog bones showed any sign of cutting, burning or chewing and the animals are presumed to have been kept as working animals and/or pets.

Relative frequency of main food species

Table 4.8 shows both the total count of identifiable fragments of each of the main food species and the estimated Minimum Number of Individuals represented by the mammal remains. Among the main food species, a consistently high percentage of between 82% (Phase D) and 52% (Phase B) of identified bone fragments is of sheep and an inverse percentage of between 12% (Phase D) and 37% (Phase B) is of cattle, showing the significance of these species in the function of the site. These figures are slightly altered when the MNI is taken as the basis for comparison between the species but the significant importance of sheep over cattle in Phase D and the greater equity between the two species in Phase B remains. The ratio of sheep to cattle in each phase is also shown in Table 4.9a and b, based on both the fragments count and on MNI estimations. The ratio varies slightly according to the calculation method employed and it is likely that the figure based on MNI estimation is more likely to be nearer the truth as it takes into account the differential fragmentation of sheep and cattle bones. This gives a ratio of 3:1 sheep:cattle in Phase D reducing to effectively 1:1 by Phase B.

There is no evidence for Red Deer being a significant factor in the diet of the occupants of RUX6, due to the paucity of bones and teeth identified from the different Phases. At first glance deer look particularly significant in Phase C but the small sample sizes there made the data statistically insignificant and no value can be placed on the figures. Teeth generally indicate animals of maturity due to eruption and wear patterns and it is always possible that deer were not deliberately targeted as food animals. This suggestion is backed by the apparent lack of spear/arrowhead evidence in the lithic collection (Torben Ballin pers.com.) and the bias on foot bones and antler in the deer remains. Collection of targeted parts of dead specimens (presumably deceased due to natural causes) would provide both antlers and metapodial bones as raw material for implement and ornament manufacture (see PART 5). The presence of a cast antler, including pedicle and brow-tine, also supports opportunistic collection.

Age of food animals at death

Age estimation of the main food mammals is unsatisfactory, largely due to the small samples available from this site. Due to the fragmentary condition of many of the maxillae and mandibles, it is generally possible only to indicate that a jaw represents an animal over or under a particular age and this causes problems of grouping and interpretation. The situation regarding tooth wear is even worse as this method (e.g. Grant 1975) is based solely on lower teeth and jaws which reduces the sample still further. Epiphyseal fusion seems to give more intelligible results but the samples are still woefully small and can only indicate the likely situation.

Table 4.10 compares the epiphyseal data for sheep by phase and shows that only in Phase D can any pattern be suggested. Here, most sheep seem to have been killed before they were 3 years old with only a few surviving to a greater age. Few neo-natal animals are represented and sheep under 12 months old are rare. This pattern of most animals being killed between 6 and 30 months old is supported by the evidence of tooth eruption in Table 4.11a which also indicates that some sheep were allowed to survive to over 40 months. Tooth eruption data is limited but shows no evidence for animals dead under 6 months. It is suggested that the evidence for sheep at the site represents a breeding flock with the gradual culling of surplus animals from lambs under 10 months of age right up to old ewes over 40 months.

Table 4.12 shows that there is little epiphyseal aging evidence for cattle in any of the phases due to the small numbers of identifiable elements which can be used in age estimation. The available tooth eruption data in Table 4.11b almost totally consists of loose teeth, giving little in the way of definitive aging and generally indicating an age bracket of 3 years' duration rather than a post-eruption boundary. It appears that most animals were killed off between 6 and 42 months, with some surviving over 42 months old and a few over 48 months. It is suggested that the evidence of cattle from the site may represent a small dairy resource with male animals killed before their second winter and female animals being overwintered and maintained as providers of dairy products and the continuance of the herd, but the small sample size makes this merely conjecture.

Species	ovi-caprid			Bos			Equus			Canis			Cervus			seal		
Age group	adult	juvenile	neo-natal	adult	juvenile	neo-natal	adult	juvenile	neo-natal	adult	juvenile	neo-natal	adult	juvenile	neo-natal	adult	juvenile	neo-natal
Phase A	4	2	1	4	1	1	1	0	0	2	0	0	1	0	0	0	0	0
modern	88	NISP		39	NISP		0	NISP		1	NISP		3	NISP		0	NISP	
	36	loose teeth		33	loose teeth		2	loose teeth		1	loose teeth		3	loose teeth		0	loose teeth	
Phase B	4	1	1	5	1	1	0	0	0	1	0	0	2	0	0	1	0	0
LBA	33	NISP		23	NISP		0	NISP		0	NISP		7	NISP		0	NISP	
	35	loose teeth		63	loose teeth		0	loose teeth		2	loose teeth		5	loose teeth		2	loose teeth	
Phase C	1	0	1	1	0	0	0	0	0	1	0	0	2	0	0	0	0	0
LBA	14	NISP		2	NISP		0	NISP		1	NISP		5	NISP		0	NISP	
	12	loose teeth		12	loose teeth		0	loose teeth		0	loose teeth		1	loose teeth		0	loose teeth	
Phase D	15	5	2	5	1	1	0	0	0	1	0	0	2	1	1	0	0	0
Late Neolithic	432	NISP		64	NISP		0	NISP		1	NISP		31	NISP		0	NISP	
	65	loose teeth		37	loose teeth		0	loose teeth		1	loose teeth		7	loose teeth		0	loose teeth	
Phase E	1	0	0	1	0	0	0	0	0	0	0	0	1	0	0	0	0	0
natural	1	NISP		3	NISP		0	NISP		0	NISP		0	NISP		0	NISP	
	1	loose teeth		0	loose teeth		0	loose teeth		0	loose teeth		1	loose teeth		0	loose teeth	

Count by context

Species	ovi-caprid			Bos		
Age group	adult	juvenile	neonatal	adult	juvenile	neonatal
Phase D	4	3	1	4	1	1
Building 1 (DJ)	166	NISP		29	NISP	
	20	loose teeth		3	loose teeth	
Phase D	13	3	2	1	1	1
Building 2 (DH)	214	NISP		22	NISP	
	303	loose teeth		17	loose teeth	

NISP Number of individual specimens

TABLE 4.8:
Relative frequency of species.

	ovi-caprid (sheep/goat)		Bos (cattle)		Cervus (deer)		totals	ratio sheep:cattle (NISP)	
Phase	NISP	%	NISP	%	NISP	%	NISP	sheep	cattle
A	88	68%	39	30%	3	2%	130	2.3	1
B	33	52%	23	37%	7	11%	63	1.4	1
C	14	67%	2	10%	5	24%	21	7.0	1
D	432	82%	64	12%	31	6%	527	6.8	1
E	1	25%	3	75%	0	0%	4	0.3	1
TOTAL	568	79%	131	18%	46	6%	745	4.3	1

Estimation of relative significance of food mammals by NISP (Number of Individual Specimens) calculation (excluding neo-natal bones and loose teeth)

TABLE 4.9a:
Ratios of food mammals.

Phase	ovi-caprid (sheep/goat)		Bos (cattle)		Cervus (deer)		totals	ratio sheep:cattle (MNI)	
	MNI	%	MNI	%	MNI	%	MNI	sheep	cattle
A	7	50%	6	43%	1	7%	14	1.2	1
B	6	40%	7	47%	2	13%	15	0.9	1
C	2	40%	1	20%	2	40%	5	2.0	1
D	22	67%	7	21%	4	12%	33	3.1	1
E	1	33%	1	33%	1	33%	3	1.0	1
TOTAL	38	54%	22	31%	10	14%	70	1.7	1

Estimation of relative significance of food mammals by MNI (Minimum Number of Individuals) calculation (excluding neo-natal bones and loose teeth)

TABLE 4.9b:
Ratios of food mammals.

Age (months)	Bone and epiphysis	ovi-caprid (sheep/goat)									
		Phase A		Phase B		Phase C		Phase D		Phase E	
		unfused	fused	unfused	fused	unfused	fused	unfused	fused	unfused	fused
Neo-natal	metapodial proximal		4		2				29		
	% unfused										
10	humerus distal radius proximal		6	1	2			2	20		
	% unfused			33%				9%			
13-16	1st phalange proximal 2nd phalange proximal	1	7	1	2			7	38		
	% unfused	12.55		33%				16%			
18-24	metacarpal distal tibia distal	4	2	1			1	7	6		
	% unfused	67%		100%				54%			
20-28	metatarsal distal							1	1		
	% unfused										
30-36	radius distal ulna proximal femur proximal	4	2	1	1			21	13		
	% unfused	67%		50%				62%			
36-42	femur distal tibis proximal tibia proximal humerus proximal	2						11	4		
	% unfused	100%						73%			
		Bos (cattle)									
Neo-natal	metapodial proximal phalange distal		8		4			3	5		
	% unfused	-		-		-		37%		-	
7-18	humerus distal radius proximal scapula proximal 1st phalange proximal 2nd phalange proximal	3	6		4			4	2		

TABLE 4.10:
Sheep epiphyseal ageing data.

	ovi-caprid (sheep/goat)				
Age (months)	Phase A	Phase B	Phase C	Phase D	Phase E
Under 6 months	0	0	0	0	0
%					
6-18	1	4	1	2	0
%	33%	40%	20%	10%	
6-30	2	3	2	4	0
%	66%	30%	40%	20%	
18-30	0	1	1	5	0
%		10%	20%	25%	
Over 30	0	1	1	3	0
%		10%	20%	15%	
over 40	0	1	0	6	0
%		10%		30%	
TOTAL	3	10	5	20	0

Table 4.11(a)

	Bos (cattle)				
Age (months)	Phase A	Phase B	Phase C	Phase D	Phase E
Under 6 months	0	0	0	0	0
%					
Under 18 months	0	1	0	0	0
%		12%			
6-42	2	3	1	3	0
%		38%	100%	38%	
Over 30	0	1	0	0	0
%		12%			
Over 42	1	0	0	4	0
%				50%	
over 48	1	3	0	1	0
%		38%		12%	
TOTAL	4	8	1	8	0

Table 4.11(b)

TABLE 4.11a and 4.11b:
Tooth eruption data (sheep and cattle).

		Bos (cattle)									
Age (months)	Bone and epiphysis	Phase A		Phase B		Phase C		Phase D		Phase E	
		unfused	fused	unfused	fused	unfused	fused	unfused	fused	unfused	fused
Neo-natal	metapodial proximal phalange distal		8		4			3	5		
	% unfused	-		-		-		37%		-	

TABLE 4.12:
Cattle epiphyseal ageing data.

Age (months)	Bone and epiphysis	Bos (cattle)									
		Phase A		Phase B		Phase C		Phase D		Phase E	
		unfused	fused	unfused	fused	unfused	fused	unfused	fused	unfused	fused
7-18	humerus distal radius proximal scapula proximal 1st phalange proximal 2nd phalange proximal	3	6	4				4	2		
	% unfused	33%		-		-		67%		-	
24-36	metacarpal distal metatarsal distal tibia distal				1			3	4		
	% unfused	-		-		-		53%		-	
36-42	calcaneum										
	% unfused	-		-		-		-		-	
42-48	femur proximal femur distal tibia proximal humerus proximal radius distal ulna proximal		1						1		
	% unfused	-		-		-		100%		-	

TABLE 4.12 continued:
Cattle epiphyseal ageing data.

Size of food animals

Measurements of sheep bones are shown in Figure 4.5 and Figure 4.6, compared to representative examples from modern Soay and Shetland breeds which are generally considered to be more representative of 'unimproved' breeds. The RUX6 examples appear to be closer to the Shetland than the Soay examples and consideration of the RUX6 material against the sheep bones from the later pre-historic wheelhouse contexts at the Udal South shows no significant alteration in the stature of these small, slender animals throughout pre-history (Finlay 1984) Figure 4.7.

There are insufficient measurable cattle bones from RUX6 to graph the results, but consideration of the RUX6 material against the cattle bones from the later contexts at the Udal appears to show no significant size change between periods and suggests a small, stocky breed, similar in skeletal proportions to an example of a 'Celtic Ox' (RSM 1905-46) in the reference collection of the National Museums Scotland.

Health of the main food animals

There is no evidence of pathology (deformity or disease) evident in the bone record at RUX6, suggesting that the animals kept were generally well-nourished and healthy.

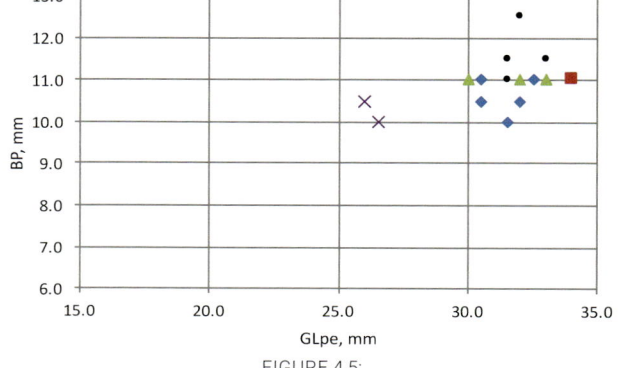

FIGURE 4.5:
Comparative size of RUX6 sheep – 1st phalanges.

Purple cross = Modern Soay   Blue diamond = RUX6
Red square = Udal RUX3   Black circles = Modern Shetland
Green triangle = Northton Neolithic

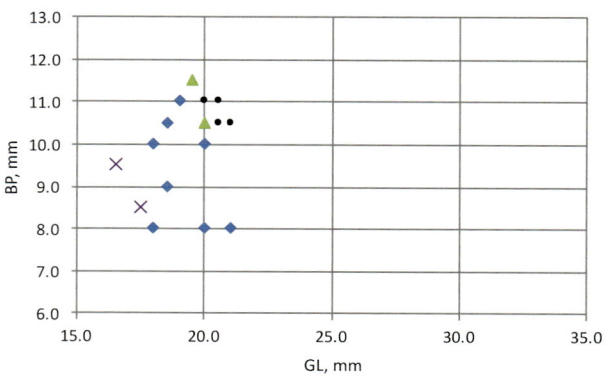

FIGURE 4.6:
Comparative size of RUX6 sheep – 2nd phalanges.

FIGURE 4.7:
Modern day Shetland sheep. © BBS.

Butchery practices

Much of the identified cattle and sheep remains – and indeed much of the material not identifiable to species but clearly long-bone fragments – show signs of butchery and a study has been made of the degree of fragmentation of bone for sheep and cattle. Conclusive results are not possible due to the small sample size but the data supports the hypothesis that the larger (cattle) bones are liable to be broken into more pieces than the smaller (sheep) bones. In the study the main meat bones (humerus, radius, femur and tibia) were compared to the bones most suitable as potential raw material (metapodials) and the number of complete bones was expressed as a percentage of the total identified bones within each of the two categories. Table 4.13 shows this fragmentation data for sheep and cattle. The degree of fragmentation was similar for meat bones and for metapodials for sheep but it is recognised that the presence of worked metapodials in the artefacts recovered from the site (see PART 5) shows that the data is flawed as suitable bones as raw material will probably have been extracted intact from the carcase. Very few intact cattle bones were recovered, reflecting their size and the need to process the carcase into manageable portions for the pot or spit.

	ovi-caprid (sheep/goat)						Bos (cattle)					
	Humerus, femur, radius, tibia			Metapodials			Humerus, femur, radius, tibia			Metapodials		
Phase	complete bones	NISP	% complete	complete bones	NISP	% complete	complete bones	NISP	% complete	complete bones	NISP	% complete
A	1	24	4%	0	15	0%	0	3	0%	0	4	0%
B	0	8	0%	0	7	0%	0	25	0%	0	18	0%
C	0	1	0%	0	1	0%	0	0	0%	0	0	0%
D	2	99	2%	2	66	3%	0	0	0%	0	0	0%
E	0	0	0%	0	0	0%	0	0	0%	0	0	0%
TOTAL	3	132	2%	2	89	2%	0	28	0%	0	22	0%

NB. Neo-natal bones have been excluded from these figures.

TABLE 4.13:
Fragmentation data.

TABLE 4.14:
Butchery cuts

Cut type	ovi-caprid (sheep/goat)														
	strip	strip	strip	chop	strip	strip	chop	chop	chop	chop	chop	chop	chop	longitudinal
Phase	1	2	3	4	5	6	7	8	9	10	11	12	13	split
A	-	1	-	-	1	-	-	2	1	1	-	-	2	11
B	-	-	-	1	-	-	-	-	-	-	-	-	-	3
C	-	1	-	-	-	-	-	-	-	-	-	-	-	-
D	-	-	-	-	-	-	-	-	3	1	-	-	1	23
E	-	-	-	-	-	-	-	-	-	-	-	-	-	-
TOTAL	0	2	0	1	1	0	0	2	4	2	0	0	3	37

Cut type	Bos (cattle)														
	strip	strip	strip	chop	chop	strip	strip	chop	chop	strip	chop	chop	strip	chop	longitudinal
Phase	1	2	3	4	5	6	7	8	9	10	11	12	13	14	split
A	-	1	-	-	-	-	-	-	-	-	-	-	-	-	10
B	-	-	-	-	1	-	-	-	-	-	1	-	-	-	2
C	-	-	-	-	-	-	-	-	-	-	-	-	-	-	-
D	-	-	-	-	1	-	-	-	-	-	1	-	-	-	26
E	-	-	-	-	-	-	-	-	-	-	-	-	-	-	1
TOTAL	0	1	0	0	2	0	0	0	0	0	2	0	0	0	39

RUX6 bones were also examined for chop marks where the bone had been split or the joint separated and for cut marks indicating the cutting of tendons or skinning activity. In general, long bones were split at joints, and/or at mid-shaft in a medio-lateral direction and/or split sagitally to facilitate marrow extraction. The overall impression is that carcases were reduced to fairly small portions, presumably to fit into a cooking pot. These were compared with the pattern of butchery which was established for the nearby Iron Age Wheelhouse site at Sollas (Finlay 1981) and the two were found to tally closely. Neolithic sheep bones from RUX6 were more often broken mid-shaft but whether for marrow extraction or meat removal is impossible to determine. In general, both sheep and cattle bone from RUX6 falls into the same pattern of butchery as that from Sollas, although the number of bones exhibiting cut marks is much fewer. Table 4.14, Figure 4.8 and Figure 4.9 detail the results.

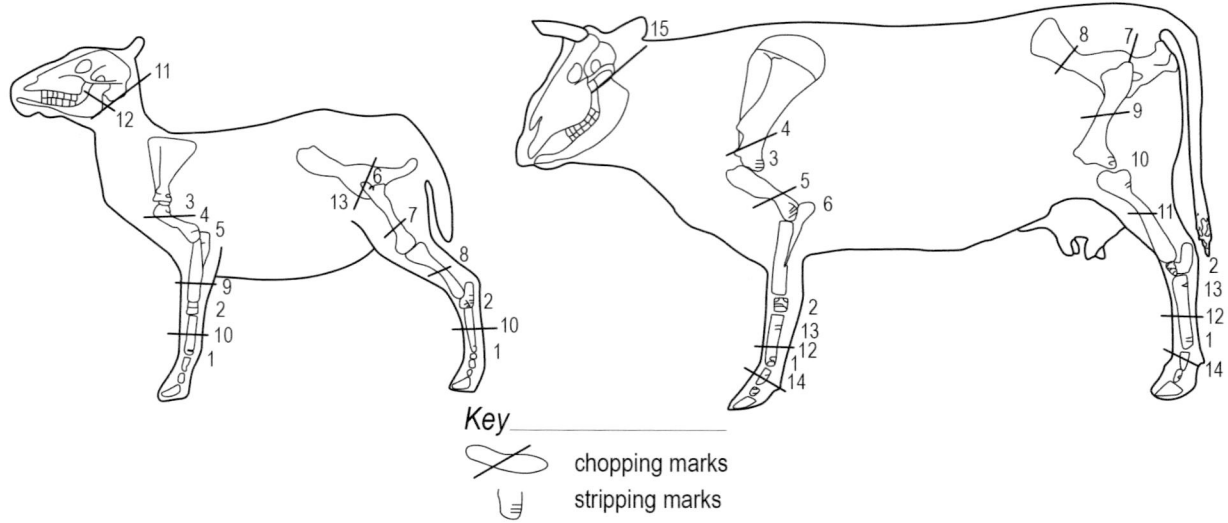

FIGURE 4.8:
Diagrammatic skeleton indicating butchery marks likely to be represented on sheep bones from RUX6 (after Finlay 1981).

FIGURE 4.9:
Diagrammatic skeleton indicating butchery marks likely to be represented on cattle bones from RUX6 (after Finlay 1981).

There was evidence from Phases A (sheep), B (sheep) and D (sheep and red deer bone) of gnawing in the form of 'puncture' and 'furrowing' marks, both of the form created by dogs and described by Binford (1981: 44-49). The importance of studying modification of bone by canine animals is that reduction of the bones by chewing will greatly alter the number and character of the fragments recovered in the faunal assemblage. Thus the results of the fragmentation study above (Table 4.13 and Table 4.14) are in reality dependant not only on man's choices of butchery, cooking and tool-making preferences but also on the degree and nature of chewing of the discarded bone by dogs. Unfortunately, it is not possible to estimate the impact of this bias, any more than it is for most of the other constraints on archaeological interpretation.

Animal husbandry

The ratios of sheep to cattle detailed in Table 4.9 suggest (albeit tentatively) that sheep is consistently the most important food animal represented at RUX6 but also that cattle increases in importance in Phase B. Consideration of the RUX6 data in comparison with material from the later periods of prehistoric occupation at the Udal and from other prehistoric sites in the Outer Hebrides (Finlay 1984) suggests that the sheep remains from RUX6 represent a small breeding flock with the gradual culling of surplus animals from lambs under 10 months right up to old ewes over 42 months. It is difficult to discern any husbandry pattern from the scanty cattle aging data but it seems likely, again from comparison with other prehistoric sites, that what is represented at RUX6 is a basic subsistence strategy involving the maintenance of a few mature breeding cows (and presumably a bull) and the slaughter of surplus animals before 1-year-old. This can be postulated as the early winter culling of the calves whose births would be necessary to start cows lactating but whose continued upkeep gave an insufficient meat return to justify their continued living when artificial methods of stimulating lactation could be employed (Serjeantson et al. 2005).

Wild resources

There are only a few teeth, bone fragments or antler of red deer in each phase.

Marine mammals are represented only by two seal teeth of an undetermined species in Phase B, although it is known that both grey and common seal are represented throughout the later sites at the Udal (Finlay 1984), and by a vertebra from a small whale (SF 26888).

Mice/rat/vole/shrew bones were recovered from Phases A to D and since the site was extensively tunnelled through by rabbit burrows, it is assumed that most if not all of these are intrusive. Rabbits are certainly intrusive as they were not introduced to the Hebrides until medieval times.

Birds were found in Phases A to D, including through dry and wet sieving and include land species such as starlings and blackbirds alongside shore species such as redshanks and oyster catchers and marine species such as puffins and great auks. All are still common in the area today, with the exception of the great auk, extinct since the mid-nineteenth century. Species which are present on the island in winter only, those present throughout the summer only, and those resident all year round, are evenly represented (see Table 4.15).

A diverse range of fish were also found in Phases A to D and include *Gadidae* (cod, ling, saithe and rockling), *Rajidae* (rays), *Pleuronectic* flatfish (probably dab), mackerel, conger eel and ballan wrasse (Listed in Table 4.16). All are still found in the coastal waters of North Uist today and have been recorded historically as exploited for food by humans. The species represented suggest exploitation of a range of coastal waters as species such as the ballan wrasse prefer rocky coasts, others such as rays and flatfish prefer sandy areas. Some of the species, such as conger eel, can often be caught in rock pools when the tide is low, others, such as rays, can be taken with rod and line, or those such as mackerel, in nets, or, in the case of flatfish, by spear. It is noticeable that there is no evidence for specialised and intensive fishing at the site, unlike later (Iron Age) phases at the Udal where Gadoids predominate.

Distribution across the site by phase and context

Detailed consideration of the representation of animal species by time (Phase) and space (Context) was undertaken to aid in the interpretation of human activity at RUX6 and tabulated in Table 4.7.

Phase A represents a very long period of time (post-Bronze Age to nineteenth century) and much potential disturbance and this is shown by the presence of every type of animal in the deposits, with the exception of marine mammals!

SF No	Species	Bone	Comments
Phase A			
17442	?	?	
23029	NID	radius distal	NID
	NID	scapula	NID
23038	cf. *Alle alle*	sternum	cf. Little Auk
	cf. *Alle alle*	ulna	cf. Little Auk
23244	NID	radius distal	NID
23500	NID	NID frag	NID
23810	*Alca impennis*	1st phalange digit 2	Great Auk
23224	cf. *Motacilla alba*	C/MC	cf. Pied Wagtail
Phase B			
23151	*Sturnus vulgaris*	ulna	Starling
	Sturnus vulgaris	radius	Starling
	Tringa totanus	C/MC	Redshank
23178	*Tringa totanus*	humerus	Redshank
23387	NID	NID frag	NID
25121	*Sula bassanus*	quadrate	Gannet
25220	cf. *Sturnus vulgaris*	sternal frag	cf. Starling
25299	*Alca impennis*	beak frag	Great Auk
	NID	2 x NID frags	NID
25341	*Alca impennis*	beak frag	Great Auk
Phase B BG24 and cist			
25472	cf. *Alauda arvensis*	humerus	cf. Skylark
26219	cf. *Larus marinus*	T/T distal frag	cf. Gt Black-backed Gull
26344	NID	NID frag	NID
26383	NID	T/T frag	NID
26877	?	?	
Phase C			
23167	NID	ulna distal	NID
SF No	Species	Bone	Comments
Phase C			
23194	NID	2 x NID frags	NID
23228	*Calidris alpina*	sternum	Dunlin
23413	*Fratercula arctica*	sternal frag	Puffin
23429	NID	NID frag	NID
23476	NID	NID frag	NID
25456	*Alca impennis*	beak frag	Great Auk
Phase D Building 1 (DJ)			
23620	cf. *Arenaria interpres*	T/MT	cf. Turnstone
23817	*Haematopus ostralegus ostralegus*	beak frag	Oystercatcher
Phase D Building 2 (DH)			
27210	?	?	

TABLE 4.15:
List of identified birds by phase.

SF No	Total no. frags	Species	Description	Comment
Phase A				
23597	1	*Limanda limanda*	vertebra	Dab
23795	2	flatfish	interoperculum, vertebra	flatfish
23912	1	Conger conger	Vertebra.	Conger eel
23873	4	Conger conger	Vertebrae	Conger eel
25032	1	NID	frag	
25060	3	*Gadus morrhua*	pre-caudal vertebra	Cod 7th winter
26075	1	*Pollachius virens*	super operculum	Saithe
23577	1	*Scomber scombrus*	caudal vertebra	Atlantic Mackerel - 6th autumn
24806	1	NID	Very small.	
Phase B				
23152	18	*Motella tricurrata*	jaws, vertebrae, etc	small Rockling
23176	2	cf. Gadoid	vertebra	
		NID	frag	
23185	1	?	skull	
23904	1	NID	frag	
23941	1	flatfish	hyomandibular,	flatfish
	2	cf. Euselachii	vertebral frag	cartilaginous fish - shark/ray?
	2	NID	frags	
25084	3	NID	Plus some frags; incl. 1 ?tooth.	one frag with cut marks
25101	1	*Gadus morrhua*	operculum frag	Cod
25132	1	NID	Fine vertebra.	
Phase B				
25163	1	*Molva molva*	pre-caudal vertebra	Ling - 8th winter
25228	1	*Limanda limanda*	os anale	Dab
25289	1	*Gadus morrhua*	otolith	Cod
25293	1	NID	frag	
25325	1	flatfish	Vertebra.	flatfish
25390	1	*Labrus bergylta*	pharyngeal	Ballan wrasse
25437	2	NID	Broken - now three frags	
25555	1	NID		
26072	85	*Gadus morrhua*	Numerous - jaws, vert, etc.	small Cod
26073	1	NID	vertebra frag	
26387	1	?	skull	
Phase B BB3 cist				
23116	2	NID	2 x frags	
Phase B BG 24				
26345	1	NID	Rib	
Phase C				
25213	1	NID	Now two frags.	

TABLE 4.16:
List of identified fish by phase.

SF No	Total no. frags	Species	Description	Comment
Phase D Building 1 (DJ)				
23658	1	cf. Euselachii	Disc.	cartilaginous fish - shark/ray?
23712	1	cf. Euselachii	vertebral disc	cartilaginous fish - shark/ray?
23818	1	*Molva molva*	epibranchial	Ling
23855	1	NID	frag	
25676	1	*Gadus morrhua*	ceratobranchial	Cod
Phase D Building 2 (DH)				
25605	1	*Labrus bergylta*	dentary	Ballan wrasse
25992	5	NID	3 ribs + 2 frags.	Wet-sieved 5 mm mesh.
26074	2	NID	2 frags	
27220	10			Extracted from 27206; wet-sieved 5 mm mesh - lost
27228	8	cf. Euselachii	vertebral discs	cartilaginous fish - shark/ray?
27229	1			Extracted from 27206; wet-sieved 5 mm mesh - lost
Phase D				
23983	1	NID	?Rib.	
25767	1	*Labrus bergylta*	dentary	Ballan wrasse

TABLE 4.16 continued:
List of identified fish by phase.

Phase B is described as a Bronze Age phase, containing two significant contexts.

- The kerbed cairn and cist produced only few loose teeth and limb elements of sheep, cattle and deer, plus a few unidentifiable fish bone fragments.

- From the BG4 structure was recovered a few loose teeth and bones of sheep, a single deer tooth, a single dog tooth and a bone from a greater black backed gull. Cattle remains included, in addition to a few loose teeth and foot bones, several left fore and rear limb elements and some skull elements. Unfortunately, the bones are largely shaft fragments and it is not possible to age the animals represented in this deposit, beyond noting that the teeth are moderately worn, but there is no obvious reason why the bones could not have been deposited as an assemblage, possibly as articulated limbs.

- *Phase C* includes evidence interpreted as Bronze Age ploughing and pits but the only faunal material recovered is a few loose teeth of sheep, cattle and deer, and two unidentifiable fish bone fragments.

- *Phase D* includes two late Neolithic structures, Neolithic pits and Neolithic 'ritual structures' and produced the largest assemblage of faunal remains from RUX6.

- Building 1 (DJ) produced 166 identifiable sheep bones and 20 loose teeth, estimated as representing at least eight individual animals, and 29 identifiable cattle bones and three loose teeth, estimated as representing at least six individual animals. For both species (and the few red deer bones here), the majority of the bone elements were from lower limbs – metapodials, ankle and toe bones, suggesting some activity taking place which required specific anatomical elements to be brought to the structure. Tool manufacture using the metapodials is a possibility here (see Bone Artefacts, this volume), as is work with skins and hides where the carcase is skinned, leaving the feet attached until further specialist work on the hide detaches them. The only other species represented in this context are a single bone of an oystercatcher and a few fish fragments of ling, cod and a cartilaginous fish such as ray or shark.

- Building 2 (DH) produced 214 identifiable sheep bones and 303 loose teeth, estimated as representing at least 18 individual animals and 22 identifiable cattle bones and 17 loose teeth, estimated as representing at least three individual animals. Again, many of the identifiable elements are from lower limbs, particularly for sheep and deer remains. However, among the sheep assemblage from this building are some significant deposits of bone, very different in quantity from the recovered material from Building 1, and suggesting a different activity may have been taking place here. For example, from a single Small Finds Number are identified at least three left and three right sheep pelvis, at least three left and one right femurs, at least two left and two right humerus, at least two right tibia, at least one each of left and right ulna, at least two left and one right radius, plus many foot elements. This argues that whole carcases were being brought into Building 2 for processing and that the remains from the activity were not removed from the scene but left to decay in situ. The only other bones recovered were a single loose tooth and a bone fragment from a medium-sized dog, around the size of a large terrier, a single unidentifiable bird bone fragment, two jaws from ballan wrasse and an unidentifiable fish bone.

- Neolithic pits contained only a cattle leg bone fragment and a worked bone point, made from a sheep metacarpal shaft
- The Neolithic ritual structures contained only two sheep loose teeth, some cattle shaft fragments and some red deer fragments. The cattle and deer shafts had been split longitudinally, presumably either to extract the marrow and/or to manufacture into artefacts

The faunal remains, taken at face value, suggest that the two buildings identified from Phase D may have had very different functions. Building 1 suggests a specialist production area, given the number of sheep foot bones recovered, possibly for hide processing or, since most of the metapodials recovered from this context are either proximal or distal elements, for artefact manufacture, although no worked bone was recovered from this structure. Cattle foot bones from Building 1 are almost all split longitudinally, which could be indicative of tool manufacture, or could be to extract marrow. Building 2 contained a similar number of sheep metapodial elements (although out of a greater total number of sheep bones so a smaller percentage) and three worked points made from sheep metapodials. The presence of a fairly wide range of sheep skeletal elements here (some from very young lambs), which could suggest whole carcase processing and retention of the bone waste, makes it difficult to put forward a convincing argument for the building's function from the faunal material alone. The miniscule quantity of remains from the ritual contexts indicates no relationship between faunal remains and activity or purpose.

Phase E represents the lowest deposits at RUX6, resting on bedrock and has yielded only a few loose teeth and small limb extremity elements from sheep, cattle and deer. This is the only Phase from which there no rabbit or rodent bones were recovered.

Subsistence and economy: the site in context

When the faunal material was first identified and evaluated in the early 1980s, few prehistoric site in the Hebrides had been excavated scientifically and even fewer had been published. The history of exploration, rescue and research in the Western Isles has been usefully summarised by Sharples (2015), who points out the lessons learnt by the late 1980s and the efforts over the past 30 plus years to capture evidence for the prehistory of the islands in a planned, resourced and reported manner. At the present day there are a range of published prehistoric coastal excavation sites, on island machair soils which had local potential for growing barley or oats, as well as stock-raising. These are useful to provide a context against which to compare the RUX6 data.

The Neolithic husbandry pattern at RUX6 appears to be mainly focused on pastoralism – the maintaining of sheep and, to a lesser extent, cattle to sustain the human population in meat, milk and raw materials. This is similar to the evidence excavated from Northton, Harris (Finlay 2006) which suggests a shepherding economy with a few cows, presumably for dairy products, and a kill-off pattern for both ruminant species of young animals under a year old killed off to maximise winter feeding resources. There is no evidence to suggest seasonal occupation of either site. Wild resources at RUX6 include only a few elements of red deer, of resident birds and of common sea fish. The paucity of wild remains is surprising, given the systematic wet and dry sieving undertaken at the site and the generally excellent preservation of bone, and is taken as a true reflection of the unimportance of wild resources to the Neolithic settlers there. The same pattern is evident in the Neolithic phases at Northton also, but contrasts with contemporary remains from Links of Noltland, Westray in Orkney, where the inhabitants are identified as primarily cattle farmers who also kept some sheep and dogs, and who accessed abundant wild resources of bird, fish, deer and marine mammals.

The small amount of faunal material recovered from the Bronze Age layers at RUX6 is statistically insignificant for any kind of detailed analysis and shows merely that sheep and cattle continued to be kept. Loose teeth are more commonly recovered in these contexts than bones, which is more likely to be due to taphonomic factors than human activity as teeth survive better through exposure on the ground surface, and this phase includes ground disturbance interpreted as ploughing. There is no evidence for an increase in focus on wild resources, apart from the presence of two seal teeth in the deposits, in themselves no guarantee of deliberate targeting of the species. In contrast, the limited evidence from the early prehistoric phases at Northton, Harris (Finlay 2006) suggests a distinct shift in economic emphasis from the Neolithic period which is dominated by sheep, to the early Beaker period where a greatly increased percentage of deer bone and antler is associated with a reduction in sheep, an increase in cattle and the introduction of pigs (whether wild or domesticated). Horse appears in the later Beaker level at Northton. The Northton husbandry pattern is mirrored at Kilellan, Ardnave, Islay, where Serjeantson (2005: 151) quoting Burgess describes the Bronze Age phase as 'essentially pastoralists herding cattle and to a lesser extent sheep'.

Evidence of cultivation from Phase C at RUX6 and from other sites in the Outer Hebrides such as Rosinish, Benbecula (Sharples 2009) and Ardnave 2, Islay (Harman 1983) is in the form of excavated ard marks and cultivation soil sprinkled with Beaker pottery sherds, perhaps as part of midden scatter for manuring. In addition, the identification of barley and emmer from these periods at sites such as Rosinish and Dalmore, Lewis suggest the development of a mixed economy in some areas.

In conclusion, the inhabitants of RUX6 seem to have practiced a pattern of animal husbandry common to the Hebrides in the third and second millennia BC, maintaining variable proportions of sheep and cattle, with the introduction of pig and horse in some areas, such as Cill Donnain, South Uist (Vickers *et al.* 2014) and Barvas, Lewis (Harman 2010). Dogs, represented throughout the area in prehistory, were presumably kept as working animals and/or pets and came in different sizes, presumably reflecting the impact of selective breeding. Wild resources such as red deer, birds and fish were readily available to the inhabitants of the Outer Hebrides throughout prehistory, but some communities, including those at RUX6, apparently chose not to make hunting, snaring and fishing a significant part of their resource strategy.

Non-worked marine shell

By Catherine Smith

'Dèan maorach fhad 's a tha tràigh ann'

(a Gaelic proverb translated as 'collect shellfish when the tide is out', that is, strike when the iron is hot)

Introduction and method

A large quantity of marine mollusc shells formed a substantial part of the environmental assemblage from RUX6. Precise details of the method by which they were retrieved is unfortunately not known but the site archive indicates that soil samples were meticulously sieved on site and that the shell collection resulted from that process. Some samples, but not apparently all, are noted as having been 'wet-sieved with a 5 mm mesh'. Samples were bagged with a unique small finds number which could be associated with context descriptions, although identifying context numbers were not used. It would appear that the shells were roughly sorted as part of this process, since each small find number corresponded with a particular grouping.

An example may help to illustrate this: SF 27213, listed as coming from Neolithic Building 2, layer XI.03, contained mainly limpet fragments, with a few wulks, flat wulks, mussel fragments and gastropod fragments. This sample had an identical context description to SF 27222, but the latter contained mainly wulk, flat wulk, dog wulk, netted dog wulk, topshell, mussel fragments, blue-rayed limpet and terrestrial mollusca. It seems fairly obvious that these are sorted samples which came from the same archaeological context; during processing limpets and wulks must have been roughly sorted into separate bags and given different finds numbers. This process had not been exact and some wulks were to be found in the bags containing limpets, and vice versa. The minor species were mainly, but not exclusively, bagged with either limpets or wulks.

For this reason, the initial identifications had to be considered as a rough guide to the contents of the bags rather than an exhaustive descriptor.

In determining which samples were to be analysed for the current report, prior sorting had to be taken into account in order to avoid artificially biasing the results towards one species with respect to any other. In other words all the samples with the same description had to be located and analysed together. Further, due attention was paid to whether the deposits from which the samples came were likely to have been disturbed by prior coastal erosion, wind blow or animal burrowing. These deposits were not recorded in detail. However, the preliminary database prepared as part of the current post-excavation process records the weights of shell in each SF number and is thus a rough guide to the contents. All small find (SF) numbers are listed in the site archive (see RUX6 dbase) and a list of the SF numbers which formed the selected samples is provided as Appendix 1.

Naming, identifying and quantifying the shells

No attempt was made to identify limpets to particular species since some of the identifying characteristics, such as shell colour, in the live (or recently dead) animal do not survive burial. As is customary, all limpets (Gaelic *bàirnich*) are characterised as *Patella* sp. In the case of gastropods, the commonest species in the assemblage, *Littorina littorea* is here referred to as the 'wulk', the name by which it is commonly known on mainland Scotland; on Orkney they are 'wilks' (Fenton 1978: 542), in English terminology they are 'winkles' or 'periwinkles', while in Gaelic, wulks are *faochagan*. The name 'wulk' is used to avoid confusion with the much larger gastropod, the English 'whelk', *Buccinum undatum*, which in Scotland is known as a 'buckie'. Presumably this name derives from the Gaelic *bocaid* or *bucaid*. To add to the confusion, in common parlance the name 'whelk' is often used interchangeably with 'winkle', as it has been in the original site records.

The other gastropod species commonly found in the collection are the closely related *L. obtusata*, the flat wulk, and *Nucella lapillis*, the dog wulk. Smaller *Littorina* specimens were problematic and were thought to be merely juvenile specimens of *L. littorea*, although some may have been the much smaller *L. saxatilis* or *neritoides*. Topshells (*Gibbula* sp.) were the small species known as 'silver tommies'. Other species were given their Latin names if there are no common Scottish or English terms.

The method of quantifying the shells depends on establishing minimum numbers of individuals. In the case of limpets and gastropods, the apex of the shell represented one individual and any fragments without the apex were not counted, as recommended by Claassen (1998: 104–6). Instead, these were quantified by weighing (in grams) and roughly determining the volume (in ml) in a measuring jug or cylinder. Whole limpets and wulks were also weighed and their volume estimated.

For bivalves, only fragments in which the hinge was present were considered to represent an individual, bearing in mind each bivalve has an upper (right) and a lower (left) valve. Other fragments in which the hinge was absent were weighed rather than counted and did not contribute to the minimum number of individuals estimation.

Results

The total weight of all marine mollusc shells from the site, that is, from all finds numbers including possible disturbed features, was 131.82 kg (information from the dbase). This compares with the total weight of 63.157kg for limpets, wulks (all species) and mussel fragments identified in the selected samples, which therefore comprised just less than half of the total recovered.

Total numbers of shells, with weights and volumes of most abundant species, by phase, are quantified in Table 4.17.

The total weight of limpet shells (*Patella* sp.) quantified, including fragments, was 56.39 kg, with an approximate volume of 82.37 litres, representing at least 13,270 individuals. Wulks (*L. littorea*) accounted for the second most abundant species with 1742 individuals weighing 5.871 kg with an approximate volume of 6.97 litres. Other gastropods occurred consistently, but in far fewer quantities. Flat wulk (*L. obtusata*) was represented by 322 individuals weighing 213 g in total while the thicker-shelled dog wulk (*Nucella lapillis*) scored 77 individuals weighing 254 g.

Other edible species were retrieved only in small quantities. In the case of the mussel, (*Mytilus edulis*) this is almost undoubtedly due to the relative fragility of the shells, which are more susceptible to crushing than those of gastopods.

Species	Phase A	Phase B	Phase C	Phase D	Phase E	Total
Limpet (*Patella* sp) (apex)		5401		7869		13270
Limpet weight (g)		11475		33132		44607
Limpet volume (ml)		18115		50360		68475
Limpet fragment weight (g)		4368		7410		11778
Limpet fragment volume (ml)		4960		8940		13900
Wulk (*L. littorea*)	41	1236	2	463		1742
L. littorea (weight)	111	3539	6	2215		5871
L. littorea (volume)	150	4605		2215		6970
L. obtusata	18	146		158		322
L. obtusata (weight)	15	90		108		213
L. cf *saxatilis/neritoides*		6		34		40
Dog wulk (*Nucella lapillis*)	5	28		44		77
Nucella (weight)	12	42		200		254
Hinia sp.		1	1	6		8
Buckie (*Buccinum ondatum*)		3		10	1	14
Gastropod		110		26		136
Topshell (*Gibbula* sp)	1	28		12		41
Mussel (*Mytilus edulis*) (hinge)	1	21	1	81		104
Mussel (weight including fragments)	2	37		390	2	431
Oyster (*Ostrea edulis*)		1		1	1	3
Great scallop (*Pecten* sp)		6		8		14
Queen scallop (*Aequipecten opercularis*)				1		1
Scallop (*Chlamys* sp)	1					1
Cockle (*Cerastoderma* sp)	3	42	2	5		52
Laevicardium			1			1
Cowrie		1		1		2
Macoma		1				1
Arctica		1		6	1	8
Tellin				9		9
Venus sp				1		1
Ensis sp				1		1
Goose barnacle (*Lepas* sp)		5		5	3	13
Blue-rayed limpet (*Helcion pellucidum*)	1	3		17		21
Unidentified	2	11	1	3	1	18

TABLE 4.17:
Total numbers of marine mollusc shells, with weights and volumes of most abundant species, by phase.

Other edible bivalves included cockles (*Cerastoderma* sp.), oyster (*Ostrea edulis*) (Figure 4.10), great scallop (*Pecten maximus*), queen scallop (*Aequipecten opercularis*), clams represented by thick fragments only, thought to be *Arctica islandica*, and single valves of heart cockle (*Laevicardium*) and *Venus*. The size of the oyster shell recovered from SF 26800 is worthy of note. A single valve from Phase E weighed 288 g and probably represents an individual of advanced age. Razorshells are surprisingly few and are represented by a single hinge of *Ensis*, again an indication of the fragility of the shells.

A further unusual survivor was the goose barnacle (*Lepas anserifera*), a few of whose delicate thin shells were found in site Phases B and D. The goose

FIGURE 4.10:
External and internal views of an oyster *(Ostrea edulis)* SF 26800 from Phase D.

barnacle is considered edible and in some parts of Europe, a rare delicacy.

Discussion

Having attempted to remove from the analysis any naturally deposited shells which have accumulated through wind blow, or contaminated archaeological features affected by erosion and redeposition, the perennial question 'were the shellfish eaten by humans (and their cattle) or were they mainly used as bait in fishing?' has to be asked. Certainly ethnographic evidence from other Scottish islands indicates that limpets and mussels were particularly important as bait, up to at least the twentieth century (Fenton 1978: 542). Fenton (ibid) describes how fishermen took half-boiled limpets out from the Orkneys on their boats, chewed them into pulp before spitting them into the surrounding sea where they attracted sillocks, or young coalfish. Craig fishing, or fishing from the rocks with rod and line, also used limpets, first mashed to a pulp in holes (or cups) in the rocks and thrown into the water as a lure, then followed with chewed limpet bait on the hook (ibid, 534). In recent times, direct human consumption of limpets, and sometimes wulks, was looked down on and considered 'famine food'.

There is no direct evidence from the RUX6 shells themselves as to whether they were bait or food. However, the dominance of limpets over wulks, which at the present day are more acceptable as human food, possibly indicates that some of them at least may represent use as bait. As has been pointed out in the analysis of Mound 3, Bornais, South Uist (Sharples 2005: 159) limpets used as bait would be far more likely to have their shells discarded near to the fishing area, rather than being brought from the shore to the site, then back to the rocks again.

Some of the limpet shells show damage, mainly to the posterior part of the shell where a roughly rectangular section has been removed, possibly consistent with removal from the rock with an implement, but this only shows how the shells were collected, not to what use they were put. The large quantity of fragmented limpet shells may however indicate that both limpet flesh and shells were pounded together to produce bait. Although this cannot be proven, it seems odd that the intact limpet shells are well-preserved and in some cases, robust and large in size. Mashing shells complete with contents to a pulp might explain the presence of the broken fragments.

The quantity of wulks recovered was, as noted, much smaller than that of limpets. In the recent past, wulks have been considered a more acceptable food than limpets and can still be bought in Scottish fish shops, unlike the limpet. According to Scottish Fisheries statistics, 225 tonnes of periwinkles (i.e. wulks) were landed in the Fishery District of Stornoway in the year to 1986 (Boyd and Boyd 1990: 362). No limpets were landed commercially.

F Marian MacNeill, an authority on historic Scottish food, recommends that a 'small pail-full' is needed to make a good soup, as eaten in the Hebrides prior to 1929 (1974: 126). Anecdotal evidence suggests that a pint of wulks, eaten with bread and butter, would make a supper for one hungry person or a snack for two people (David Bowler, Director, Alder Archaeology, Perth, pers comm).The entire RUX6 sample of 6.97 litres of wulks, approximately 12.3 Imperial pints, would probably therefore have fed only about 24 hungry people!

Dog wulks (*Nucella lapillis*) made up a small but significant part of the RUX6 sample. Their shells are much thicker than those of the *Littorea* species, therefore it is interesting that some of them have been broken. While a purple dye may be obtained from this species, the quantity of shells needed to produce it is very large, in comparison to the numbers retrieved from the site. It seems most likely that they were probably eaten but nowadays they are removed from gathered shell collections by the fishmonger. The species' carnivorous habits and its predilection for attaching itself to carrion may account for the fact that it is considered 'unclean' and not usually eaten at the present day. However, this was not always so, and it was certainly consumed at several prehistoric shell midden sites in the north of Scotland studied as part of the Scotland's First Settlers Project (Milner 2004; 2007). Ray Mears, a noted eater of wild foods (2007: 64) has also observed that 'they make good eating' and that future experimental archaeology might identify the best way to prise the meat out of the thick shells.

The preceding species are inhabitants of the upper and middle shore. The lower shore was also exploited though certainly not to the same extent, as the small numbers of sand- or mud-dwelling molluscs attest. A very small number of oysters, clams, scallops and other bivalves are indicators that although available, and in the case of the oyster valves, well-grown and large it may have been too troublesome to collect by dredging or digging when limpets, wulks and mussels were easily available on nearby rocks. It is however surprising that so few razorshell fragments were found as this species is considered an edible delicacy; however the shells are thin and fairly fragile therefore may not have survived crushing weights. Known as 'spoots' because of their water-spouting habit, they may be found in the sand of the lower shore at low tide.

A resource which may have become available after storms is the goose barnacle. This species is occasionally found washed ashore attached to wreckage and driftwood. Botanical analysis (Ramsay, this volume) showed that wood from larch (*Larix*) and spruce (*Picea*) were present in Phases B and D and were considered to have originated from driftwood.

Distribution of marine shell across the site

As mentioned above, the majority of marine shell sample was from floors and features, whose stratigraphy was largely undisturbed, and where there was a possibility that they had been deliberately collected and brought back to the site.

The Phase A marine shell samples were not examined further as they would not have aided the understanding of the prehistoric activities that took place on the site.

The structural activities in Phase B, at the start of the early Bronze Age, indicated that marine shell played little part in activities associated with the construction of the cist and the cairn in the western part of the site. To the east, the story is a little different. The burnt depression of structure BG24, with its occasional pits was identified as having several episodes of use, including the accumulation of three floor deposits. The numbers of limpet apices (c. 5900) and fragments was high and mainly distributed on the eastern side of the floors, nearest the shore, also with wulks, dog wulk, gastropods, *Gibbula* sp., and the occasional cockle, mussel and great scallop. This distribution could have been the result of use and the deliberate collection of shell for activities associated with the area, combined with natural wind-blown accumulation during inactive periods, and also disturbance of the Phase C accumulation of sand and shell. Although two pits were noted as having shells within their fills, the majority of marine shell was associated with the floors. The layers beneath the skeleton placed in the adjacent cist (BM1) also contained limpets, wulks and fragments of other species but it is considered unlikely that the shell was deliberately placed there prior to the reception of the human remains. The cist suffered from coastal erosion and some of the shell could have been the result of deposition after coastal scouring. The presence of articulated crustacean remains within the cist confirms this interpretation (see Crustacea, below).

Phase C was not sampled as it was the natural sand accumulation that led to the abandonment of the buildings at the end of the late Neolithic. It is only identified with ploughing and pit digging as the sand became consolidated by vegetation.

Marine shell (limpets, wulks, flat wulks, dog wulk and mussel), in the (Phase D) late Neolithic DJ building, was found in a variety of features - a soakaway/drain, a possible posthole and post-setting, but only in small numbers, indicating their accidental or natural occurrence. The small increase of limpets from around the hearth might imply that these were deliberately collected for bait or even famine food, along with species such as mussel, cockle and other gastropods, which were found in small numbers with them.

Interestingly the largest number of species and greatest number by weight and volume came from the last surviving floor of the building (Floor 1). A large number of limpet apexes (503), wulks (111), with flat wulks (22), dog wulk (20) and mussel (12) were retrieved from the floor, although precisely where is not known. This distribution of marine shell can be interpreted in a number of different ways. Admittedly, some of these species could have been bought back to the building as bait for fishing or for food. However, the cessation of occupation of the building and the abandonment of the structure occurred due to the rapid accumulation of a 0.2 m to 0.7 m thick layer of blown sand. As shell is relatively light, a storm wind and even high tides could have brought them along with sand onto the site to accumulate on the floor of the abandoned building.

The distribution of marine shell in the adjacent building DH was similar. A pit in the floor of the structure contained a large number of limpet apices and fragments (4606) with a small number of wulks and mussel fragments and the occasional shells from a few other species, implying that this was a deliberate cache or deposit. The top of the last floor and the area around the hearth were dominated by limpet apices and fragments (5701), with wulks (146), dog wulk (45), mussel fragments (24) and blue-rayed limpet (12), but again this could have been due to the accumulation of wind-blown sand, as possible sea incursion or the preparation of bait and food before the building was abandoned.

Samples from around these two buildings and the shaft of bedrock (DC) exposed to the east of the buildings, indicated only natural occurrences of marine shell in windblown sand.

The upper layers of the natural geology underlying the site (Phase E) were disturbed by anthropogenic activity, and a number of species (*Buccinum*, mussel fragments, oyster, clam, goose barnacle, and unidentified) appeared mixed into these layers in very small numbers, with only topshell (*Gibbula* sp.) dominating. They are interpreted as occurring naturally.

The perforated mollusc shell

By Catherine Smith

Numerous marine mollusc shells appear to have been recovered from the excavations at the Udal, North Uist. This report deals with a selection of pierced or perforated shells 'extracted from the RUX6 assemblage.'

Two species of shell were found to bear small sub-circular perforations: limpets (*Patella* sp) and flat wulks or winkles (*Littorina obtusata*). Perforations in limpet shells were located in four main positions on the shell. *Apical* refers to the point on the dome of the limpet; *anterior* to the 'front' of the shell, where the head of the live animal is found, *posterior* to the 'back' of the shell, where the most of the fleshy foot of the animal protrudes and *lateral* to either side of the shell, at right angles to the antero-posterior axis.

Most of the limpet perforations occurred singly but in some cases two holes were present. Expressed as a percentage, the location of the holes was 23% anterior, 42.1% posterior, 15.1% lateral and 20% apical or almost apical. Magnification of the area surrounding the holes showed that in most cases the edges of the perforation were rough, with delamination of the nacreous material on the inner aspect of the shell. There were no apparent traces of use-wear around the holes on any of the limpet shells examined and the most likely explanation for their presence is predation by other sea creatures. A variety of other organisms predate limpets, including starfish, crabs and other mollsca, but the predator which most probably caused the damage is the dog wulk. *Nucella lapillus* is a common gastropod which feeds on a variety of other molluscs, including limpets (Crothers 1985: 2). The method of attack is for the dog wulk to settle on the surface of the limpet or other shell, extrude a rasping chitinous organ called a radula, with which it bores through the shell of the limpet, helped along by secretions of a shell-softening chemical (Carriker 1981). After boring the hole, the predator secretes digestive enzymes which soften the muscle of the limpet and finally the liquefied flesh is sucked out through the perforation by the dog wulk's proboscis. It can be seen that the position in which the holes are found probably reflects the relative ease of attachment of the predator. An apical position on the dome of the shell is possibly harder to negotiate and maintain than one on a sloping surface. The fact that the posterior slope is less steep than the anterior may be one good reason for the higher percentage of posterior holes in the RUX6 sample (42.1% posterior, compared with 23% anterior) although shell thickness at the boring site may also play a part.

Limpets in the RUX6 assemblage which had delaminated into 'rings' were suspected to have either disintegrated naturally, have been damaged by human agency or had possibly been attacked by the pincers of crabs. Three crab species were indeed recognised within the crustacean assemblage.

All of the perforations found in the flat wulks (*L. obtusata*) were located on the lateral surface of the shell. In these shells, the outer surface showed irregular scratch marks around the hole, within a possible area of flattening. As with the limpets, these marks are entirely consistent with predation by molluscs. However, in order to test this, some unstratified shells were deliberately rubbed against, firstly, an emery board coated with abrasive sand, and secondly, a piece of local gneiss retrieved from the sieved samples. In both cases, it was the work of a few minutes to produce a flat platform, with irregular striations. Continued rubbing of the shells against the stone eventually thinned the shell until a small hole appeared on the lateral surface. Shells treated in this way are almost indistinguishable from the archaeological specimens, save for the fact that the striations in the Neolithic shells are more abraded. It is thus entirely possible that these perforated shells could have been strung together through the perforation into the aperture and suspended from a string as jewellery. Interestingly all the perforated flat wulks came from features associated with Building DH in Phase D.

At the Mesolithic site of Sand, cowries with two perforations were identified as having possibly had some artefactual purpose although predators were also suspected (Hardy 2009). Possibly flat wulks may have been used in the same way.

The crustacea

By Catherine Smith

The remains of crustaceans were recovered from RUX6 in association with the molluscan shells. Crustacea are segmented arthropods, with an exoskeleton of chitinous material strengthened by calcium carbonate, hence survival of crab remains at a machair site where the soil is both shelly (containing much calcium carbonate) and sandy (a suitable pH for preservation), is good.

The main identifiable parts of the crustacean body found in archaeological deposits are the cheliped (claw) and carapace (body). The walking legs, or pereiopods, are slender and less commonly found, if at all, and none were identified in the material from RUX6. Table 4.18 further classifies crustacean chelipeds (claws) as the component parts of pollex, dactylus or unspecified pincer. In some cases, the claw may be partially articulated and include a further segment known as the carpus. A total of 589 fragments was identified.

Identification to species is difficult and has only been attempted where fragments are diagnostic. For example, carapace fragments of the partan or edible crab (*Cancer pagurus*) can be recognised from their distinctive pie-crust edges. Similarly claw fragments of the partan are also relatively easy to recognise where large fragments survive.

It is thought that three crab species were present: the partan or edible crab (*Cancer pagurus*), the green shore crab (*Carcinus maenas*) and the harbour crab or other relative from the family of swimming crabs (*Liocarcinus cf. depurator*), the last identified by its characteristic grooved and knobbled reddish claws.

Crab habitats abound on rocky shores, the shore crab in particular being easily found under the algal covering of rocks and in rock pools. Juvenile partans are also abundant on rocky shores (Hayward *et al.* 1996 164), which perhaps accounts for the higher proportion of smaller pincers in the assemblage. In previous centuries in the Northern Isles, partans were caught by hand or pulled from their hiding places with a hook (Fenton 1978: 542).

The partan is the species most likely to have been eaten by humans although it could also have been used as fishing bait. Some of the partan remains

Species	Phase A	Phase B	Phase C	Phase D	Total
partan	✓	✓			
cf. partan	✓	✓		✓*	
harbour crab	✓	✓			
cf. harbour crab		✓	✓	✓	
shore crab		✓			
Identified parts					
Pollex	18	11	2		31
Dactylus	2	3	1		6
Chela (pollex & dactylus)	11	13	1		25
Carpus	4	8	1		13
Chela and carpus	4	5			9
Unspecified pincer	10	28		8	46
Carapace	4	39			43
Approx no. unidentified fragments	172	240	4		416

✓ = present
*Includes blackened and ?burnt fragments and also conjoined fragments.

TABLE 4.18:
Phasing and identification of crustacea.

were of small or medium individuals and it might be wondered whether these were worth cooking and eating. Similarly the remains of shore crabs were of small size and it is possible that these undersized animals were the remains of bait. In Shetland, crabs and limpets intended for bait were pounded into 'soe' in cup-shaped rocks on the beach (Nicolson 1990: 117). However this would not account for the larger claw fragments at RUX6, as producing pulped bait in this way would inevitably pulverise the shells into unrecognisable fragments.

Also notable was the fact that some of the partan claws were semi-articulated. Their survival seems curious given that they contain useful meat. Perhaps in the case of the smaller claws it seemed like too much trouble, or only the white meat found within the carapace was utilised, as it is today.

Milner (2009) has pointed out that crustacean remains, while sometimes surprisingly plentiful at archaeological sites, are rarely given much consideration and their contribution to the site economy is therefore often overlooked. At the site of Sand (ibid) the author suggests that crustacean remains should be considered as more than just a peripheral resource, used only in times of food scarcity. Through their collection by people of all ages and genders, the hunting, cooking and consumption of crabs may contribute to the complexity of social interactions at the site as well as providing a nutritious and enjoyable foodstuff. The results of this brief analysis of the Udal crabs should be considered alongside that of the other available foods resources, from shellfish to birds and large mammals as well as plant-based foods, both wild and domestic, in order to gain a picture of what life may have been like at the site.

Distribution of crustacea across the site

The most recent layers (Phase A) contained 172 fragments of crustacea, which derived from contexts affected by late human and animal disturbances to the site, the erosion of the coastal edge and the redeposition of material after erosion. These events also affected the Phase B cairn, which partly collapsed and also the stratigraphy of lower phases. In this mix of material fragments of partan and cf. partan, cf. harbour crab, dominated, with small fragments possibly indicating the effects and power of wind and rain. Some of the occurrence of crustacea fragments may also be the result of gulls nesting or roosting on the site and taking crabs for food.

The number of crustacea fragments (240 pieces), from the early Bronze Age (Phase B) is the highest recorded from the site. Of that number, the majority (201) came from the fill of the cist containing the buried remains of human skeleton SF 17642. The cist remained open for a period of time before it was capped and sealed, but the skeleton may have been partly covered over with layers of pale and mauve-coloured sand. The upper half of the cist remained open to be filled naturally with sand and fragments of small stone, but it also led to the intrusion of rodents and birds. The rodents may have been responsible for the accumulation of partan, cf. partan, shore crab and harbour crab remains, including carapaces, into the layers of sand just above the skeleton. The open cist with its protruding sides, and partly sheltered interior would have provided an opportunistic gull with both nesting and roosting sites, the large number of partan and harbour/shore crab fragments attesting to its preferences. Once the lid of the cist was placed in position the crustacea fragments were sealed in-situ and remained undisturbed until excavated in 1974.

Phase C contained little artefactual or environmental evidence, except in this instance four crustacea fragments including two of cf. harbour crab.

The late Neolithic occupation (Phase D) of the site produced a total of 8 pincer fragments of cf. partan and cf. harbour crab from the floors and pits of two buildings and an external feature (Table 4.18). One of the cf. partan pincers comprised two conjoining fragments, and another was probably burnt. Although slight, the evidence suggests that crabs at least were being brought back to the buildings, most likely for consumption. The paucity of evidence is likely to be the result of cleaning of floors of the buildings and the removal of debris to other areas.

Terrestrial snail assemblages

By Ruby Ceron-Carrasco

Introduction

The terrestrial snails from the RUX6 were mostly retrieved during the sieving of bulk samples: three samples processed by flotation through 1 mm mesh produced thousands of micro terrestrial mollusca remains.

The broad aims of the terrestrial mollusca analysis was to identify and record the terrestrial mollusca assemblage, to identify the environment of the various archaeological contexts where these were recovered and try to reconstruct the environmental history of the site and its surroundings. Finally, to look at any changes of species present and identify any spatial variation in the environment during the phases of occupation represented.

Analytical procedures followed Evans (1972), and the nomenclature by Kerney and Cameron (1996). The molluscs were identified using the author's own reference collection and by reference to Kerney and Cameron (ibid.)

The result of the identifications is given as a catalogue per phase and context. A summary of the species representation is given as Table 4.19; this was based on species characteristic i.e. 'Dominant' where significant amounts of shell were recovered and as 'Indicators' to show where these were recovered in smaller numbers. Fifteen taxa were identified in the catalogue (Table 4.20).

Results

Almost all the specimens of terrestrial snails recovered from bulk samples contained the shells of the white-lipped snail *Cepaea hortensis*; some remains amongst those of *C. hostensis* may be shell of *Arianta arcustorum* but this has not been possible to determine as these have lost the colouring and brands. *Trichia striolata* were also found in smaller quantities.

The flot samples produced large amounts of micro gastropod species: *Vallonia costata*, *Valonia pulchella*, *Pupilla muscorum*, *Collumbella edentula*, *Cochlicopa lubrica*, *Cochlicopa lubricella*, *Aegopinella nitidula*, *Vitraea crystallina*, *Helicella itala*, *Vertigo pygmaea* and *Cochlicella acuta*.

In general, the analysis demonstrates the presence of terrestrial mollusca sensitive to vegetation and moisture, and to the lack of vegetation in areas where wind-blown sand may have been deposited. A few of these terrestrial snail species are sensitive to the deposition of fresh domestic organic waste, i.e. middening. Careful investigation of the way in which molluscan faunas vary within archaeological deposits allow an insight into the way that local environments and land use patterns may vary across a site and how this variation changes through time. In turn these patterns constitute the site formation processes that result in the archaeological deposits.

Discussion

The anthropogenic and natural environments and environmental change

The varied and distinctive character of the terrestrial mollusca assemblage recovered at RUX6 has great

Species	Dominant	Indicator	Habitat
Cepaea hortensis	D	I	Dunes in wet and cold places, grassland, woods
Arianta arcustorum		I	Meadows and herbage, woods, always in wet places
Vitrea crystallina		I	Moist grassland, damp places
Helicella itala		I	Dunes and calcareous grassland, dry exposed habitats
Vallonia pulchella	D	I	Sand dunes, moist meadows, open calcareous places, wet places
Vallonia costata	D	I	Dry open calcareous places, stone walls, sand dunes, occasionally open woods
Aegopinella nitidula	D	I	Moist places, woods, herbage, amongst rocks, often in humanly disturbed habitats
Cochlicopa lubricella	D	I	Dry, limestone and in calcareous sand dunes
Cochlicopa lubrica	D	I	Moderately damp places of all kinds, grassland, woods
Vertigo pygmaea	D	I	Dry calcareous grassy places, sand dunes
Columella edentula		I	In moderately damp places and calcareous lowland places, woods and marshes
Pupilla muscorum		I	Dry exposed places
Cochlicella acuta		I	Maritime. Usually found in dunes and coastal grassland, occasionally calcareous ground inland
Trichia striolata		I	Woods, hedges and waste ground; widely spread by man

TABLE 4.19:
Summary of terrestrial mollusc species representation defined by characteristic and habitat.

potential as environmental indicators. In general terms, there may have been minor change from woodland to various types of grassland before the site was built.

Analysis of the stratigraphy indicates that *Cepaea hortensis* was the commonest species found in all phases of the site. Its natural habitat of dunes and grassland indicate that throughout all phases, the Udal peninsular was predominantly wet and cold. During the late Neolithic (Phase D), a total of 88 *C. hortensis* were found in a pit in Building 2 (DH), related to its last floor and hearth. This information suggests that the floor of the building could have been damp and possibly that the water table was high. Only seven specimens of the same species were found on/in the floor of the adjacent Building 1 (DJ), the only terrestrial molluscs to have been found there.

In the wall tumble around these two buildings further specimens of *C. hortensis* were noted indicating that the cold and relatively damp conditions between the stone walls was suitable for its survival. Additional evidence of the environment around the buildings came from SF 27022, where more than 600 molluscs from seven different species (*Cepaea hortensis, Cochlicopa lubrica, Cochlicopa lubricella, Pupilla muscorum, Vallonia pulcella, Vallonia costata* and *Columella edentula*) were identified. Together these species indicated mixed conditions of dry calcareous sand dunes in open and exposed lowland areas with grassland/meadows, contrasting with damp to wet dunes, cold places and stone walls. It is also possible that some of these species were brought on to the site in possibly grassland turf from the surrounding area to form packing between the outer and inner skins of stone walls. Crawford (1980) considered the mauve coloured sand he found there to be decayed turf.

The terrestrial molluscs found outside the buildings was consistent with those found in the following Phase C, and therefore they also reflect the changing conditions at the end of the late Neolithic with the accumulation of a thick deposit of windblown calcareous sand and the development of the machair. More than 6300 molluscs were identified from this phase, some in connection with the fills of pits that were interpreted by Crawford as having been quickly dug and rapidly filled in. The occurrence of *Aegopinella nitidula* in the assemblage of this phase also indicates the disturbance of habitats caused by ploughing and pit digging.

The prevalence of the machair environment, of cold, often damp and dry conditions in the calcareous sand with its covering of meadow/grassland herbage is noted in the early Bronze Age (Phase B) with the occurrence of species noted in Phases D and C, especially in the eastern part of the site where over 6500 molluscs were identified. In the western part of the site, species were largely restricted to *C. hortensis* due to the alteration of habitat the construction of a large stone cairn. The most notable addition to the assemblage at this time, but probably as a result of modern disturbances in Phase A, where it was also found or introduced, was *Trichia striolata*. Its presence indicates disturbed and waste ground - the digging of stone-robbing pits into the Bronze Age cairn, and the digging of a long trench used for a saw-pit. The increasingly wetter weather conditions of the last half century is noted in the occurrence of *Arianta arcustorum* in the top layers of Phase A, which prefers meadows and herbage in wet places.

Functions and processes

Some of the contexts may have filled up naturally because there are distinctive species, some through anthropogenic activity such as waste disposal, this is supported by the large presence of *Vertigo pygmaea*. There are also qualitative differences between various contexts, some with high *Vallonia costata* for example, indicating a damp environment and stable conditions associated with short-sward grazed grassland influenced by the grazing requirements of sheep and/or cattle (Law and Thew 2015; Thew 2003).

Conclusion

The terrestrial snails recovered at RUX6 have allowed for evidence of spatial variation as demonstrated by the species present, these have shown how the formation processes may have developed. Terrestrial snails are sensitive to local environmental changes such as changes in vegetation cover. At RUX6 such processes included associated buildings i.e. walls where the cold and dump conditions were suitable for the survival of certain mollusca species; whilst others show the use of the machair environment with grassland habitat suitable for grazing domesticated animals.

SF No	Phase	Level	Context	Species	Quantity - total numbers of molluscs
Phase A					
27128-9	A	I/II/VIII	Sand layers	*Cepaea hortensis* cf. *Arianta arcustorum Trichia striolata*	64
27123	A		Ditch infill	*Cepaea hortensis Trichia striolata*	157
27124	A		Ditch infill	Terrestrial molluscs	14
All	A	1-VIII	All other disturbances and features	*Cepaea hortensis*	365
Phase B					
	B	VIII/IX/X	Trench extension	*Cepaea hortensis*	53
27078	B	VII/VIII/IX.1	East of saw pit	*Cepaea hortensis Trichia striolata*	38
23396	B	IX.1	East of saw pit	*Cepaea hortensis Cochicopa lubrica Cochlicopa lubricella Pupilla muscorum Vallnia pulcella Vallonia costata Columella edentula Aegopinella nitidula*	>5000
27073	B	VIII/IX.1	East of saw pit	*Cepaea hortensis Trichia striolata*	22
All	B	VIII/IX	East of saw pit	*Cepaea hortensis*	1525
All	B	VIII/IX	West of saw pit	*Cepaea hortensis*	27
26267	B	IX.31	Area around standing stone	*Cepaea hortensis Cochlicopa lubricella Vallonia costata*	14
	B	VIII/IX	BG24 floors & pits, cist capping	*Cepaea hortensis*	88

TABLE 4.20:
Terrestrial snails catalogue.

Phase C						
26249	C	IX.3	Pits and level	Cepaea hortensis Cochicopa lubrica Cochlicopa lubricella Pupilla muscorum Vallnia pulcella Vallonia costata Columella edentula Aegopinella nitidula	> 5000
23220	C	IX-XI	Across the area, east and west halves, pit infills	Cepaea hortensis	1309
Phase D						
27022	D	XI.2	East half external to buildings	Cepaea hortensis Cochlcopa lubrica Cochlicopa lubricella Pupilla muscorum Vallonia pulcella Vallonia costata Columella edentula	>600
27066	D	XI-XI.2	External to the buildings, west and east halves, small stone tumble	Cepaea hortensis	55
26723	D	XI.2	East of saw pit; facade mural packing	Cepaea hortensis	3
Phase D - Building 1 (DJ)						
27136	D	XI.01/02	Interior and hearth	Cepaea hortensis	7
Phase D - Building 2 (DH)						
27031	D	XI.01/03	Floor 1, shell pit, around hearth,	Cepaea hortensis	88
26826	D	XI	Hearth area	Terrestrial molluscs	1
Phase E						
26691	E	XIc	East of saw pit	Cepaea hortensis	3

TABLE 4.20 continued:
Terrestrial snails catalogue.

PART 5 Exploitation of natural resources and the uses of artefacts

Introduction

By Beverley Ballin Smith

The site's artefacts in bone, stone (flint, pumice, quartz and other stone) as well as pottery vessels have a long and somewhat chequered history of assessment, (re)analysis and reporting, which has involved many specialists over two decades. They are also well-travelled materials that seem to have diminished a little since they were excavated, probably in connection with the move of the whole Udal collection from Cambridge to Dumfries and Galloway, where Crawford had gone to live sometime in the early 1990s. The loss of the objects did not stop there, as further samples of flint, stone and other materials were sent to specialists for different analyses up and down the UK in the late 1990s, even after they had been reported on previously. Whether they were all returned is uncertain.

It was always Crawford's intention to further the post-excavation programme of the entire Udal collection, but the writing up of RUX6 and the start of specialist analysis of materials was strongly encouraged by the former Scottish Development Department (SDD), later Historic Scotland and now Historic Environment Scotland, in the mid-1990s, as it had funded the rescue excavation of the site. It was during that time that Crawford must have visited the National Museum's Artefact Research Unit in Edinburgh, to discuss post-excavation analysis of some of the finds. The results were the preliminary analyses of the stone and quartz by Ann Clarke and the flint by Caroline Wickham Jones. Crawford was always reluctant to impart information, least of all stratigraphic information, and his relationship broke down with the Artefact Research Unit. The type of scientific analyses of materials that he expected may not have been able to be carried out, but financial matters were probably a major consideration too.

Crawford had always maintained a relationship with Edinburgh University since the time he was employed at the School of Scottish Studies, and in 1994-95 he engaged Andy Dugmore (now Professor) and later Anthony Newton (now Dr) to undertake scientific analysis of the pumice. Crawford's link with the University of Glasgow was tenuous but he agreed to Robert Squair, a PhD student at the Department of Archaeology, borrowing the RUX6 pottery assemblage at the beginning of 1997 for his research. After the pottery assemblage was returned to Crawford, and Squair delivered his report and attained his PhD, all further post-excavation analysis of assemblages from RUX6 ground to a halt.

The assessment of the whole Udal collection took place between 2010 and 2012 and it became clear during that exercise that previous specialists did not necessarily get access to the whole of an assemblage, and that many finds, especially the larger stone (samples and artefacts) and the quartz, had either disappeared, disintegrated or mysteriously, more samples had joined them. Small assemblages like flint and pumice where straightforward to reconcile with the earlier reports, but others were more difficult or unsuccessful, and assemblages have had to be re-examined, re-categorised, re-analysed, re-written and updated. As mentioned in other sections, the passage of time has not been kind to the collection.

With the re-analysis of the quartz assemblage by

Torben Ballin in the light of the site stratigraphy and the database, problems emerged mirroring the missing environmental evidence seen in PART 4. Although there is no record of the two late Neolithic buildings being treated differently from the point of view of soil and sample sieving, it is quite obvious that there is a large amount of fine detail and data missing from Building 1 (DJ), which has affected the analysis of artefacts, and has probably skewed the understanding of the activities that took place there.

The antler and bone artefacts were originally examined by Imogen Crawford in 1997, but they have been subsequently re-examined with the rest of the faunal collection by Judith Finlay Aird (PART 4), with an input from Catherine Smith, to determine what was worked and what was unworked, or what was the product of taphonomic conditions on the site.

Time has also not been kind to the pottery assemblage, even though Squair mentions the large quantity of very small fragments and dust, more of it has accumulated. It has also been extremely difficult reconciling the pottery sherds with Squair's descriptions and interpretation due to continued abrasion in spite of careful bagging and labelling. For this publication, his description and discussion of the assemblage have been shortened, edited and updated to bring out the important aspects of his work and that of the collection in relation to the structures.

Flaked flint and other fine-grained lithic materials

By Caroline Wickham-Jones with a contribution by Torben Bjarke Ballin

General composition of the assemblage

The assemblage includes 158 pieces, recovered from the five phases of the site (Table 5.1). The main artefact categories are defined as in the quartz report (Ballin this volume). Regular flakes are defined as likely tool blanks (pieces with more than 10 mm of straight acute edge), whereas irregular flakes are perceived as likely waste (pieces with less than 10 mm of straight acute edge) (Wickham-Jones 1990: 58).

Phases	A	B	C	D	E	Total
Regular flakes	11	8	1	31	2	53
Irregular flakes	7	7	2	45	1	62
Indeterminate pieces				5		5
Pebbles and split pebbles	1			6	1	8
Cores	1			8		9
Retouched pieces	4	2	2	12	1	21
Total	24	17	5	107	5	158

TABLE 5.1:
Composition of the assemblage by main category and phase

Raw materials

The assemblage is composed primarily of flint, although there are pieces made of other materials (Table 5.2).

Raw material	Total
Flint	144
Mylonite	5
Pseudotachylite	2
Siltstone	2
Calcite	2
Chalcedony	1
Shale	1
Unknown	1
Total	158

TABLE 5.2:
Raw materials

The flint is all greyish-white and it is mostly heavily corticated (*sensu* Shepherd 1972), probably a post-depositional effect. There are eight pebbles or split pebbles, and the surviving external surfaces of the debitage and tools suggest that throughout the history of RUX6, the knappers collected small, rolled pebbles of flint as raw material. It is most likely that these came from the local beaches. There has not been any quantitative work carried out regarding the availability of flint on the beaches of North Uist, but research on Islay and Barra indicates that it may have been relatively plentiful in the southern parts of the Inner Hebrides and the Western Isles (Sinclair and Finlayson 1989; Dickens 1990). The main primary flint sources are found in this direction, on and off the Antrim coast (Wickham-Jones and Collins 1978), and it is possible that there was a fall-off in both quantity and quality towards the north, although local offshore sources of Cretaceous and non-Cretaceous flint/chert may also have been available (Ballin 2014b). All the evidence from RUX6 indicates that the knappers only had access to very small pebbles, and as they also made stone tools of other materials, such as quartz and gneiss (see Ballin this volume), flint only played a small part in their flaked lithic repertoire. Most flint types have excellent flaking properties, and produce durable working-edges, but this had to be weighed up against the small size and possible rarity of the available beach pebbles. The stone-workers at the Udal therefore made use of a variety of lithic raw materials to fulfil their needs.

FIGURE 5.1a and b:
The two opposed faces of flakes SF 25223 and SF 25743 (photographed by BBS).

The other lithic raw materials listed in Table 5.2 were probably also procured locally. Mylonite (a notably stripy type of silica: SF 23383, 23432, 23592, 25036a, 25305b) and pseudotachylite (a black flint-like glass SF 25223, 25743, Figure 5.1a and b, by some geologists referred to as 'flinty crush-rock'; Higgins 1971: 21) are related lithic ('cataclastic') materials formed in the area around the Outer Hebrides Thrust Zone, where two tectonic plates ground against each other and altered some of the local country rock (Fettes et al. 1992, 135). All pieces of mylonite from RUX6 are missing.

Most of the assemblage is relatively fresh. Only eleven burnt pieces were recognised, and in addition there were seven heavily abraded pieces (see Ballin this volume and forthcoming a for discussion of local 'sandblasting' of artefacts in machair environments). The burnt artefacts were scattered through the assemblage, and it is likely that many burnt pieces remain unrecognised. Research elsewhere has indicated that the majority of burnt pieces are not altered significantly (Finlayson 1990: 54). The abraded pieces were mainly from Phases A and D.

In addition, three pieces appear to have been re-flaked. It is likely that they were discarded at some stage and later picked up by another knapper and reused, but it is not possible to quantify the period between the different use phases. They were all recovered from Phase D and are discussed in more detail below.

Technology

Primary knapping

The assemblage includes relatively little waste from knapping (47% of the total, including irregular flakes, indeterminate pieces, pebbles and split pebbles, as well as cores). The small number of pieces and the long time span covered by on-site activity, (even if not continuous), make it difficult to extract information on the applied reduction techniques. The analysis of knapping techniques depends on the recognition of trends among the detachment characteristics of an assemblage, and at RUX6 the individual phases into which the flint assemblage is divided, are too numerically small for reliable trends to be discerned. Nevertheless, some general comments may be made.

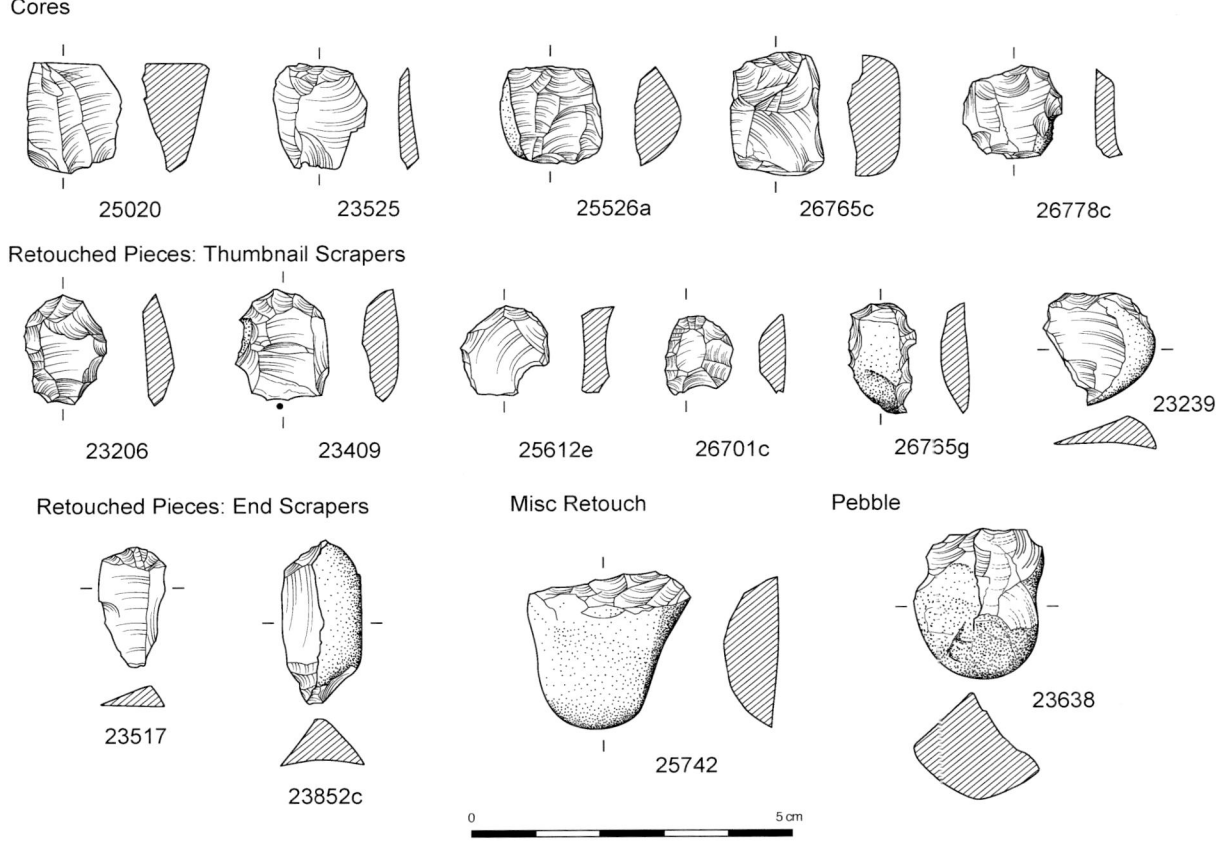

FIGURE 5.2:
Cores SFs 25020, 23525, 25226a, 26765c and 26778c, thumbnail-scrapers SFs 23206, 23409, 25612c, 26701c, 26765g and 263229, end scrapers SFs 23517 and 23852c, miscellaneous retouch SF 25742 and split pebble SF 26368.

The flint industry of the site is characterized by the exclusive production of flakes; there are no true blades, nor any true blade-cores in the assemblage. Hard percussion as well as bipolar techniques were applied in the production of the flakes. Although only one platform core was recovered, it is likely that others were worked-down by bipolar flaking before the cores were abandoned (Figure 5.2). Bipolar flaking (hammer-and-anvil technique), is particularly suited to the knapping of small pebbles, such as those recovered from the site, and would have helped the knappers to make the most of their limited resource (Finlayson 2000).

The collection's split pebbles demonstrate that small pebbles were first opened by bipolar technique, and some were then flaked further by the application of the same approach (Figure 5.2). Platforms were formed on other pebbles, but after the production of some platform flakes, cores which became too small for free-hand reduction were then exhausted completely by the use of bipolar knapping. There is, however, no difference in size between the bipolar cores and the collection's solitary platform core. All were reduced to the same minuscule size (Table 5.3).

SF no	Length	Width	Thickness	Type
2520	18	13	10	Platform
25526a	17	15	8	Bipolar
23525	16	15	3	Bipolar
23706d	20	13	8	Bipolar
26606	17	12	8	Bipolar
26765c	21	15	10	Bipolar
26778c	15	16	5	Bipolar
27232a	16	14	10	Bipolar

TABLE 5.3:
Dimensions of the complete cores (mm)

As might be expected, the flakes include pieces made by the application of platform technique as well as bipolar technique. This, no doubt, reflects the success of bipolar knapping as a technique to manufacture regular flakes on small pebbles. The large number of irregular platform flakes (waste flakes) indicates the general trimming work that is necessary to maintain a platform core (thus producing more small and irregular flakes). The assemblage is small (115 flakes in all), and it is difficult to make generalisations, but, on the whole, the regular flakes are more commonly bipolar, whereas the irregular flakes are mainly struck from platform cores (Table 5.4).

	Bipolar	Platform	Total
Regular flakes	45%	55%	53%
Irregular flakes	23%	78%	62%

TABLE 5.4:
The various techniques applied to produce the site's flakes

Although the regular flakes are of roughly the same size, the platform flakes tend to be slightly larger than the bipolar flakes (Figure 5.3), which lends support to the hypothesis that the cores were first worked by the application of platform technique, and once they had been reduced below a certain size, by bipolar knapping.

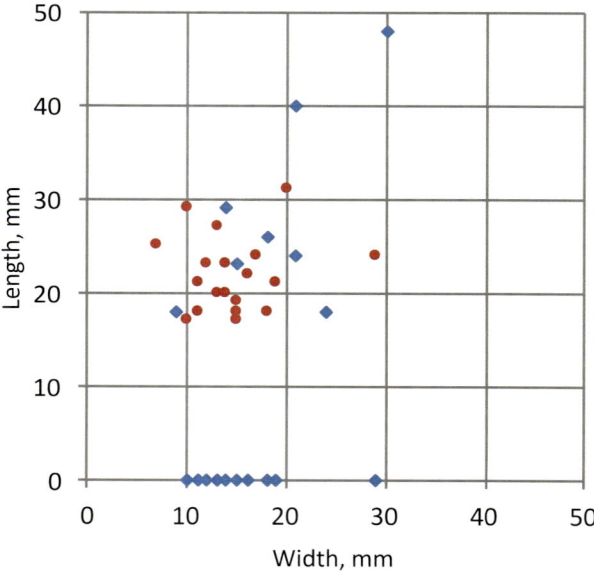

FIGURE 5.3:
Length:width ratio of all intact regular flakes; blue: platform flakes, red: bipolar flakes

Many flakes would have been suitable for a variety of uses without further alteration. The examination of the lithic assemblage did not include detailed use-wear analysis, but it was noted that only two flakes showed signs of macroscopic edge damage. This may be a result of use, but the pieces may have been damaged in other ways, such as post-depositional pressures, and the absence of edge-damage is not necessarily indicative of a piece not having been used. Nevertheless, a number of pieces were modified, mostly into small thumbnail-scrapers.

Secondary knapping

In total, the assemblage includes 21 retouched pieces, all of which are flint. They were modified by fine, steep, pressure-flaking, and the working-edges are generally highly regular. Most of these pieces are small scrapers, with tiny, round, thumbnail-scrapers (Figure 5.2) being the most common (Table 5.5).

Phase	A	B	C	D	E	Total
Thumbnail-scrapers	3	1	1	9	1	15
End-scrapers	1			1		2
Truncated scrapers			1			1
Resharpening flakes				1		1
Pieces w bifacial retouch		1				1
Pieces w other retouch				1		1
Total	4	2	2	12	1	21

TABLE 5.5:
Retouched pieces by type and phase

The scrapers are based on small flake blanks, of which inner flakes are more common than cortical ones. Most have one retouched edge, usually at the distal end, and occasionally extending along one side. On most scrapers, the retouched edge is flanked by two relatively straight lateral sides, perhaps to facilitate hafting. The small size of these pieces would certainly make them hard to use without a haft.

The two end-scrapers, SF 23517 and 23852c (Figure 5.2), are made on long triangular blanks (one inner and one secondary flake), and they have both been retouched at the proximal end. The retouch is slightly longer than that of the thumbnail-scrapers. The truncated scraper (26202b) is similar in size and shape to a thumbnail-scraper, which it may originally have been. Most of the scrapers have some macroscopic edge-damage along the retouched edge, although in some cases it is very slight.

In addition, there is one resharpening flake (SF 25612c). The edge is much abraded, but it is impossible to tell from what type of tool it was struck. The angle of the edge suggests that it may have been detached from a scraper. There is also one piece SF 25742 (Figure 5.2) with more extensive, straight, acute retouch across the distal end, and this piece may be an expedient scale-flaked knife. It has been classified as 'a piece with other retouch'. A piece with bifacial retouch (SF 27084) is thought to be a fragment of an arrowhead. Like the mylonite pieces (above), this tool is missing. In the original catalogue, the raw material of this piece was classified as 'shiny and black', possibly 'shale', but it is possible that this piece is in fact pseudotachylite, like one black arrowhead from the Barabhas machair (the Elliott Collection; Ballin forthcoming a).

Chronology and distribution

Phase A

Twenty four flint artefacts were recovered from Phase A (Table 5.1). The finds include mainly regular flakes (11 pieces), and there is also much waste from knapping (pebbles, cores, and irregular flakes) (Table 5.6). In addition, there are four retouched pieces: three thumbnail-scrapers, and one end-scraper. The flakes reflect the use of both bipolar and platform flaking, with slightly more emphasis on platform flaking (Figure 5.4).

Phase	A	B	C	D	E
Knapping waste	37%	41%	40%	60%	40%
Regular flakes	46%	47%	20%	29%	40%
Retouched pieces	17%	12%	40%	11%	20%
Total (%)	100%	100%	100%	100%	100%
Quantity	24	17	5	107	5

TABLE 5.6:
Main artefact categories by phase

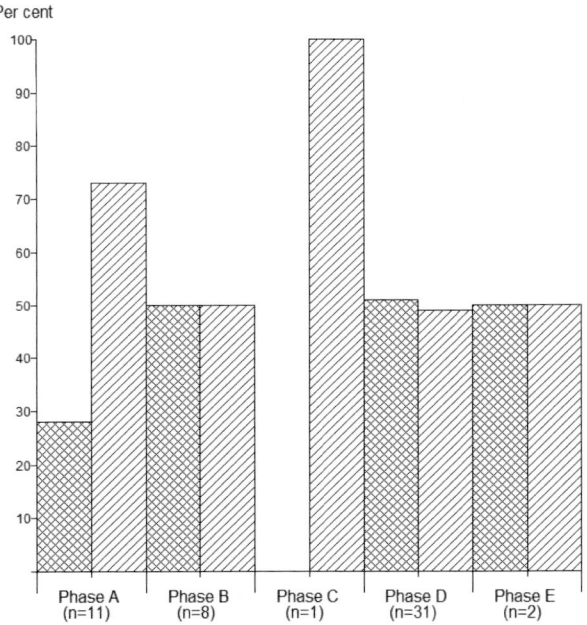

Figure 5.4:
The ratio of bipolar:platform flakes within each phase; hatched bars = platform flakes, cross-hatched bars = bipolar flakes

Phase B

The assemblage from Phase B includes seventeen flint artefacts (Tables 5.1 and 5.6). There were only two retouched pieces, namely one thumbnail-scraper, and one broken bifacial arrowhead, and the rest of the pieces were all flakes, seven of which are irregular whereas eight are regular. There were no cores, or pebbles. Both bipolar and platform flaking were used (Figure 5.4). One piece, SF 26309, a regular secondary flake of chalcedony, was recovered from floor 1/2 in the building BG24, which may, judging from the composition of its quartz assemblage (Ballin this volume), be a domestic building.

Phase C

Only a small number of flint artefacts (five pieces) were recovered from Phase C (Table 5.1). Two of these are retouched pieces, namely one thumbnail-scraper, and the small truncated scraper, and there are two irregular flakes and one regular flake (Table 5.6). All of the flakes appear to have been struck from platform cores (Figure 5.4).

Phase D

The largest number of flint artefacts (107 pieces) was retrieved from Phase D (Table 5.1). This assemblage includes much knapping waste (60%; Table 5.6), such as eight bipolar cores and the only indeterminate pieces to be found in the assemblage (five in all). In addition to the regular flakes, twelve pieces were retouched, namely nine thumbnail-scrapers, one end-scraper, one resharpening flake, and one piece with other retouch. Both bipolar and platform flaking were used (Figure 5.4).

Several of the pieces were associated with the Neolithic buildings 1 and 2. Building 1 (DJ) yielded 16 pieces, all of flint. This assemblage includes one bipolar core (SF 23706d), but it consists mainly of regular flakes and retouched pieces (only 37% was knapping waste, Table 5.7), reflecting an assemblage primarily derived from the use of flaked stone tools.

Building 2 (DH), in contrast, yielded more pieces and much more evidence for the manufacture of tools. This assemblage includes two bipolar cores and two pebbles, and 72% of the assemblage is knapping waste (Table 5.7). The finds from Building 2 also include five retouched pieces, such as the resharpening flake (SF 25612c) and the piece with other retouch, which may be a scale-flaked knife (SF

25742). The greater number of pieces from Building 2 may have been augmented by the fine sieving of the soil near the hearth, which certainly increased the number of tiny irregular flakes.

	Building 1	Building 2
Regular flakes	7	10
Irregular flakes	5	31
Indeterminate pieces		4
Pebbles and split pebbles		2
Cores	1	2
Retouched pieces	3	5
Total	16	54
Debitage (% of total)	37%	72%

TABLE 5.7:
The lithic artefacts from Neolithic Buildings 1 and 2

Phase E

Five artefacts were recovered from Phase E (Table 5.1). There were no cores, with the assemblage including two pieces of waste (one irregular flake and one pebble), two regular flakes, and one retouched piece, a thumbnail-scraper (Table 5.6). There is evidence for both bipolar flaking and platform flaking (Figure 5.4).

Discussion

The flint assemblage (158 pieces) is small compared to the numerical size of the site's quartz assemblage (9,811 pieces; Ballin this volume). Nevertheless, when considered by itself, it does contain some interesting features.

Although the material derives from five phases which differ in terms of date and activities, there is remarkably little variation between the sub-assemblages from the different phases. The individual artefacts are almost identical: a flake from Phase E is more or less identical to a flake from Phase A and the flakes seem to have been knapped by the application of the same percussion techniques. The retouched pieces do not vary across the different phases. Phase A contains the only platform core of the assemblage, but there is plenty of evidence for the knapping of platform cores at lower levels. However, none of the individual phases yielded enough flint artefacts for meaningful generalisations to be drawn.

However, the variation of the numerical size of the individual sub-assemblages may itself be interesting.

Phases C and E, for example, each yielded very low numbers of flint artefacts. This variation has to be considered in relation to the use of other lithic materials in each phase (e.g. the quartz; see Ballin this volume), and to the interpretation of the features and contexts of the different phases. Phase C, for example, represented activities that might therefore not be expected to yield much lithic material. Phase D, on the other hand, which represented settlement activity, produced the largest assemblage of flint artefacts, and as shown in Table 5.7 the assemblages from the two Neolithic buildings are clearly composed differently. The material from Building 1 (DJ) appears to reflect the *use* of stone tools, while that from Building 2 (DH) may reflect the *manufacture* of tools (see discussion of the quartz assemblages from the two buildings, Ballin this volume). However, this picture may have been affected by the loss of sieved residues from Building 1 (DJ) (see PART 4) in contrast to the survival of finds from Building 2 (DH).

Throughout the life of the site, the knappers were working small pebbles of flint to produce flakes and a limited range of implements. The lack of formal variation among retouched pieces was touched upon above, and it merits further discussion. Such uniformity in a group of modified pieces would be noteworthy in itself had all the artefacts been recovered from contexts of one period. At the site, tiny thumbnail-scrapers were apparently being made from the time of its earliest occupation up to the final stone-using levels (cf. composition of the quartz implements; Ballin this volume). These scrapers dominate the assemblage, only supplemented by a few pieces with edge-retouch and scale-flaking/invasive retouch (Table 5.5).

Most likely, the preponderance of this tool type reflects its usefulness in relation to the dominance of one specific activity throughout the 'life' of the site, such as the processing of certain materials. Scrapers are a ubiquitous tool type in prehistory, and they dominate all Neolithic and early Bronze Age lithic assemblages from the Western Isles (Ballin this volume; also see Ballin 2008b). They have been linked to many different environments and tasks all over the world. Without detailed functional analysis it is impossible to be more specific about their use at RUX6.

The flint assemblage as a whole is not only unusual in terms of the composition of the retouched pieces, it is also unusual in terms of the almost complete absence of waste material from knapping throughout all the site's phases. The manufacture of stone tools is a reductive process that leaves behind much debris. Usually, most of the waste is tiny, and apart from that recovered from Neolithic Building 2, this debris may have been missed or lost, whereas larger pieces of waste may have been removed in connection with site maintenance (Binford 1983). Nevertheless, there usually remains a quantity of irregular debris (classified according to size – larger and smaller than 10 mm – as indeterminate pieces and chips), as well as cores, split pebbles, and irregular flakes. The assemblage from RUX6 does include artefacts belonging to the latter three categories, but there were few indeterminate pieces and no chips. This is unusual and suggests that knapping could have taken place elsewhere, away from the excavated areas, unless it entirely reflects the loss of sieved residues from most of the site. In the present case, the artefacts recovered from the site may reflect the use of formal and informal tools (retouched pieces and regular flakes).

Finally, there are three pieces which show signs of recycling (two flakes, SF 23622a and 26765b, and a flaked pebble, SF 25612b). They come from Phase D (including both buildings). All are corticated, and SF 26765b is very abraded. These pieces most likely represent RUX6 knappers who scavenged the debris left by previous generations of knappers.

Conclusion

The flint assemblage from RUX6 includes 158 artefacts of various materials, mainly, if not entirely, related to the use, as opposed to the manufacture, of tools. Cores and irregular flakes are present, but there is little of the more irregular and smaller waste from knapping, such as chips and indeterminate pieces. Interestingly, the only area of the site which contained mainly evidence for the manufacture of lithic artefacts is Building 2 (DH) in Phase D. Building 1 (DJ), in contrast, contained little knapping waste that survived. However, this may to a degree, reflect the loss of sieved residues.

The knappers applied bipolar as well as platform flaking to reduce small pebbles, probably collected from a local beach. The small size of the unworked and split pebbles in the assemblage suggests that manufacture may have been restricted by the available raw material, and there is evidence for the recycling of flaked material left by previous generations of knappers. Flint only forms a small part of the lithic assemblage from RUX6 with quartz artefacts dominating heavily (see Quartz, this volume).

Despite the fact that the site reflects use through much of early prehistory, there is little variation in the composition of the assemblage. There is no evidence for the alteration of the manufacturing techniques through time; production is focused on the manufacture of small flakes and thumbnail-scrapers, which dominate the assemblage. It seems likely that flint, a scarce resource at the site, was used mainly for one specialised task, which may relate to the processing of one or more specific materials.

The quartz assemblage

By Torben Bjarke Ballin

Introduction

The excavation of RUX6 yielded almost 10,000 quartz artefacts, Artefacts in flint and other stone objects recovered during the excavation were characterized and catalogued separately by Wickham-Jones (flint), and Ballin Smith et al. (other stone and pumice). A preliminary report on the quartz was prepared by Clarke (1997).

The purpose of this analysis was to characterize the quartz artefacts in detail, with special reference to raw-materials (quartz types) and typo-technological attributes, and to date and discuss the finds. The examination of the quartz is based upon a detailed catalogue (an Access database) of the finds referred to by their catalogue number (CAT no.).

Key elements of the discussion are the procurement and reduction of the quartz (operational schema[s]), intra-site artefact distribution (vertically as well as horizontally), on-site activities, dating of the site, and comparison with other Western Isles quartz assemblages. As part of the analysis, the present collection is compared first and foremost with the finds from Barabhas in northern Lewis (Ballin 2010a-e; forthcoming a), as these collections were recovered from the same environment (machair) as those from RUX6 and therefore exposed to some of the same post-depositional effects (e.g. aeolian activity/sand-blasting; see below). The Elliott Collection (Ballin forthcoming a) has been chosen as the main focus of the comparative analysis as this assemblage is numerically comparable with RUX6 (6,883 quartz, flint and stone artefacts, 5,702 of which are quartz/quartzite), and has been characterized the same way, in terms of terminology and nomenclature.

The quartz collection from RUX6 is dominated by finds from the Neolithic period, but supplemented by artefacts from the early Bronze Age, whereas the collection from Barabhas (Elliott Collection) is dominated by early Bronze Age material, but supplemented by finds from the Neolithic. Barabhas and RUX6 are both dominated by domestic activity, supplemented by some burial/ritual activity, indicated at Barabhas (the Elliott Collection), by the recovery of several well-executed, but re-deposited, arrowheads, and at RUX6 by features.

Definitions of the characterization of the assemblage (and abbreviations)

The definitions of the main lithic categories are as follows:

Chips: All flakes and indeterminate pieces the greatest dimension (GD) of which is ≤ 10 mm.

Flakes: All lithic artefacts with one identifiable ventral (positive or convex) surface, GD > 10 mm and L < 2W (L = length; W = width).

Indeterminate pieces: Lithic artefacts which cannot be unequivocally identified as either flakes or cores. Generally the problem of identification is due to irregular breaks, frost-shattering or fire-crazing. Some refer to these as *chunks*, but this practice should be avoided as many indeterminate pieces (for example thermal specimens) are not 'chunky' (see below).

Blades and microblades: Flakes where L ≥ 2W. In the case of blades W > 8 mm, in the case of microblades W ≤ 8 mm.

Cores: Artefacts with only dorsal (negative or concave) surfaces – if three or more flakes have been detached, the piece is a core, if fewer than three flakes have been detached, the piece is a split or flaked pebble.

Tools: Artefacts with secondary retouch (modification).

GD: Greatest dimension.

Av. dim.: Average dimensions.

Quartz is notoriously problematic to analyse due to the intricate ways most quartz types flake and fracture. It has therefore been difficult to develop a terminology which accurately describes quartz debitage and cores, and due to the different ways specialists apply the available terms, many reports on Scottish quartz assemblages are not directly comparable. It was therefore decided in this section to discuss some of the more problematic terms, not least as it is highly likely that the complex flaking and fracture patterns of the RUX6 quartz (see Raw Material section) without careful use of the quartz-related terminology would have resulted in different specialists producing very different reports on the same assemblage.

As mentioned above, the traditional use of the term 'chunk' to describe pieces of debitage, which cannot be defined more precisely is unhelpful and the term 'indeterminate pieces' is preferred, as this is what they are. In most cases, the difficulties involved in the more precise definition of these pieces are based on the specific fragmentation of them, usually due to their fracturing along internal fault planes or due to thermal action, such as exposure to frost or fire. A flake, for example, which as a result of thermal action has shed its ventral face cannot be identified as a flake, and is moved to a lower level in the lithic classification hierarchy – it becomes an indeterminate piece, and it is not necessarily 'chunky'.

In connection with the inspection of other (usually older) assemblages, the author noticed that there was a tendency to frequently characterize 'chunky pieces' according to their general appearance, rather than according to their specific technological attributes. If a quartz fragment displays even the smallest bit of a ventral face, or an identifiable bulb of percussion, or bipolar terminal, it is a flake - chunky or not. If the RUX6 assemblage had not been classified using technological attributes it would have included substantially fewer flakes and substantially more 'chunks'. Occasionally the term 'angular shatter' is used as a replacement for the term 'chunks', but it is an equally imprecise term, as it is primarily based on appearance and not technological principles.

Occasionally, quartz reports use the expression 'amorphous cores' (or irregular-, multi-platform- or multi-directional cores) to describe slightly larger 'chunky' pieces, but the definition of these pieces should be based on technological attributes and not appearance. If for example, a large chunky piece

FIGURE 5.5a and b:
Four refitted orange-segment flakes from the Norwegian site Lundevaagen 21, SW Norway, a) refitted and b) individual flakes. © Torben Bjarke Ballin 1999.

has the smallest surviving part of a ventral face, or other attributes such as percussion bulbs or bipolar terminals, it is a large flake and not a core. If a large chunky piece is a core, it should be classed primarily according to its number of striking platforms and their position in relation to each other (see Butler 2005).

Waste products from bipolar (anvil-based) primary production are difficult to deal with, and there are two main problems. Firstly, how can bipolar cores be distinguished from bipolar flakes and secondly, how can different types of bipolar flakes be distinguished? Both problems relate to the specific way lithic raw materials tend to flake and break when reduced on an anvil, not least the way many pebbles tend to split radially into so-called orange-segment flakes, leaving no apparent core at all (Figure 5.5a and b).

The four pieces in Figure 5.5b could have been characterized as unifacial bipolar cores, but they are technically flakes (as they have one or more ventral faces) which, due to their cross-sections, are referred to as orange-segment flakes. In terms of defining bipolar cores, the same definitions apply as those used to define platform cores (above), in that they have no ventral (convex) face, and all cortex-free faces are concave faces (unless they are internal fault planes).

Splitting a pebble according to the 'orange-segment principle' creates some terminological problems, such as which faces of the orange-segment flake are dorsal (outer) and which are ventral (inner)? And subsequently, when is a cortical orange-segment flake a primary flake (with full outer cortex) and when a secondary flake (with partial outer cortex)? These questions are relevant to the calculation of a number of ratios and the production of some tables (e.g. Tables 10 and 13).

Usually, a standard platform flake would have one dorsal face (with ridges separating the scars left by previously detached flakes or cortical areas), and one ventral face (where the flake was attached to its parent core), but as shown in Figure 5.6, a bipolar orange-segment flake with a triangular cross-section would have two ventral (inner) faces and one convex dorsal (outer) face. The fact that many quartz pebbles opened by the application of bipolar technique tend to split according to the 'orange-segment principle' also explains the tendency of many bipolar blanks to be elongated flakes or even metrical blades (see Figure 5.8).

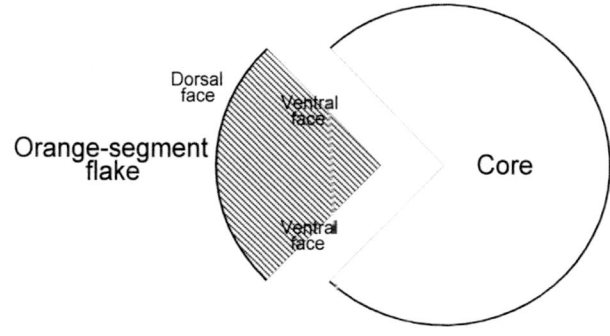

FIGURE 5.6:
The elements of a bipolar orange-segment flake.

During the characterization and cataloguing of the worked quartz from RUX6, there was another terminological problem. The assemblage included many more 'blades' (flakes where L ≥ 2W) than expected. However, it was noted in many cases that one lateral side would be sharp, whereas one would be blunt, and in the latter case, attributes (such as the directionality of the ripples of that 'lateral' side) would indicate that this was not an actual 'lateral' side, but a break facet. In other words, many of these pieces are not elongated blades, but the lateral fragments of broader blanks i.e. flakes. The lateral sides of all elongated pieces (potential blades) from RUX6 were therefore examined, and in many cases it was possible to determine that they were simply fragments of broad flakes, and the collection's number of blades were brought down to the expected level (based on comparison with other contemporary quartz assemblages from the Western Isles e.g. Barabhas [Elliott Collection], Barabhas 3, Dalmore, Calanais, Barpha Langais, Rosinish, and Point Braighe (Ballin 2000b; 2002b; 2008a; 2010a-e; 2015a; 2016; forthcoming a;).

The assemblage

During the excavations at RUX6, a total of 9,811 quartz artefacts were recovered (Table 5.8); their distribution across the phases is mentioned in the descriptive sections, but discussed in the distribution and dating sections below.

In total, 96% of the assemblage is debitage, 3% are cores, and 1% are tools (Table 5.9). These ratios are approximations, as the abrasive or sandblasting effect of the wind and sand, as well as the specific character of the RUX6 quartz, made it difficult to identify edge and surface modification as well as other attributes, and some flakes and blades may be sandblasted cores and tools. Due to the size of the collection, only the greatest dimensions of the flakes are recorded, but all three dimensions of blades/microblades, cores and tools are noted (see below). Table 5.9 shows that the composition of the quartz artefacts does not vary notably across the main phases, but the low numbers of tools (1%) is discussed below.

Debitage	
Chips	1,865
Flakes	6,679
Blades	221
Microblades	35
Indeterminate pieces	620
Crested pieces	16
Total debitage	9,436
Cores	
Split pebbles	28
Single-platform cores	37
Single-platform/discoidal cores	5
Opposed-platform cores	8
Cores w two platf at angle	4
"Flaked flakes"	2
Irregular cores	28
Atypical cores	2
Bipolar cores	148
Core fragments	5
Total cores	267
Tools	
Backed knives	1
Scale-flaked knives	1
Discoidal scrapers	2
Blade-scrapers	1
Short end-scrapers	39
Double-scrapers	2
Side-scrapers	4
End-/side-scrapers	2
Atypical scrapers	2
Scraper-edge fragments	7
Piercers	6
Pieces w retouched notch(es)	1
Denticulated pieces	2
Pieces with edge-retouch	25
Points	4
Percussoirs	1
Percussoirs/anvils	1
Hammerstones	3
Hammerstones/split pebbles	1
Hammerstones/anvils	1
Pounders	2
Total tools	108
TOTAL	9,811

TABLE 5.8: General artefact list

	Quantity					
Phase	A	B	C	D	E	Total
Debitage	1,444	2,212	529	4,720	94	8,999
Cores	64	58	16	112	3	253
Tools	14	39	7	46		106
Total	1,522	2,309	552	4,878	97	9,358
	Per cent					
Phase	A	B	C	D	E	Total
Debitage	95	96	96	97	97	96
Cores	4	2	3	2	3	3
Tools	1	2	1	1		1
Total	100	100	100	100	100	100

TABLE 5.9:
The main artefact categories by phase.

Raw materials – types, sources and condition

The quartz is almost exclusively white milky quartz, with saccharoidal forms being virtually absent. The quartz is mostly pure, but some pieces are characterized by grains or, in some cases, sheets of mica and/or hornblende (an amphibole). Some rare larger pieces contain much feldspar, occasionally in the form of up to 30 mm long crystals. As feldspar, due to its cleavage planes and poor flaking properties, is quite useless as a toolstone, these pieces were generally abandoned at an early stage. Feldspar-bearing irregular core CAT 26212/001, for example, had a GD of 155 mm when it was discarded, and single-platform core CAT 25740/006 was almost 60 mm long when it split along the long axis due to the presence of a 20 mm long feldspar crystal.

These different impurities indicate the local rock formations the quartz came from. Pegmatites appear to be absent on North Uist, but several of the gneisses and granites in the area contain 'other' minerals in addition to the quartz, namely mica, hornblende, and feldspar, and several of the gneisses and granites contain feldspar in the form of large lenticular crystals or 'augen' (Fettes et al. 1992: 14-16, 44-47). The North Uist bedrock generally consists of gneiss, but outcrops of intrusive granite may be found throughout the island.

Although the quartz is generally of good quality, a relatively large proportion of it is a form of 'reconstituted' quartz, consisting of compressed grains of quartz of considerable size, frequently measuring several millimetres across. Although these pieces have been packed so tightly together that they mostly appear to form a massive entity, it is possible on close inspection to identify the fine fault-lines between the individual grains. These fault-lines, in conjunction with the frequent occurrence of thin sheets of mica and/or hornblende makes this type of quartz flake in an irregular manner when struck. This, may to a degree, explain the low tool ratio of the assemblage (1%), as unmodified edges of this form of quartz tend to be somewhat irregular and jagged, or naturally denticulated, therefore making it very difficult to safely identify actual secondary retouch. The 'reconstituted' character of this form of quartz also means that when modified edges (e.g. scraper-edges) are damaged, large chunks occasionally break off and remove notable proportions of the original working-edge. In addition to this specific RUX6 problem, quartz is generally considered a difficult raw material to analyse, in the sense that its reflective surfaces makes it difficult to identify secondary retouch and thereby tools (Ballin 2008b: 73).

	Quantity		Per cent	
	Flakes	Blades	Flakes	Blades
Primary pieces	526	19	8	8
Secondary pieces	960	34	14	13
Tertiary pieces	5,193	203	78	79
Total	6,679	256	100	100

TABLE 5.10:
Reduction sequence of all unmodified and modified flakes and blades.

In Table 5.10, almost 80% of the flakes and blades are tertiary (inner) material, with c. 20% being cortical material (a piece was defined as tertiary if it had a cortex-free dorsal face but a cortical platform remnant; see core and technology sections). The low cortex ratio, in conjunction with the fact that many inner pieces are characterized by coated, rust-coloured fault-planes, suggests that some of the quartz was quarried from local veins, whereas the abraded surfaces of most of the cortical pieces (as well as pieces with cortical platforms) indicate that the majority of the quartz may have been procured by combing the beach adjacent to the site (Ballin 2008b: 56).

A total of 8% of the quartz artefacts from the site were defined as sand-blasted that is abraded by aeolian activity in the machair environment. Sandblasting varies between light to heavy abrasion, and some of the latter pieces have been so heavily affected that they are on the verge of becoming pebbles or cobbles again. However, the 'blast ratio' of the five Phases A to E varies notably, with the ratio of Phase A being 38.2%, and the ratios of the following phases 4.5%, 0.9%, 0.5% and 2.1%, respectively (Figure 5.7). The ratio of Phase E is based on a sample of less than 100 pieces and may be affected by random statistical fluctuation. It is thought that the 'blast ratio' may basically be an 'erosion ratio', representing the degree to which the various phases were disturbed by aeolian activity and the repeated formation of deflation and conflation areas (Barber 2011: 45).

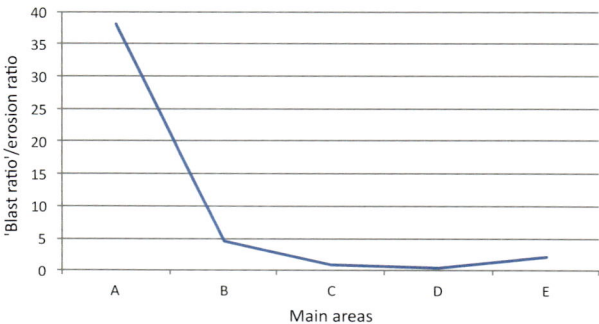

Figure 5.7:
The ratio of sand-blasted pieces by phase (the 'blast ratio'), probably indicating the level of erosion the main phases were exposed to.

In comparison, the c. 6,000 pieces of worked quartz and other lithic artefacts collected by Mark Elliott from the Barabhas dunes (Ballin forthcoming a) had a 'blast ratio' of 78%, but this collection only includes surface collected material, whereas many of the lithics from RUX6 had been recovered by excavation of deeper levels, which were almost entirely unaffected by aeolian activity. Interestingly, several pieces are only sandblasted on one face, demonstrating that the opposed face was protected in the ground.

Phase	A	B	C	D	E	Total
Burnt no.	36	248	7	67	0	358
Total per phase	1,522	2,309	552	4,877	97	9,357
Burnt ratio	2.3	10.7	1.2	1.4	0	4

TABLE 5.11:
Burnt quartz ratio by phase

In total, c. 4% of the assemblage has been affected by exposure to fire but, as shown in Table 5.11, the burnt quartz ratio varies significantly from phase to phase – where Phases C and D have burnt quartz ratios just above 1% that of Phase B is almost 11%. This could possibly reflect different activity patterns, with Phase B representing Bronze Age burial/ritual activities and Phase D Neolithic settlement, but it is also possible that these differences reflect changing recovery policies during the excavation, with little sieving taking place during the initial phases, and significant and meticulous sieving taking place in connection with the investigation of Phase D (particularly Neolithic building 2 and its knapping floor; see distribution section). The sieving of Phase D resulted in the retrieval of thousands of tiny quartz chips (see the debitage section) which were so small that they were defined as 'not burnt', and may therefore artificially have lowered the burnt quartz ratio of Phase D. Numerous pieces had ashy sand attached to their surfaces, but pieces were only defined as burnt, if exposure to fire had altered their physical appearance (i.e. if there were colour changes, weight loss, light crazing, deep fissures, etc.; Ballin 2008b: 39).

Debitage

In total, 9,436 pieces of debitage were recovered from the site, which includes 1,865 chips, 6,679 flakes, 221 blades, 35 microblades, 620 indeterminate pieces, and 16 core preparation flakes (all crested pieces). As shown in Table 5.12, the chip ratio varies significantly across the site's main phases, from 1% in Phase A to 37% in Phase E. It cannot be ruled out that these differences reflect differences between settlement (much primary production and many chips) and burial or ritual activities in Phase B (no primary production and few/no chips). As indicated by Figure 5.7, Phase A is heavily affected by erosion and the digging of stone-robbing holes and a saw-pit into lower levels that introduced at least some chips into this level. Most likely, the differences in terms of chip ratio mainly reflect differences in recovery policies, with Phase D probably being the only phase consistently sieved.

Phase	A	B	C	D	E	Total
Chips, number	7	83	34	1,725	7	1,856
Total debitage per phase	1,444	2,212	529	4,720	94	8,999
Chip ratio (per cent)	1	4	7	37	8	21

TABLE 5.12:
The site's smallest pieces of debitage (chips) and their distribution across the phases

In order to compare like with like Table 5.13 was produced, which shows the composition of the debitage within the phases. The debitage included in the table is all relatively large pieces, whose recovery should not be affected by the absence of sieving in some phases.

	Quantity					
Phase	A	B	C	D	E	Total
Flakes	1,272	1,879	432	2,652	79	6,314
Blades	35	49	13	108	2	207
Microblades	2	3	2	26		33
Indeterminate pieces	123	197	48	203	5	576
Crested pieces	5	1		6	1	13
Total	1,437	2,129	495	2,995	87	7,143
	Per cent					
Phase	A	B	C	D	E	Total
Flakes	89	88	87	88	91	89
Blades	2	3	3	4	2	3
Microblades	trace	trace	trace	1		trace
Indeterminate pieces	9	9	10	7	6	8
Crests	trace	trace		trace	1	trace
Total	100	100	100	100	100	100

TABLE 5.13:
The composition of the debitage within the main phases (less chips)

	Quantity		Per cent	
	Flakes	Blades	Flakes	Blades
Soft percussion	0	0	0	0
Hard percussion	1,987	15	52	11
Indeterminate platform technique	37	4	1	3
Platform collapse	171	1	4	1
Bipolar technique	1,622	117	43	85
Total	3,817	137	100	100

TABLE 5.14:
Applied percussion techniques (technologically definable unmodified and modified flakes and blades)

Across the site, flakes dominate heavily (80%), followed by indeterminate pieces (8%). Blades only make up 3%. Taking into account that the debitage assemblage from Phase E is numerically very small, and therefore possibly affected by random statistical fluctuation, the only notable difference between the other phases is the fact that more blades were recovered from Phase D than from any of the later phases (5% compared to 2-3%). This corresponds well with the suggested dates of the different phases (see dating section), with Phase D dating to the late Neolithic, and Phases A-C (in terms of the quartz) to the early Bronze Age. In Scotland, the Neolithic period is usually associated with blade-producing industries and the Bronze Age with flake-producing ones.

As shown in Table 5.14, there are considerable differences in terms of the percussion techniques applied to produce the site's flakes and its blades. Where flakes were manufactured by the application of hard percussion (pronounced bulb) and bipolar technique (crushed terminal) almost in equal measure, most blades are bipolar, with 11% having been produced by the application of hard percussion. However, the fact that many platform-flakes display crush-marks at their distal ends, suggest that hard percussion may also frequently have involved the use of an anvil (see Figure 5.24).

The fact that most blades are bipolar pieces probably reflects the tendency mentioned above of bipolar orange-segment blanks frequently becoming somewhat elongated – a tendency further supported by a comparison between the general flake:blade ratio (96:04) and the flake:blade ratio of the bipolar material (83:17). The bipolar material therefore includes notably more blades than the site as a whole. Figure 5.8 also clearly shows that the bipolar blades are more elongated than the platform blades.

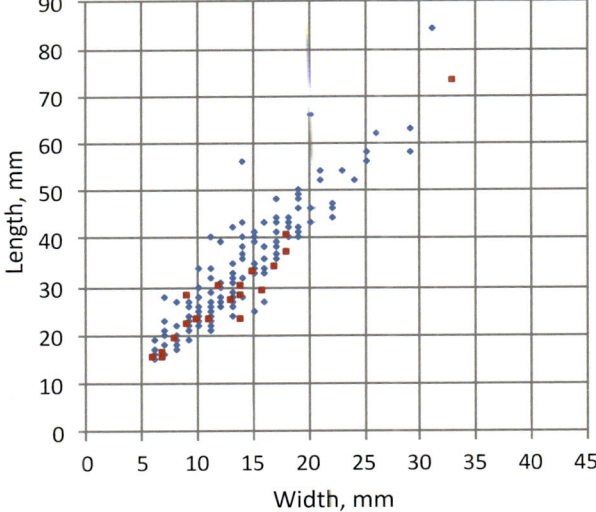

FIGURE 5.8:
The main dimensions of all intact blades/microblades in quartz; hard-hammer pieces (red), and bipolar pieces (blue).

However, RUX6 also yielded a small group of broad blades, some of which were characterized as either 'regular' (three pieces), 'quite regular' (two pieces), 'unusually regular' (three pieces: CAT 25139/12-13, 25143/9), or 'perfectly regular' (one piece: CAT 26004/54, Figure 5.9). These nine blades form a group of their own of aesthetically pleasing and technologically well-executed pieces and they are easily distinguished from the vast majority of blades included in Figure 5.8. Only five of these pieces are technologically definable, with two being hard percussion pieces, and three display platform collapse. Considering that these pieces are generally characterized by parallel lateral sides and dorsal arrises, they are most likely to have been manufactured by the application of platform technique, as bipolar blades tend to have somewhat curved lateral sides and criss-crossing dorsal arrises (cf. Ballin 1999).

FIGURE 5.9:
Intact blade SF 26004/54.

One piece was recovered from Phase A, two from Phase B, and six from Phase D, and given that Phases A and B are affected by erosion (Figure 5.7), the well-executed platform blades are predominantly associated with Phase D. All 'unusually regular' and 'perfectly regular' pieces derive from this Phase (see distribution section).

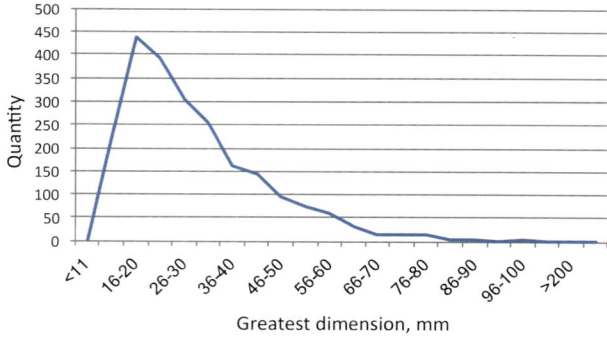

FIGURE 5.10:
The greatest dimension of all intact quartz flakes (2,237 pieces).

The flakes vary considerably in size (Figure 5.10), and the average greatest dimension of 2,237 intact flakes is 30 mm. However, some flakes (218 pieces) are as small as just over 10 mm (separated from the chips by the metric definitions of the two categories), whereas some pieces reach a greatest dimension of c. 100 mm. The peak in Figure 5.10 is at 16-20 mm (437 pieces). It is highly likely that the clustering of flakes towards the left side of the diagram is due to the smaller size categories including much debris from the preparation of the cores (decortication, trimming, etc.), whereas most of the flakes from the central and right side parts the diagram are likely to include larger proportions of actual 'target blanks', i.e. intentional tool blanks.

As shown above (Figure 5.8), the blades include hard percussion (19 pieces) as well as bipolar specimens (158 pieces). The average dimensions of these two categories are 28.6 by 13.2 by 7.6 mm (hard percussion blades) and 33.2 by 13.8 by 9.7 mm (bipolar blades), and although these two sets of values may not seem to differ much from each other, the resulting length:width ratios do. They are 2.1 (hard percussion blades) and 2.4 (bipolar blades), showing that (as mentioned above) the bipolar blades are relatively more slender than their platform counterparts.

The average dimension of the indeterminate pieces is 23 mm. They are fairly small and vary in size between 11-100 mm. Many of these pieces are fragments of cores which, due to the presence of internal fault planes or impurities, disintegrated during reduction. However, the fact that as many as 13% of the indeterminate pieces are burnt (where only 4% of the assemblage as a whole shows signs of exposure to fire) suggests that a large proportion of these pieces may be fragments of quartz artefacts which either accidentally fell into domestic fires during production or use, or were discarded and secondarily used as pot-boilers (e.g. some hammerstones).

Phase	A	B	C	D	E	Total
Cortical platforms	119	140	35	173	5	472
Total per phase	228	278	64	330	13	913
Cortical platform ratio	52	50	54	52	39	52

TABLE 5.15:
Flakes and blades with cortical platform remnants (only intact pieces or proximal fragments).

As flake production was usually initiated without preparing a striking platform (that is, most early-stage single-platform cores have cortical platforms; cf. Figures 5.22 and 5.23), a substantial proportion of the flakes and blades also have cortical platforms (Table 5.15). The presence of blanks with cortical platforms was also noticed in connection with the analysis of the Barabhas material (Elliott Collection), although in this case such platforms were not quantified ('... *and practically all hard-hammer blanks have cortical platform remnants* ...; Ballin forthcoming a). At RUX6 52% of all intact or proximal fragments of hard percussion flake and blade blanks have cortical platforms.

Due to the preference for cores with cortical platforms amongst the RUX6 knappers (see core and technology sections, below), the preparation flakes are all crested pieces, with platform rejuvenation flakes being entirely absent. A total of 13 examples out of the 16 crested pieces are unilateral, with three pieces being bilateral (that is, with small flakes and chips detached to one or both sides of a central dorsal crest) (Figure 5.11). Seven intact pieces have average dimensions of 46 by 23 by 12 mm, and 13 of the 16 intact and broken crested pieces were defined as blade-based. Approximately half of the category was recovered from Phase D, and where crested pieces from the upper levels include both flake- and blade-based specimens, those from Phase D are all blade-based (compare with the stratigraphy of the site's well-executed blades, above).

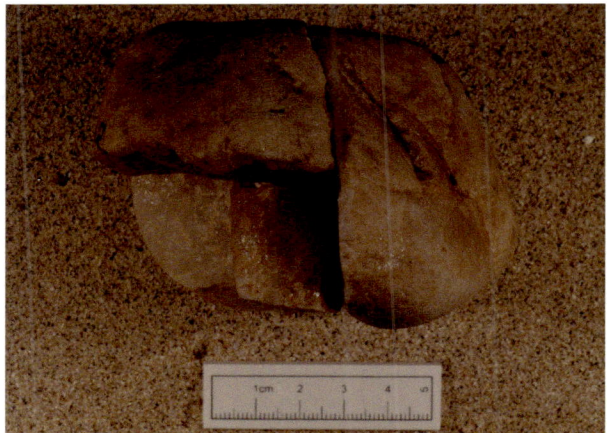

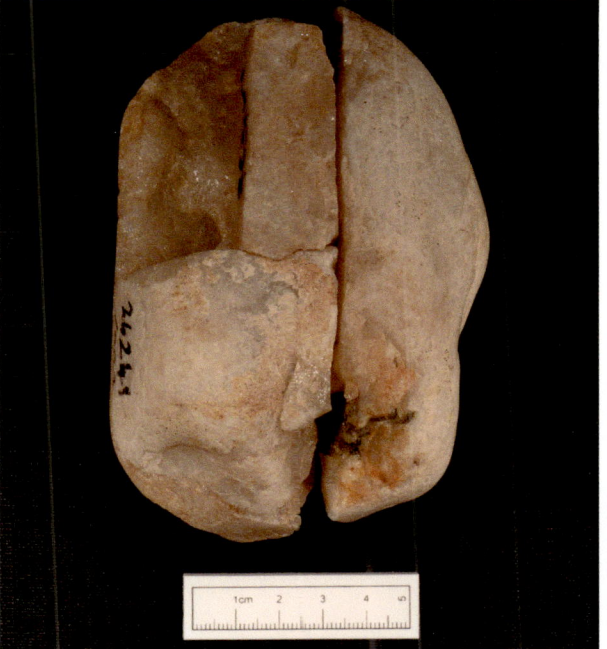

FIGURE 5.12a and b:
Split pebble SF 26241/1-4, refitted, a) top view and b) side view.

FIGURE 5.11:
Crests SFs 26310/37, 23786/58 and 23678/4.

Cores

During the excavation 267 cores were retrieved (Table 5.8): 28 split pebbles (including one that was refitted; Figure 5.12), four conical cores, 33 single-platform cores, five single-platform/discoidal cores, eight opposed-platform cores, four cores with two platforms at an angle, three 'flaked flakes', 28 irregular cores, 149 bipolar cores, and five core fragments.

The dimensions (L by W by T) of cores are measured in the following ways: in platform cores, the length is measured from platform to apex, the width is measured perpendicular to the length with the main flaking-front orientated towards the analyst, and the thickness is measured from flaking-front to the often unworked/cortical 'back-side' of the core. In the case of bipolar cores, the length is measured from terminal to terminal, the width is measured perpendicular to the length with one of the two flaking-fronts orientated towards the analyst, and the thickness is measured from flaking-front to flaking-front. More 'cubic' cores, like cores with two platforms at an angle and irregular cores, are simply measured in the following manner: largest dimension by second-largest dimension by smallest dim.

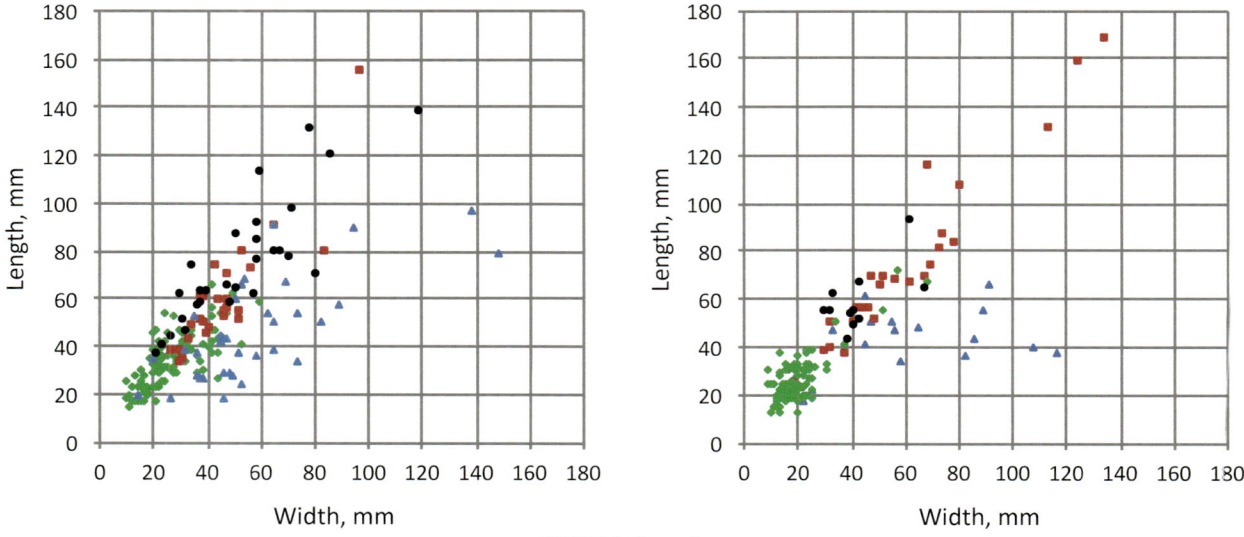

FIGURE 5.13a and b:
The dimensions of the four main core types (intact pieces) recovered at a) RUX6 compared with the cores recovered by Elliott's at b) Barabhas: split pebbles (black), single-platform cores (blue), irregular cores (red), and bipolar cores (green);

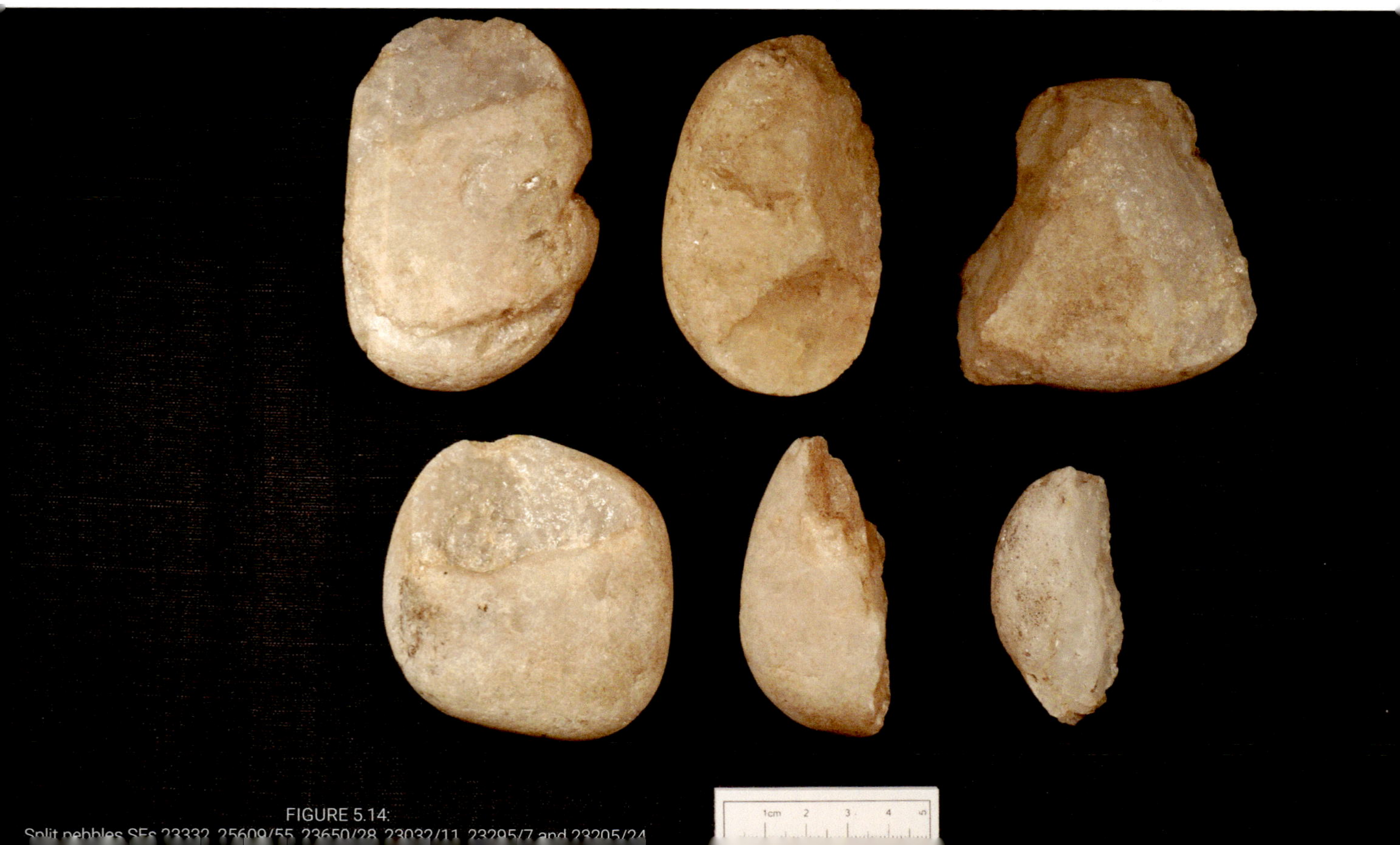

The main dimensions of the four most numerous core types recovered are shown in Figure 5.13a: Split pebbles (early-stage bipolar cores or broken hammerstones), single-platform cores, irregular cores and bipolar cores. Figure 5.13b shows the dimensions of the same core categories from Barabhas (Elliott Collection; Ballin forthcoming a). Comparison of Figure 5.13a and b shows that there are considerable differences between the sizes of the cores from the two sites. The split pebbles (Figure 5.14) from RUX6 are notably larger at *c.* 40-140 mm than those from Barabhas at c. 40-70 mm with one outlier. The single-platform cores from both sites are fairly broad, but those from RUX6 are less so than the exaggeratedly broad pieces from Barabhas. Most single-platform cores from RUX6 have L:W ratios of *c.* 1:1 to 1:1.5, whereas those from Barabhas vary between *c.* 1:1 and 1:3. The irregular cores from RUX6 are mostly medium-sized (*c.* 30-90 mm with one outlier), whereas those

FIGURE 5.14:
Split pebbles SFs 23332, 25609/55, 23650/28, 23032/11, 23295/7 and 23205/24

from Barabhas include some considerably larger specimens at c. 40-170 mm. The bipolar cores from RUX6 are generally notably larger at c. 15-65 mm than those from Barabhas of c. 10-40 mm. It is thought that these considerable variances may largely reflect differences in chronology between the two sites, where RUX6 appears to be dominated by late Neolithic material, supplemented by some early Bronze Age artefacts, Barabhas (Elliott Collection) is dominated by early Bronze Age material, supplemented by Neolithic artefacts. These differences are discussed further in the following sections.

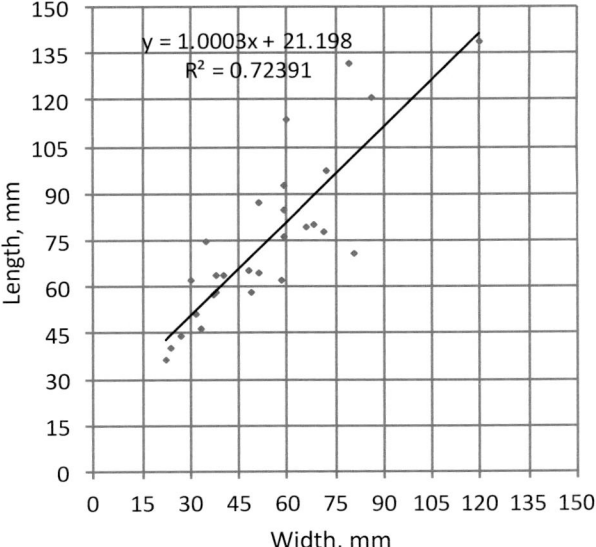

FIGURE 5.15:
The dimensions of the split pebbles.

Split pebbles: A total of 28 split pebbles were recovered from the site. As shown in Figure 5.15, more than half of these are so large that they are technically cobbles (i.e. with a GD exceeding 64 mm; definition according to Hallsworth and Knox 1999: Figure 13). The average dimensions of these pieces are 75 by 64 by 38 mm.

Generally, split pebbles are perceived as a formally and technologically simple category, defined by having had too few flakes or bits detached to be classified as cores (see terminology section). However, as shown in Figure 5.16 this artefact group is considerably more varied than commonly thought, and inspection of size and formal variation revealed a number of sub-types. Figure 5.16 shows the main split pebble sub-types identified in the RUX6 assemblage. In some cases, flakes were detached, in other cases cubic bits broke off along internal fault planes, planes of weakness, or impurities. Most split pebbles are defined by one or more flakes or bits being detached from one end of the pebble, probably where it was exposed to the most pressure. Flakes or bits may be detached from one or both faces, and in rarer cases from both ends. The detached flakes and bits may be slightly smaller or larger than indicated by the idealized sub-types shown in Figure 5.16, and a small number of pieces belong to rarer sub-types. Type 8 is relatively rare and was formed by being struck either at the centre of one lateral side or at the centre of one face.

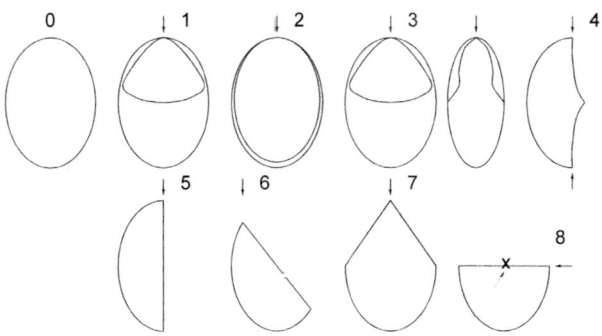

FIGURE 5.16:
Main sub-types of split pebbles. Upper row: flakes detached; bottom row: cubic bits detached. 0) Undamaged oval pebble; 1) one half-length flake detached from one terminal, one face; 2) one full-length flake detached from one terminal, one face; 3) two half-length flakes detached to either side of one terminal; 4) two approximately half-length flakes detached, one from either terminal, one face; 5) the pebble split along the central long-axis; 6) diagonal split; 7) two corners split off either side of one terminal; 8) the pebble broke across; and 9) other forms of limited flaking or splitting.

Split pebbles are usually perceived as early-stage bipolar cores, and the relatively small specimens retrieved at Barabhas (Elliott Collection) may be just that (Figure 5.13). However, the split pebbles from RUX6 (Figure 5.13) are generally considerably larger, and the site's fist-sized pieces may actually be damaged hammerstones (which would also explain the low number of hammerstones recovered from the site). Generally, the damage to damaged early-stage bipolar cores and damaged hammerstones would probably be the same, but several of the split pebbles from RUX6 have notable hammer marks at one or both ends (Figure 5.14), indicating hammering that goes beyond what would usually be expected from early-stage bipolar cores. One piece (SF 26587/1) has a weakly developed crushed ridge at one end and notable peck-marks at the centre of one lateral side, and this piece was defined as a damaged hammerstone (below). Pieces of Type 8, which split due to pressure to the centre of one face, may be damaged anvils (e.g. SF 24974/31). As shown in Figure 5.13, the site's bipolar cores are all shorter than 60 mm, corresponding to the smallest of the split pebbles.

FIGURE 5.17:
Single-platform cores (conical) SFs 26020/28, 26327/7, 25091/14 and 25129/7.

Single-platform cores: The 42 single-platform cores recovered from the site include a number of sub-types, the most important being specimens with a well-defined conical shape (four pieces: CAT 25091/14, 25129/7, 26020/28 and 26327/7, all Figure 5.17), 'standard' single-platform cores (33 pieces), and flat hybrid specimens with attributes usually associated with single-platform and discoidal cores (five pieces: 25009/70, 26006/1(Figure 5.18), 26549/2-3 (Figure 5.19), 26748/4). The dimensions of the single-platform cores are shown in Figure 5.20a and b, where the cores are subdivided according to their morphology (Figure 5.20a), and according to the character of their platform (Figure 5.20b). These diagrams were produced to allow the various sub-types to be precisely defined, not least as some of these may be diagnostic (see technology and dating sections).

Figure 5.20a shows that the conical cores are generally considerably smaller (no larger than 40 by 40 mm) than other single-platform cores. In terms of the cores' L:W ratio, the conical cores are generally within the band 2:1-1:1 indicating they are elongated, whereas most standard single-platform cores are within the band 1:1-1:2, which is relatively broad, and the L:W ratio of the single-platform/discoidal cores approaches 1:2, which is notably broad/flat. Figure 5.20b shows that cores with cortical platforms tend to be within the band 1:1-1:2 of relatively broad to broad, whereas cores with plain and faceted

FIGURE 5.18:
Single-platform/discoidal core SF 26006/1.

platforms cluster around a trendline corresponding to L:W = 1:1. Two of the conical cores appear to be blade-cores (26020/28 and 26327/7), whereas the other single-platform cores may be flake cores.

As shown in Figure 5.13, some single-platform cores from Barabhas (Elliott Collection) are exceptionally broad/flat, and in connection with the characterization and discussion of these pieces, two different types of broad/flat single-platform cores

FIGURE 5.19:
Single-platform cores (broad/flat) SFs 26549/3, 23690/13, 26199/1 and 23470/105.

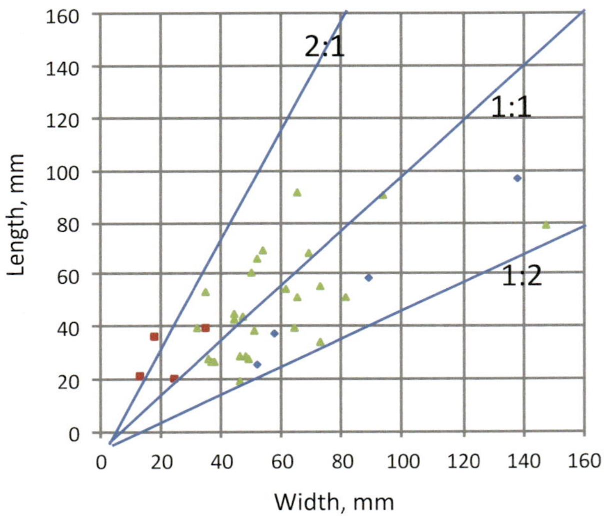

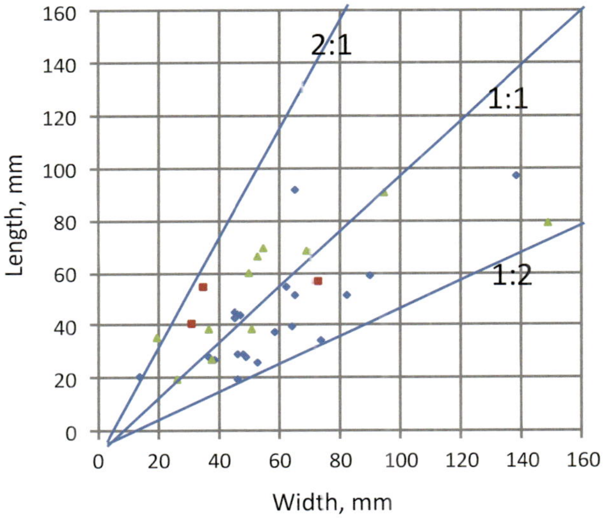

FIGURE 5.20a and b:
The dimensions of the intact single-platform cores. In a), the cores are subdivided according to their morphology (red conical; green: 'standard' single-platform cores; and blue: single-platform/discoidal cores), and in b) according to the character of their platform (blue: cortical; green: plain; and red: faceted).

were defined (Figure 5.21). The broadest/flattest single-platform cores found at Barabhas are the flake-based ones, where flakes were detached from the circumference of large parent flakes (Figure 5.21b), and the absence of pieces like this in the RUX6 collection explains the lack of very broad/flat pieces in Figure 5.13. However, most of the RUX6 single-platform cores are broader than what would be expected on the basis of how this type of cores appears in most other lithic assemblages.

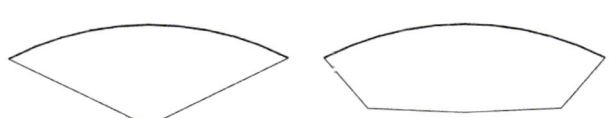

FIGURE 5.21a and b:
Single-platform cores sub-types defined amongst the quartz cores from Barabhas (Elliott Collection). a) shows the cross-section of CAT 5167 – these pieces are simply very flat cores with very acute flaking-angles; b) shows the cross-section of CAT 216, 4698, 5526, and 5869 – these cores are based on thick flakes from which smaller flakes were detached along the entire circumference, giving some of them a scraper-like appearance. The thick line indicates the cortical platforms of the pieces.

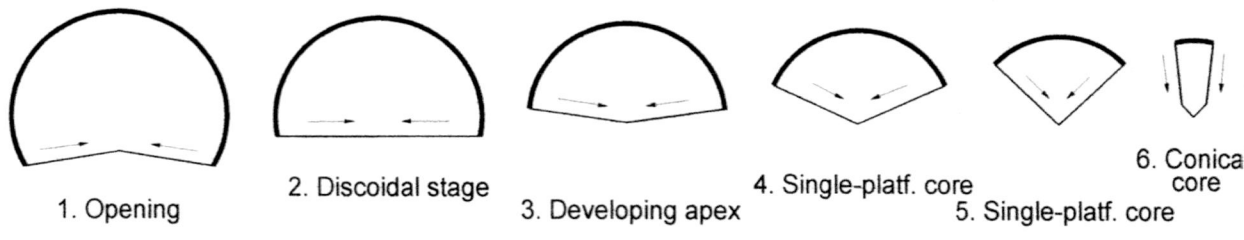

FIGURE 5.22:
The operational schema of the main platform core type identified at RUX6.

During the characterization and cataloguing of the RUX6 assemblage, the larger single-platform cores were subdivided into 'standard' single-platform cores (33 pieces) and single-platform/discoidal cores (five pieces) (Figure 5.20a). As there are more single-platform cores at RUX6 (42 pieces) than at Barabhas (Elliott Collection) (21 pieces). The RUX6 collection included pieces discarded at all stages of the operational schema, and it was possible to define the operational schema applied in some detail (Figure 5.22). In contrast, the fewer cores available at Barabhas did not allow this, and it is thought that the core Figure 5.21a may correspond to the RUX6 core illustrated as Figure 5.22-4. Once all the cores from RUX6 had been characterized and catalogued, it became clear that the single-platform/discoidal cores form the first stages of this operational schema (Figure 5.22-1-3), and may formally be defined as hybrids between traditional single-platform cores and discoidal cores.

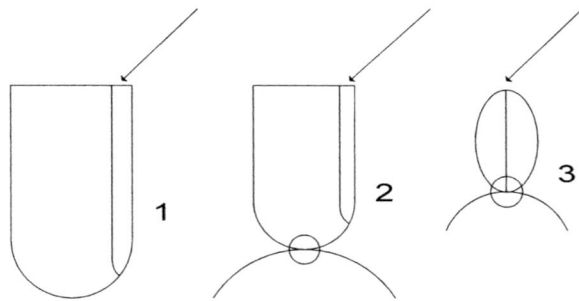

FIGURE 5.23:
1) Free-hand hard percussion; 2) hard percussion on anvil; and 3) bipolar ('hammer-and-anvil') technique; after Callahan 1987: illus 97.

This operational schema is defined as:

0 - A large quartz cobble is opened by 'gnawing' an opening in the cortex of one side or end.

1-3 - Flakes are detached by hard percussion along the entire circumference of this opening and at these stages it is not possible to recognize a well-defined apex (single-platform/discoidal cores).

4-5 - As the produced hard percussion flakes (detached by striking the cores at rather acute angles) are quite thick at the proximal end and thin at the distal end, they tend to be approximately wedge-shaped (when seen from the side), and through the reduction process, the sides of the cores grow increasingly steep, and a more notable apex develops ('standard' single-platform cores).

6 - At the end of this operational schema, the cores may have quite steep sides and a well-defined apex, and they can now be termed conical cores. As mentioned above, some finer conical cores from RUX6 appear to be specialized blade-cores, but some may also be the final products of this formal sequence.

Platform-edge trimming was rarely applied. The presence of crush-marks at the distal end of many platform-flakes suggests that anvils may generally have been used in connection with the manufacture of hard percussion flakes (Figure 5.23-2). Platforms were generally not prepared and as a result as many as 52% of all intact or proximal fragments of hard percussion flake and blade blanks have cortical platforms.

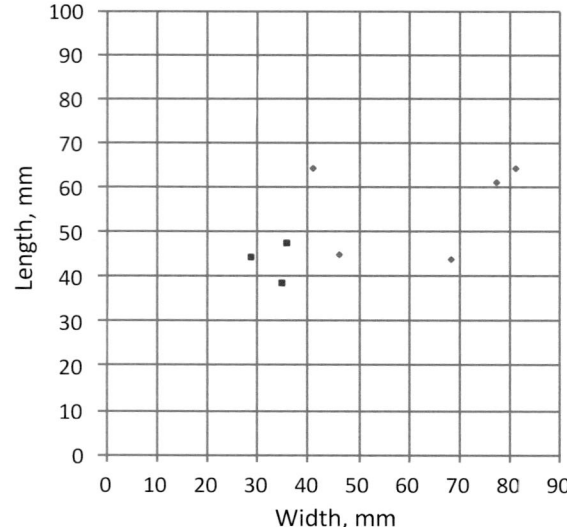

FIGURE 5.24:
The dimensions of the opposed-platform cores (intact pieces). Red: approximately cylindrical pieces; blue: broad unifacial pieces.

FIGURE 5.25:
Opposed-platform cores SFs 25204/6, 23470/104, 25724/6, 25148/162, and 26710.

Opposed-platform cores: This category includes two subtypes (Figure 5.24): broad unifacial pieces and approximately cylindrical pieces, which tend to have been reduced along the entire circumference. The former (CAT 23470/104 (Figure 5.25), 25057/97, 25077/27; 25204/6 (Figure 5.25), 25254/55, are fairly large, plain pieces with average dimensions of 56 by 63 by 38 mm; L:W 0.9:1), from which flakes were struck, whereas the latter (CAT 25148/162, 26710/45 and 25724/6; Figure 5.25), are smaller, more regular pieces with average dimensions of 43 by 33 by 24 mm (L:W 1.3:1), from which elongated flakes or blades were removed. The larger cores tend to have cortical platforms and cortical 'backsides', whereas the smaller ones tend to have platforms defined by parallel fault planes and they were usually worked along the entire circumference. These forms appear to be dual-platform versions of the 'standard' single-platform cores and the conical cores. Platform-edge trimming was rarely applied.

Cores with two platforms at an angle: The small group of four cores with two platforms at an angle are mostly fairly irregular pieces (three pieces), with one piece being quite neat (CAT 23332/28). The former three pieces are crude objects with highly denticulated, untrimmed platform-edges from which irregular flakes were detached. They vary in size between GD 55 mm (CAT 26020/25) and GD 21 mm (CAT 27215/1582). CAT 23332/28, on the other hand, has two more regular flaking-fronts, one on each opposed face, and both flaking-fronts are characterised by neatly trimmed platform-edges, from which it was attempted to detach more elongated flakes. This core measures 65 by 57 by 42 mm. All these cores have at least one cortical platform.

Irregular cores: The collection includes 28 roughly cubic irregular, or multi-platform/multi-directional, cores. They are defined as having been worked from three or more different directions. As shown in Figure 5.26, the irregular cores are amongst the largest recovered from the site, only comparable in size to the larger single-platform cores (Figure 5.20a and b). They measure on average 60 by 47 by 39 mm.

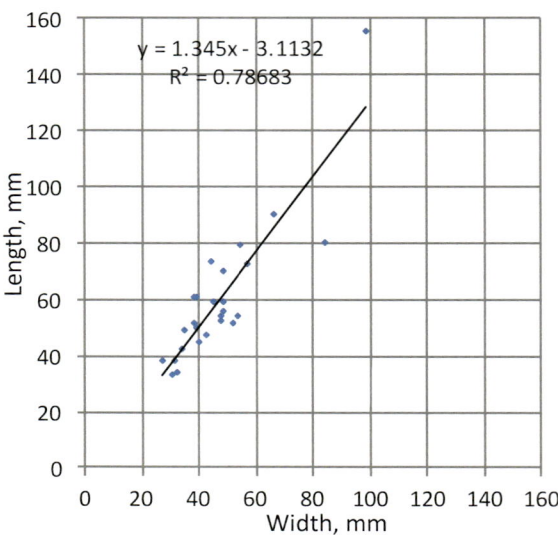

FIGURE 5.26:
The dimensions of the irregular cores (intact pieces).

A number of irregular cores have attributes which show the history of these pieces. Although some irregular cores may have been worked in an unschematic manner from the detachment of the first flake, it is generally thought that many irregular cores may represent the final stage of reduction sequences which began with single-platform cores, and where secondary and tertiary platforms were added as they were required. CAT 26681/8 may be the redefined remains of a single-platform/discoidal core. CAT 23559/17 (Figure 5.27) may be the remains of a 'standard' single-platform core; and CAT 25201/17 may be the exhausted remains of an opposed-platform core. CAT 26212/1, the largest irregular core, contains large amounts of big feldspar crystals and mica sheets, which may explain why this piece was abandoned at an early stage, after attempts to improve the reduction process by adding more and more new platforms proved ineffectual.

'Flaked flakes' (cores): This category (see Ashton et al. 1991) includes three specimens: two larger ones (CAT 23107/012, 25222/041) based on robust hard percussion flakes (GD 73-83 mm), and a smaller one (CAT 16501/004) based on a flake fragment (GD 49 mm). In all cases, the ventral face of an original flake or flake fragment was used as a flaking-front, from which smaller flake blanks were detached. However, where CAT 16501/004 and CAT 25222/041 only had a small number of flakes detached by striking one lateral side of the original flake, CAT 23107/012 had flakes detached by striking several points along the 'flaked flake's' circumference.

FIGURE 5.27:
Irregular cores SFs 23559/17, 16785/2, 23267/6, 26881/2, 26681/9,

FIGURE 5.28:
Bipolar cores SFs 26626/74, 26427/20, 25321/68, 23992/7, 23198/5, 24006/3, 23044/88, 26235/12, 26626/78, 26004/52, 26809/19 and 26468/42.

Bipolar cores: The numerically largest core category is clearly the bipolar cores and a representative selection is illustrated (Figure 5.28). A total of 149 bipolar cores were recovered from RUX6, whereas the two second largest core categories – single-platform cores and irregular cores – only include c. 30-40 pieces each. As mentioned above, some of the split pebbles may be early-stage bipolar cores, although some of the larger ones are thought to be damaged hammerstones (see below). It is thought that some of the bipolar cores may represent pebbles which were reduced entirely by the application of hammer-and-anvil, but many are likely to be the heavily exhausted remains of platform cores which were gradually transformed following the formula: Collected pebble/quarried block (GD up to c. 200 mm) ⇒ single-platform core ⇒ dual-platform core ⇒ irregular core ⇒ bipolar core. This is supported by several bipolar cores with surviving platforms along one lateral side or at one end (e.g. CAT 23168/13, 25000/74, 26226/16). However, a small number of bipolar cores are also based on thick hard percussion flakes (e.g. CAT 26626/02, 25139/22).

The average dimensions of 118 intact bipolar cores are 34 by 26 by 15 mm, but the size of the individual pieces vary within a number of sub-categories (Figure 5.29a and b). It is possible to subdivide the bipolar cores into two sets of sub-types: unifacial vs bifacial specimens and pieces with one or two reduction axes (one or two sets of opposed terminals). In previous reports (e.g. Ballin forthcoming a), it has been suggested that generally bifacial cores may represent later stages in the reduction process than unifacial ones due to the fact that they have been more extensively reduced. The pieces with two reduction axes may represent later stages than pieces with one reduction axis, having been in need of being re-orientated, although the size of the original raw material pebbles or blocks and their flaking properties may also have played a part (ibid.). Figure 5.29a and b and Table 5.16 were produced to show the general size of the bipolar cores but also to test the likely position of the various subtypes of bipolar cores in the reduction process.

	Number	%	Av. dimensions		
			Length	Width	Thickness
Unifacial	20	17	40	33	18
Bifacial	98	83	33	24	14
1 axis	104	88	34	25	15
2 axes	14	12	34	30	15
Total	118	100	34	26	15

TABLE 5.16:
The distribution of intact bipolar cores across unifacial and bifacial pieces and across pieces with one and two reduction axes

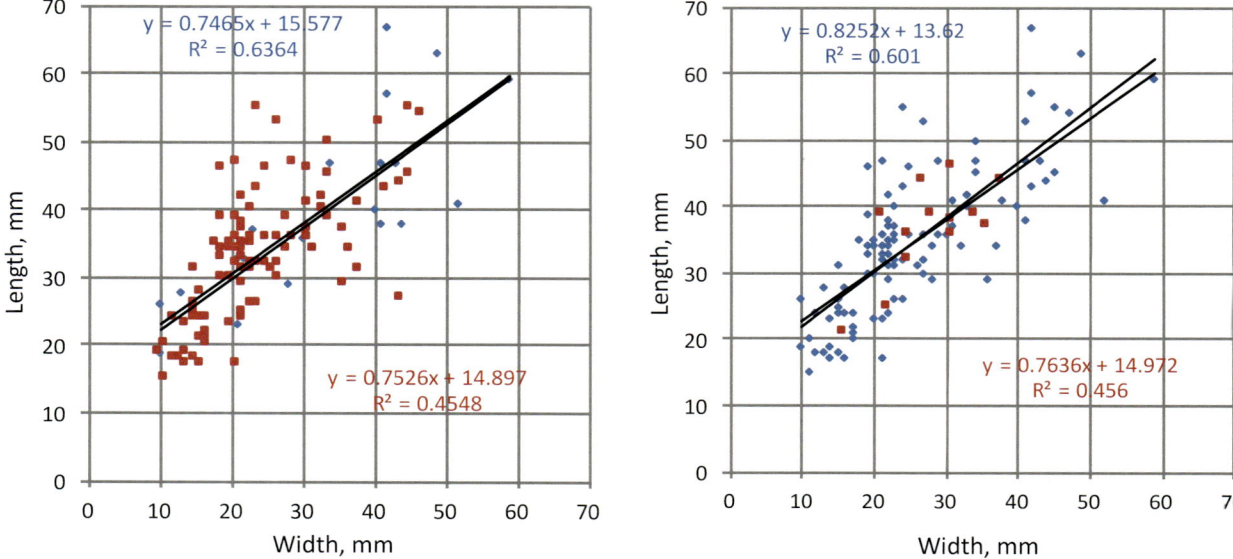

FIGURE 5.29a and b:
The dimensions of the bipolar cores (intact pieces). In a), the cores are subdivided according to whether they are unifacial (blue) or bifacial (red), and in b) according to their number of reduction axes (sets of opposed terminals) (one axis: blue; two axes: red).

Figure 5.29a and Table 5.16 show that the unifacial cores are notably larger the bifacial ones, supporting the suggestion that the bifacial cores have indeed been more extensively reduced and they are therefore likely to represent later stages in the reduction process. In terms of the cores with one and two reduction axes, Table 5.16 indicates that there may not be any size difference between these categories in terms of *average* dimensions. However, Figure 5.29b shows that most cores with two axes are cores of medium size, and that all the site's largest cores are pieces with one axis.

Core fragments: Five indeterminate core fragments were retrieved in connection with the excavation. Due to the degree of fragmentation and in conjunction with sandblasting, it was not possible to characterize these pieces more precisely. They vary in size between GD 44-57 mm.

Tools

The tool typology applied on this site generally follows Butler (2005) and Ballin (1996; 2010a; forthcoming a). The assemblage includes 108 implements, which may (following the typology developed in connection with the analysis of the lithic and stone objects from Barabhas, Ballin 2010a; forthcoming a) be subdivided into 'fine' tools (95 pieces) based on flaked blanks and blades, and 'coarse' tools (13 pieces) which tend *not* to be based on flaked blanks, although there are exceptions (see below). The latter group may again be subdivided into implements based on raw pebbles/cobbles and identified by their robust use-wear (e.g. most hammerstones and pounders), and those shaped by the flaking, pecking or polishing (or a combination of these actions) of pebbles/cobbles or, in some cases, very large flakes (e.g. points).

The 'fine' tools include: two knives, 59 scrapers, six piercers, one notched piece, two denticulates, and 25 indeterminate implements with various forms of retouch. The 'coarse' tools include: four points, two percussoirs, five hammerstones, and two pounder fragments.

Compared to the Barabhas material (260 implements from the Elliott Collection), the basic composition of fine (89%) and coarse tools (11%) is exactly the same. However, there are also notable differences, such as the fact that the finer pieces from Barabhas include large numbers of arrowheads (36 pieces) and the RUX6 assemblage none, and that the coarse implements from Barabhas include relatively large numbers of pounders (14 pieces), whereas the RUX6 material only includes two fragments. If it is accepted that many of the larger split pebbles from RUX6 are in fact damaged hammerstones, the collection includes considerable numbers of hammerstones (possibly 15-20 pieces) against the nine hammerstones from the only slightly less numerous Barabhas collection.

Knives: The knives embrace two forms: a backed knife (CAT 26146/26) and one scale-flaked knife

(CAT 23803/7) both Figure 5.30. Neither is a particularly well-executed piece, and both could be defined as expedient.

FIGURE 5.30:
Knives SF 23803/7 and SF 26146/26.

The former is a hard-percussion flake with full blunting retouch of its right lateral side, measuring 52 by 34 by 13 mm. The unmodified left lateral side would have functioned as a cutting-edge. The latter is a large, elongated indeterminate flake measuring 67 by 40 by 23 mm. Its fully cortical left lateral side would have protected the user's hand, whereas the right lateral side has been sharpened by scale-flaking. It appears that the cutting-edge may have been modified by scale-flaking from both faces, making this edge partially bifacial.

Scrapers: In total, 59 scrapers were recovered from RUX6. They include two discoidal scrapers, one blade-scraper, 39 short end-scrapers, one double-scraper, four side-scrapers, three end-/side-scrapers, two atypical scrapers, and seven scraper-edge fragments. In the analysis of the scrapers from Dalmore on Lewis (Ballin 2002b), a maximum GD of 23 mm was suggested for Western Isles thumbnail-scrapers, but as shown in Figure 5.31b, at Barabhas a GD for this entirely size-based sub-category of 25 mm would probably have been more appropriate. Only three to six scrapers (are small enough to fall into this category (c. 7% of all scrapers), whereas at Barabhas 70 scrapers (or 53%) were defined as thumbnail-scrapers (Figure 5.31b) and most of the small scrapers are in flint. Inspection of the flint scrapers from the RUX6 shows that most of those are also thumbnail scrapers. The very small scrapers found at Western Isles prehistoric sites – some of which are quartz, but with most being in flint and mylonite – are therefore more likely to be a result of the size of the raw material pebbles and blocks, rather than a diagnostic feature.

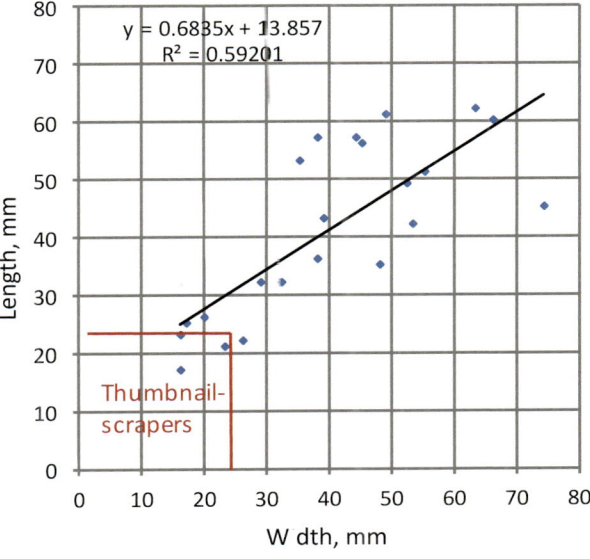

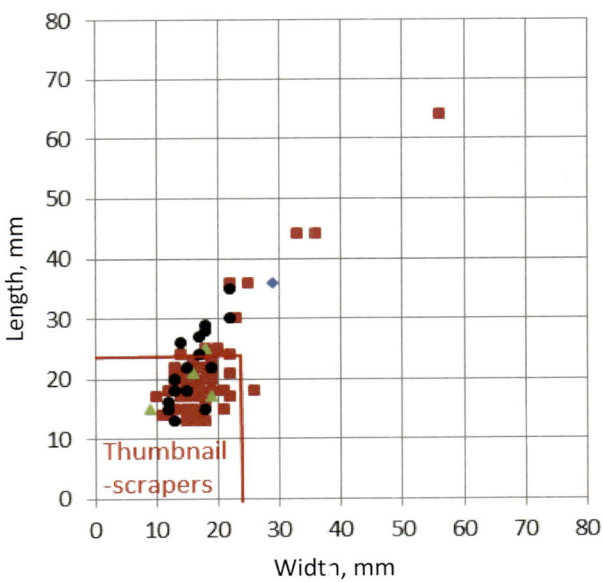

FIGURE 5.31a and b:
Comparison between the short end-scrapers of a) RUX6 and those of b) Barabhas, the Elliott Collection (red).

Two scrapers are based on blades (3%; only one is a proper blade-scraper, see below), 44 on flakes (75%), one on an exhausted bipolar core (2%), and 12 are on indeterminate blanks (20%). In comparison the figures relating to the scrapers from Barabhas (Elliott Collection) are: blades (1%), flakes (85%), exhausted bipolar cores (13%), and indeterminate blanks (1%). One notable difference between the two assemblages is the more frequent use of exhausted bipolar cores as scraper blanks at Barabhas, whereas at RUX6 it was more common to use robust flakes and tabular indeterminate pieces as scraper blanks.

Only one proper blade-scraper, CAT 23516/28 (Figure 5.32) was recovered. Where the other blade-based scraper recovered at the site is a side-scraper based on a robust bipolar blank, this piece is based on the distal fragment of a crested blade, and it has a convex, steep scraper-edge at the distal end. It measures 45 by 22 by 17 mm.

FIGURE 5.32:
Blade-scraper SF 23516/28.

The collection includes two discoidal scrapers (Figure 5.33), which differ considerably in terms of size. One (CAT 16688/7) is 66 mm long and 24 mm thick, which split along the long axis. The other (CAT 23350/5) is thumbnail-sized and measures 18 by 15 by 11 mm. Both are defined by having a steep scraper-edge running along the entire circumference, and both have a cortical lower face. They were both manufactured by the application of inverse retouch (Ballin 2000a), and are therefore formally related to the end-scrapers with inverse retouch described below.

FIGURE 5.33:
Discoidal scrapers SF 16688/7 (split) and SF 23350/5.

Short end-scrapers are defined as end-scrapers which are too short to be defined as blade-scrapers (above). This category is the most numerous of the scraper forms found at RUX6 (66%). Most are based on hard-hammer flakes, but three are on bipolar flakes and four are based on indeterminate pieces. The end-scrapers from this site are generally difficult to identify, as the raw material used ('reconstituted' quartz) makes the scraper-edges slightly less durable. In connection with heavy-duty use, for example on hard materials, large cubic bits may break off the working-edges, making them difficult to recognize.

In many cases, the short end-scrapers were manufactured on which ever blank was deemed suitable (rather than specially selected or manufactured scraper blanks), and a steep straight to convex working-edge was subsequently formed at one end. Due to this fact, most of the end-scrapers are fairly expedient and relatively irregular specimens. However, it *is* possible to identify a number of categories which may correspond to mental templates in the mind of the prehistoric knapper and thus possibly associated with specific functions:

- Very small thumbnail-scrapers with a convex, steep scraper-edge at one end (e.g. CAT 25487/176, 26487/46, 26536/3, all Figure 5.34); average GD c. 20-22 mm.

FIGURE 5.34:
Short end-scrapers (small) SF 23470/109, 23567/21, 25045/21, 16858/5, 25167/63, 26536/3, 25487/176 and 26487/46.

- Elongated (almost reaching blade proportions), well-executed pieces with a narrow, convex, steep scraper-edge at one end; only one such piece was recovered (CAT 23141/33, Figure 5.35), which is based on an elongated bipolar blank (57 by 38 by 17 mm).

- 'Standard', well-executed, mostly flake-based end-scrapers with a neat, convex, steep scraper-edge at one end, (CAT 23023/26, Figure 5.35, CAT 23567/021, Figure 5.34, 25045/21, Figure 5.34, 25148/160, Figure 5.35, 25167/62, 26209/18, Figure 5.35). Their circumference is generally irregularly rounded-oval, and their GD varies between c. 35-60 mm. Some are thick and some are thin. CAT 23567/21 has a neat, convex working-edge at the proximal end, formed by the application of inverse retouch.

- A number of end-scrapers are based on fully cortical blanks, either flakes or indeterminate pieces. Their working-edges were formed by the application of retouch from the cortical face (i.e. inverse retouch), and some of these pieces are relatively well-executed. CAT 26748/3, for example, has a regular, narrow (21 mm), convex, steep to slightly acute scraper-edge at one end.

Double-scrapers are basically short end-scrapers with two opposed working-edges. Only one double-scraper was retrieved, CAT 23657/22, and it is probably based on the central fragment of a thick hard percussion flake. It has a slightly convex, steep scraper-edge at either end

Four side-scrapers are all relatively expedient pieces, based on robust flakes or indeterminate pieces. Two are fragmented, and two are intact (CAT 23650/29, 26020/18 both Figure 5.36). The former of the two intact pieces is based on a thick, primary, bipolar blade measuring 95 by 40 by 28 mm. It has a slightly convex, steep, denticulated scraper-edge along its left lateral side. The latter is a large hard-hammer flake with a slightly convex, steep working-edge along its right lateral side.

Two of the three end-/side-scrapers are based on thick indeterminate pieces (average dimensions: 87 by 70 by 52 mm), and they both have a lateral and an end working-edge. The working-edges are all straight to slightly convex, and have been provided with robust retouch. They were worn to a degree that left the used edges notably abraded and rounded. The group's third member, CAT 23470/108, is also

FIGURE 5.35:
Short end-scrapers (large) SFs 23141/33, 25025/7, 25148/160, 23470/107, 26020/21, 23023/26, 26209/18, 25336/17.

FIGURE 5.36:
Side-scrapers SFs 26020/18 and 233650/29, and scraper-edge fragment 25944/9.

based on an indeterminate piece, but it is smaller and thinner, and probably split during use. Its end-edge and its lateral edge are both more regular than those of examples described above, and it measures 65 by 50 by 28 mm.

Two atypical or indeterminate scrapers were also found at the site. CAT 23605/159 is the exhausted remains of a bipolar core and measures 46 by 43 by 28 mm. It had one terminal transformed into a convex steep scraper-edge, and one corner at the opposed end was transformed into a similar scraper-edge. CAT 17445/3 is the fragment (GD 64 mm) of a scraper based on a thick indeterminate piece. A surviving stretch of steep, inverse scraper-edge appears to have been either concave or nosed.

Seven scraper-edge fragments represent too small parts (in relative terms) of their parent pieces to be characterized more precisely (average GD 39 mm).

Piercers: A distinction is made between various types of piercers and points with pointed working-ends. The former are relatively small and the latter relatively large, but no firm formal or metric

FIGURE 5.37:
Piercers SFs 26020/19, 26585/4, 26585/5, 25122/38 and 25351/29.

definitions of the two types have been proposed. However, in the present case, the two types are easily distinguishable by size, as no intact piercer is larger than 56 mm and no intact point smaller than 74 mm. The piercers are generally relatively fine implements, and the points relatively crude (see below).

The piercers form a very heterogeneous group, where one broken piece (CAT 26585/5; now missing its tip, Figure 5.37) may have been a small implement with a length of c. 27 mm. Three pieces could represent a relatively well-defined formal subtype (CAT 26585/4, 25122/38 Figure 5.37, 26283/14). The former piece is intact, measures 56 by 38 by 20 mm, and is based on an indeterminate flake. A regular, robust point was formed at the distal end by neat retouch of both lateral sides, and this specimen is more or less drop-shaped. CAT 25122/38 and CAT 26283/14 are well-made broken-off tips of similar implements. CAT 25351/29 (Figure 5.37) is an indeterminate piece with an almost square outline, and a fine tip was shaped by merging the retouched left lateral side and the retouched distal edge. CAT 26020/19 Figure 5.37, is a large, broad hard percussion flake (50 by 67 by 29 mm), which was equipped with a robust piercer tip at the distal end. This working-part was made by retouching two large notches next to each other, with the spur between the two notches forming the tip.

Notched and denticulated pieces: CAT 26359/30 is a broken implement based on an indeterminate flake. It has a retouched notch in one lateral side, the chord of which measures 13 mm. It is uncertain what kind of tool the original intact piece was. Two denticulates were also recovered (CAT 17445/4 and CAT 23267/7). Denticulates seem to be particularly common on Bronze Age sites (cf. Ballin 2002a; 2011b), particularly later Bronze Age sites, but this 'type' is relatively poorly defined and understood. Some may be denticulated scrapers and others crude cores or core fragments. The modification of CAT 17445/4 is relatively regular and convex, and this flake-based piece may be a denticulated scraper measuring 28 by 35 by 9 mm. CAT 23267/7, on the other hand is a slightly larger piece at 61 by 36 by 26 mm. It has three notable spurs in one lateral side, formed by retouching a series of adjacent notches (chords 10-20 mm). This piece may be either a very crude and expedient side-scraper, or possibly a 'flaked flake' (core).

Pieces with edge-retouch: A total of 25 artefacts were defined as pieces or fragments with simple edge-retouch. The blanks are mostly flakes, but one is a blade, two are abandoned bipolar cores, whereas four are indeterminate pieces. These pieces differ considerably in shape and size (greatest dimension 13-66 mm), and it is thought that this tool group includes artefacts, or fragments of artefacts, with different functions. One intact piece with edge-retouch is based on a large hard-percussion flake measuring 61 by 55 by 25 mm, with regular steep retouch along its right lateral side (Figure 5.38). This modification may be a scraper-edge (defining the piece as a side-scraper), or it may be the blunting of a backed knife, with the opposed sharp edge being the cutting-edge of the implement.

FIGURE 5.38:
Piece with edge-retouch with modification along the right lateral side.

Points: The four points are all fairly crude implements with robust tips formed by minimal modification. They are generally shaped like very plain handaxes although their context clearly suggests a date within the Neolithic-Bronze Age. In the report of the finds from Barabhas (Elliott Collection), the following preliminary definition was suggested: they have a robust, shaped point at one end, and a 'lumpy' handle-end, allowing the pieces to be used in a robust manner to either drill or chop holes in a variety of presumably indeterminate but hard materials (Ballin forthcoming a).

The two most regular pieces (CAT 25983/8, 26241/14, both Figure 5.39) were formed in slightly different ways. The former, measuring 79 by 81 by 32 mm, is based on a primary hard percussion flake, and it was shaped by 'normal' retouch of the lateral side (i.e. retouch from the ventral face), and it has a distal tip. The latter was probably made by reduction of an indeterminate piece with one fully cortical face, and it was formed by 'inverse' retouch of the lateral side (i.e. retouch from the dorsal or cortical face). In both cases, the lateral retouch was fairly acute, almost invasive, removing large parts of the affected face. In the case of CAT 26241/14 (Figure 5.39), the modification has removed almost all parts of the original cortex-free face. The use of inverse retouch is reminiscent of the approach followed to shape the inversely retouched scrapers described above.

FIGURE 5.39:
Points SF 26241/14 and SF 25983/8.

Percussoirs: This group includes two pieces: a plain percussoir and a percussoir/anvil. The former (CAT 17272/3) is based on a small pebble which has fine peck-marks at its most pointed end. It measures 37 by 29 by 22 mm. This is probably a finer version of the hammerstones presented below, and it may have been used for finer flaking, core-edge trimming, or tool modification (cf. Inizan *et al.* 1992: illus 38). The latter (CAT 26782/1, Figure 5.40a and b, see also Figure 5.51) is based on an elongated, flat pebble measuring 57 by 28 by 16 mm, and it has two opposed modified and/or used terminals, as well as peck-marks at the centre of each face from use as an anvil. The two terminals have marked, ground facets against the faces and lateral sides, and at either end the ground surfaces meet in a ridge running from lateral side to lateral side. At one end, the ground area only shows smooth abrasion, whereas at the other end, it is possible in magnification to detect striations, which follows the long axis of the piece, as well as peck-marks at the top of the terminal ridge. The anvil pits in the two faces clearly link the piece to primary reduction – more specifically, knapping of small bipolar cores – whereas the abraded ends in functional terms relate the piece to pestle-shaped pounders, although these usually have rounded working-ends, rather than ridged ones. The peck-marks at the tip of one terminal suggests its use as a fine hammerstone or percussoir.

FIGURE 5.40a and b:
Percussoir/anvil SF 26782/1, a) main face and b) one of the tips.

It is presently uncertain whether the grinding of the two ends relate to use, as seen in connection with the pestle-shaped pounders, or whether the grinding could be a form of modification, the purpose of which was to make the ends more pointed, and to prepare the piece for use in connection with the preparation and reduction of quartz cores, or the modification of quartz tools.

Hammerstones: Five hammerstones were recovered, with three being defined as plain hammerstones (CAT 24947/17, 26531/2, 26588/1), one as a hammerstone/anvil (CAT 26009/1, Figure 5.41), and one as a hammerstone/split pebble (CAT 26587/1, Figure 5.41).

Two of the plain hammerstones are broken or disintegrating pieces (CAT 24947/017, 26588/1). The former is quite sandblasted and its wear traces have almost been entirely obscured by aeolian activity, whereas the latter has notable crush-marks at one pointed end. The third plain hammerstone (CAT 26531/2) is a regular single-platform core with a cortical 'back-side', and its apex has been used for hammering. This intact specimen measures 69 by 70 by 45 mm.

The hammerstone/anvil (CAT 26009/1) is a standard oval hammerstone based on a fist-sized rounded cobble measuring 85 by 76 by 67 mm. At its two opposed pointed ends, it has notable crush-marks,

FIGURE 5.41:
Hammerstones SF 26009/1 and SF 26587/1

but it also has a crushed, slightly deeper 'pit' in a relatively flat part of one lateral side. It is possible that this indicates use as an anvil. The hammerstone/split pebble (CAT 26587/1) is an interesting piece. At one pointed end it has a crushed ridge of the sort usually associated with the terminals of bipolar cores, but it is quite likely that a use pattern like this could also develop by using the piece as a hammerstone. At the other pointed end, the piece has a typical circular area defined by fine crush-marks, of the form characterizing the hammerstones described above. This piece measures 96 by 85 by 59 mm.

Pounders: Two pieces are broad-sides of damaged pounders of pestle type (Ballin Smith 1994: 196). They are both defined by having a pecked/ground, domed area at one end, clearly set off from the main body of the original implement by a well-defined facet. The smaller piece, CAT 25110/11 (Figure 5.42), measuring 57 by 48 by 28 mm, has split straight through the pecked/ground terminal, with the split following a natural fault plane along the central axis of the piece. The other piece (CAT 25669/14, Figure 5.42) is a hard percussion flake measuring 76 by 62 by 31 mm, where the flake's bulb of percussion is exactly at the centre of the pecked/ground terminal. The fact that both pounders split precisely through the centre of one of their terminals, suggests that these two surviving parts were detached as a result of pressure to that end through use. The widths of the two pieces indicate that the original pounders differed considerably in size, one measuring 48 mm across and the other 62 mm).

Technological summary

This technological summary is based on information presented in the raw material, debitage, core and tool sections above. As previously mentioned, the assemblage is dominated by material dating to the Neolithic (Phase D), with higher levels probably including some Bronze Age material. Although some quartz may have been procured from veins, the cortex covering many cores and blanks suggests that the majority of the quartz may have been obtained by combing the beaches adjacent to the site.

The various industries present at the site are apparently all flake industries or industries dominated by flake production (Table 5.13). Blades were produced through all phases, but most of these are elongated bipolar orange-segment flakes. However, some 'proper' (i.e. percussion rather than bipolar) blades were also produced, possibly exclusively in Phase D. Some of these are impressively long (Width L:W ratios of up to 4:1), and a number of blades were characterized as 'unusually regular' for quartz blades, or even 'perfectly regular' (CAT 26004/054). Bipolar orange-segment blades are produced when

FIGURE 5.42: Pounders SF 25669/14 and SF 25110/11

an elongated pebble is reduced by the application of hammer-and-anvil technique and the pebble splits in a radial manner (Figures 5.5 and 5.6). The flakes were manufactured by the application of hard percussion and bipolar technique in equal measure, whereas the blades were produced mostly by the application of bipolar technique, with 11% being hard-hammer blades (Table 5.14).

The main (single-platform) part of the site's operational schema is shown in Figure 5.22 in idealized form. In addition, damaged, or misshapen single-platform cores were on occasion given a new lease of life by adding new platforms, and when these cores were almost entirely spent they were in some cases exhausted completely by the application of bipolar technique, following this simplified schema: collected pebble/quarried block (GD up to c. 200mm) ⇒ single-platform core ⇒ dual-platform core ⇒ irregular core ⇒ bipolar core. Although anvils are usually associated with bipolar technique, the fact that many hard percussion flakes have crush-marks at their distal ends suggests that anvils may also have been used in connection with platform reduction (Figure 5.23).

As shown in Figure 5.21, it is thought that most of the different types of single-platform cores recovered at the site may have formed part of the same operational schema. Single-platform/discoidal cores represented the opening stage, 'standard' single-platform cores the intermediate stage, and conical cores the final stage. This succession is the result of the wedge-shaped form of most hard percussion flakes forcing the cores to gradually acquire steeper and steeper sides. However, some of the more regular, smaller cores, such as some conical and cylindrical opposed-platform cores, may represent specialized blade production. Cores with plain and faceted platforms probably represent later stages in the reduction sequence. As a result, a large proportion (52%) of the site's platform flakes and blades have cortical platform remnants.

The split pebbles are interesting, as a large proportion may be broken hammerstones, with some being early-stage bipolar cores. At Barabhas (Elliott Collection), for example, they are smaller and more likely to be exclusively early-stage bipolar cores.

Generally, large quartz cobbles were reduced with no more initial core preparation than some decortication, and the production of one or more crests (guide ridges). In addition to recovered crested flakes and blades, this is supported by single-platform core CAT 25139/23 which has a surviving initial crest. The unfinished crests on single-platform cores CAT 25487/177, and CAT 26359/31 were interpreted as attempts at rejuvenating the cores at a later stage by creating new guide ridges. Although some plain and faceted core platforms are present, cortical platforms were clearly preferred (57% of all single-platform cores have cortical platforms), and no platform rejuvenation flakes (or core tablets) were recovered. In comparison, only 52% of all intact or proximal fragments of hard percussion flake and blade blanks have cortical platforms. This slight difference probably reflects the size of the original cobbles exploited at the site, as large cobbles have considerably more inner mass than small pebbles (cf. Ballin 2016). The larger the pebble/cobble, the more inner blanks it was possible to produce from the core in relation to outer cortical ones. A small number of cores display crudely rubbed platform-edges, with the more neatly trimmed pieces being the smaller ones associated with blade production.

As mentioned above, it is obvious that cortical platforms, including those based on natural fault planes, were preferred to prepared ones. The question is whether this preference indicates that the quartz knappers had a generally expedient or 'lazy' approach to lithic reduction, or whether this choice had some benefits in terms of controlling the outcome. It is not possible to answer this question in any absolute way, but considering the relatively low quality of the quartz, not least the so-called 'reconstituted' quartz described in the raw material section, it is possible that the natural surfaces – such as the smooth, slightly domed cortical surfaces, and the flat, regular surfaces of fault planes – would be better than any surfaces a knapper could hope to produce by flaking.

It is thought that the approach defined above may also have been used at Barabhas (Elliott Collection), as indicated by the core illustrated as Figure 5.23-1, but the core illustrated as Figure 5.23-2 represents a small group of 'deviant' broad/flat single-platform cores recovered at Barabhas, which are based on the detachment of medium-sized hard-hammer flakes from the circumference of large, thick flakes. Cores of this type were not recovered from RUX6, and it is possible that these cores are diagnostic and relate to the dominating early Bronze Age component of the Barabhas assemblages, where RUX6 predominantly dates to the Neolithic period.

A number of hammerstones were used to reduce the collected pebbles/cobbles and quarried blocks, and two percussoirs were probably used in connection with finer work, such as core-edge trimming, production of crests, and modification of blanks into tools. Some of the hammerstones and percussoirs also display pecked pits from their use as anvils in connection with bipolar reduction. Compared to the numerical size of the collection, there are fewer hammerstones than one would expect, but some lightly used specimens may have been discarded during the excavation of the site, whereas others may have been included in the category of split pebbles (see above).

At RUX6 only one of the recovered pieces displays invasive retouch, the scale-flaked knife CAT 23803/007. All other implements were manufactured by the use of simple edge-modification.

Distribution and activities

In this section, the distribution of the quartz artefacts is discussed in an attempt to discover spatial patterns relevant to the interpretation of the site's chronology and the site's features, and structures (activities).

Vertical distribution - stratigraphy

As mentioned in the raw material and technological sections, the site has clearly been exposed to extensive erosion in the form of repeated deflation and conflation (Barber 2011), which left substantial proportions of the quartz assemblage sand-blasted. The sand-blasting mainly affected Phase A, and to some extent Phase B (Figure 5.3), but the recovery of small numbers of sandblasted pieces in Phase D suggests that all levels were affected to some degree.

Further disturbances were introduced when people during the Bronze Age (Phases A-C) dug pits and constructed cairns and erected other structures across the area. These activities affected the deeper Neolithic levels (Phase D) and mixed objects from these levels into the later phases. One piece, CAT 23516/28 (Figure 39) a blade scraper based on a crested piece, from Phase A is an obvious example of this mixing of finds.

There are a few obvious vertical trends (Table 5.17), with the most notable one being the fact that the Phase B (Bronze Age burial and ritual features) contains few chips (c. 100+) and Phase D (Neolithic domestic features) many chips, almost 2,000 pieces (Table 5.12). The concentrations of minuscule chips ought to indicate the presence of knapping floors, in Binford's (1983: 189) terms workshop *drop zones,* which would not have been affected by any form of *preventive* or *post hoc maintenance* ('cleaning up'), and chips (refuse from primary production) should not be expected in connection with burial and ritual features. However, the fact that only contexts from Neolithic Building 2 (DH), Phase D, yielded notable amounts of chips particularly the area immediately east of the hearth – suggests that during the excavation of RUX6, sieving may only have been carried out consistently in this part of the site, and possibly not at all in connection with the investigation of the site's upper levels.

Another, probably more valuable trend is the recovery of larger numbers of blades from Phase D than from Phases A-C (5% compared to 2-3%; Table 5.13). However, many blades are simply 'metric blades', defined entirely by their relative length and mostly produced by bipolar technique, and not 'qualitative blades' (i.e. elongated pieces with parallel sides and dorsal arrises produced by platform techniques). The most stratigraphically useful piece of information in this context is the fact that all hard percussion blades defined in the catalogue as 'unusually regular' (CAT 25139/12-13 and 25143/9), or 'perfectly regular' (CAT 26004/54), were recovered in Phase D. It is generally accepted that 'proper' blades were not produced during the Bronze Age period, whereas all parts of the Neolithic period produced 'true' blades.

Another interesting stratigraphic trend is the difference between the various phases in terms of their burnt quartz ratio (Table 5.11): Phase B has a ratio of almost 11%, whereas the other phases are characterized by notably lower ratios of c. 1-2%. It is well-known that on domestic sites some lithic waste or tools occasionally fall into hearths during knapping or tool use. The fairly high ratio characterizing Phase B may be a result of other activities taking place at this level, such as rituals in connection with burials (cf. Ballin 2012, 23). However, as the burials from RUX6 are all inhumations, and not cremations, the rituals would have to be communal feasting rather than rituals associated with funeral pyres.

Unfortunately, most of the artefacts, such as the cores and tools, are fairly plain pieces. As no 'fancy' diagnostic quartz objects (to use a term introduced

	Phases												
Debitage	A	B	B/C	B/D	B/E	C	C/D	C/E	D	D/E	E	Unstrat.	Total
Chips	7	83				34			1,725	9	7		1,865
Flakes	1,272	1,879	198	8	28	432	6	6	2,652	117	79	2	6,679
Blades	35	49	5		1	13	2		108	6	2		221
Microblades	2	3	2			2			26				35
Indeterminate pieces	123	197	28		3	48			203	11	5	2	620
Crested pieces	5	1	1						6	2	1		16
Total debitage	1,444	2,212	234	8	32	529	8	6	4,720	145	94	4	9,436
Cores													
Split pebbles	8	9	1			3			7				28
Conical cores		3							1				4
Single-platform cores	10	10				3			9	1			33
Single-platform/discoidal cores	1	2							2				5
Opposed-platform cores	1	3				3				1			8
Cores w two platf at angle		1				1			2				4
'Flaked flakes'	1	2											3
Irregular cores	8	6	2			1			11				28
Bipolar cores	33	22	1			5			80	6	2		149
Core fragments	2	1	1								1		5
Total cores	64	58	6			16			112	8	3		267
Tools													
Backed knives		1											1
Scale-flaked knives									1				1
Discoidal scrapers	1	1											2
Blade-scrapers	1												1
Short end-scrapers	6	18				3			12				39
Double-scrapers									1				1
Side-scrapers		2				1			1				4
End-/side-scrapers						1			1			1	3
Atypical scrapers	1								1				2
Scraper-edge fragments		1				1			5				7
Piercers		3				1			2				6
Pieces w retouched notch(es)		1											1
Denticulated pieces	1	1											2
Pieces with edge-retouch	1	6							18				25
Points	1	2							1				4
Percussoirs	1												1
Percussoirs/anvils									1				1

TABLE 5.17:
Distribution of all artefact categories across the various phases

Tools	Phases												Total
	A	B	B/C	B/D	B/E	C	C/D	C/E	D	D/E	E	Unstrat.	
Hammerstones	1	2											3
Hammerstones/split pebbles		1											1
Hammerstones/anvils									1				1
Pounders		1							1				2
Total tools	14	39	1			7			46			1	108
Total	1,522	2,309	241	8	32	552	8	6	4,878	153	97	5	9,811

TABLE 5.17 continued:
Distribution of all artefact categories across the various phases.

by Stephen Green in connection with his analysis of British bifacial arrowheads; Green 1980) were recovered from the site, it is almost impossible to carry out any form of stratigraphic control through analysis of the distribution of formal quartz artefacts.

Horizontal distribution – levels, contexts and features

The horizontal distribution of worked quartz was investigated through a number of specially selected contexts and features (Table 5.18) which were deemed to be of potential interest to the understanding of site chronology or activities.

In Table 5.18, the list of contexts and features assigned to Phase B are at the top, followed by those from Phases C and D. Phase A was generally so disturbed by erosion and more recent activities that it was decided not to include contexts and features from this phase in the table. The artefacts are grouped according to their potential information value in relation to the two main themes guiding this discussion, chronology and possible prehistoric activities. The analysis of the main contexts and features is hampered by the same key problem as the discussion of the vertical distribution of quartz artefacts: that few formal artefact categories were identified, and that none of these are diagnostic *sensu stricto*.

Most of the features from Phases B and C are datable to the Bronze Age period and they are generally either burial or ritual monuments or related to them. Although on occasion individual or small numbers of plain lithic artefacts were deposited deliberately in Bronze Age features (e.g. at the Skilmafilly cremation cemetery, Aberdeenshire; Ballin 2012: 23), it was also quite common in some cases to deposit well-executed and curated ('fancy') implements with the deceased, such as scale-flaked and plano-convex knives (e.g. Finlayson 1997: 309; Ballin 2006: 81; 2014a). foliate knives (Ballin 2012: 11), barbed-and-tanged arrowheads (Ballin 2016), as well as maceheads, battle axes and jet objects (cf. main text and catalogue in Clarke *et al.* 1985; also Brophy and Sheridan 2012; Downes 2012).

However, the finds recovered from the Bronze Age features in Phases B and C are all generally plain, and they include finds usually associated with primary production and knapping floors/workshops, such as numerous pieces of debitage or waste; split pebbles, which may be either early-stage bipolar cores or damaged hammerstones; platform and bipolar cores; and hammerstones. Although it cannot be ruled out that the finds from the upper phases may include material from small Bronze Age knapping floors, the fact that the finds from Phases B and C are more or less identical with those from Phase D suggests that most of the quartz artefacts recovered from the Bronze Age features may be Neolithic artefacts introduced either by erosion or by the digging and construction activities of Bronze Age people. The Phase B single-platform cores, for example, include characteristic pieces which were clearly produced by the Neolithic operational schema defined in Figure 5.22, as well as relatively sophisticated conical cores. If any lithic objects were deliberately deposited in the Bronze Age features, for example in connection with burial rituals, they would be plain pieces and it is impossible to separate these from the vast volume of residual Neolithic material.

The Phase B structure BG24 may be an exception from this scenario, as it included a high volume of quartz waste (506 pieces), including clear indicators of primary production, such as one detached core preparation flake (crest), split pebbles and cores, as well as two hammerstones. However, it also includes a number of tools, such as two piercers, one notch, and one point, suggesting use as well as production

	Phase	Debitage	Regular platform blades	Crests	Split pebbles	Single-platform cores	Other cores	Bipolar cores	Scrapers	Other tools	Total
Associated pits, cist and cairn (CE), (BB3-3.1), (BC2), (BE43)	B	94			1		2		1		98
Standing stone and surroundings (BD1-2)	B	8							1	1 hammer	10
Pits (BG2)	B	51									51
'Ritual' building floors (BG24.1-4.3)	B	506		1	2	4	1	2	1	1 point, 2 piercers, 1 notch, 2 hammers	523
Tadpole (BH)	B	10									10
Cist (BM1-2)	B	79			1	1	1	3	1		86
Pits (BN1-2?)	B/C	24									24
Various ploughed areas	B/C/D	64			1	2		1	1		69
Pits (CD), (CD22)	C	71					1			1 piercer	73
Libation pit (DB, DC) and associated areas	D	74			1			1	1	1 percussoir/anvil	77
Building 1 (DJ)	D	423	2	2		4	2	9	5	1 scale-flaked knife, 2 retouch	450
Buildings 1 and 2	D	10					1				11
Building 2 (DH) hearth area	D	132	1					3			136
Building 2 (DH) hearth area	D	2,135		2	1	3	3	14	2	1 hammer/anvil, 1 piercer, 4 retouch	2,166
Building 2 (DH) pit	D	10									10
Building 2 (DH)	D	473	2		1	2	2	8		3 scrapers, 9 retouch, 1 pounder	501
Building 2 (DH) hearth	D	25						3			28
Building 2 (DH) west of saw pit	D	48									48
Burning outside Neolithic buildings 1 and 2	D	7								1 point	8
Possible building 3	D	23									23
Total		4,267	5	5	7	17	12	44	13	32	4,402

TABLE 5.18:
Distribution of quartz artefacts across a selection of contexts and features.

of tools. This scenario offers two interpretations, namely that this structure is not a ritual building at all, but a domestic one, or that a ritual structure was erected on top of an earlier combined lithic workshop/activity area.

One particular feature (BG3) within this structure adds to its re-interpretation as a domestic one. BG3 was a small stone box in the floor towards one end of the structure, immediately next to the later and intrusive cist (3M). This small box contained a small collection of large quartz objects, such as one large cobble (GD *c*. 120mm) split into four smaller refitting fragments (26241/1-4); parts of another large cobble which consists of two refitting indeterminate pieces (26241/12-13) and where the attributes of the quartz suggest that three further flakes (26241/8-10) may also have formed part; four individual flakes (26241/5-7, 11); as well as a large point (26241/14). As it is unlikely that the point was involved in knapping (it may have been used for chopping or piercing), this collection is probably not a 'knapper's tool-kit', but more likely an individual's personal cache of raw material and possibly a favoured tool (Figure 2.43).

The so-called 'libation pit' from Phase D presents a scenario similar to that of the Bronze Age features, in that a likely ritual feature contained relatively large amounts of debris associated with primary production, as well as one scraper and a percussoir/anvil. Although this feature is of Neolithic date, like most other finds recovered from this phase, the lithic objects in it should probably be interpreted in the same manner as those recovered from the Bronze Age burial/ritual features. The quartz could have entered the feature either as a result of natural (erosion) or anthropogenic agents (human activities). In this case, the lithics probably entered the feature with the infill, and they may either be contemporary with, or predate, the feature.

The 'libation pit' contained one of the most aesthetically pleasing quartz artefacts recovered - the small percussoir/anvil (CAT 26782/001). Although this neat little implement appears to be a curated and cared for tool, it may still simply be a piece relating to the site's primary or secondary production, rather than a deliberately deposited piece.

The two Neolithic buildings (DH and DJ) both share characteristics which supports the interpretation of them as domestic structures. The investigation of Building 2 (DH) clearly involved the consistent use of sieving, and the bulk (possibly three-quarters) of the 2,135 pieces of debitage found there is chips. These chips almost certainly indicate the location of the (in Binford's terminology) drop-zone of a dense knapping floor, which also yielded two core preparation flakes (crests), one split pebble, three single-platform cores, three other cores, 14 bipolar cores, and one hammerstone/anvil, which would have been involved in the production of this waste material. Almost all of this knapping debris was found immediately east of the hearth, supplemented by 135 pieces recovered west of the hearth. Two scrapers, one piercer and four pieces with edge-retouch were also retrieved from this scatter, suggesting that, although some use of tools may have taken place here, most tool-using activities probably took place elsewhere within or out-with the building. As shown in Table 5.18, three scrapers, nine pieces with edge-retouch and one pounder were recovered from other features in the building.

The composition of the finds from Building 1 (DJ) corresponds roughly to that of Building 2 (DH), apart from the fact that the former was almost devoid of chips. This supports the impression that consistent sieving may have only taken place in connection with the investigation of Building 2 and its main knapping floor, or that the sieved residues were lost. As the number of larger pieces of debitage recovered from Building 1 (423 pieces) corresponds roughly to that recovered from Building 2, it is quite likely, indeed almost certain, that the use of consistent sieving in connection with the investigation of this building would have yielded an equally large amount of chips, and that the scenario may be almost identical, with a knapping floor within the building, supplemented by some tool use. The tools from Building 1 include five scrapers, one scale-flaked knife, and two pieces with edge-retouch. The recovery of a number of highly regular platform blades from both buildings supports the identification of them as Neolithic domestic structures.

Other contexts and features did not yield enough worked quartz to allow any opinion to be formed as to their likely date or activities taking place there.

Activities

The interpretation of site activities on the basis of the recovered quartz is hampered by a number of factors, such as erosion and other forms of disturbance of the site; inconsistent recovery policies; and the fact that it is generally difficult to identify retouched edges on quartz surfaces, not least when the quartz flakes as poorly as some of the examples from RUX6 (e.g. the so-called 'reconstituted' quartz) and frequently breaks up when used, and when some quartz is 'sand-blasted'.

However, it is possible to make the following general observations:

- The assemblage is the product of two different sets of activities, Neolithic domestic activities, and Bronze Age burial/ritual activities. Some quartz artefacts, such as those from Neolithic Buildings 1 and 2 (Phase D), are associated with in-house knapping floors, and this material must be considered of certain Neolithic date. It is not possible to date any Phase A-C quartz artefacts to the Bronze Age period with certainty, and it is suggested that most of the finds from Phases A-C may be redeposited Neolithic pieces. However, the unusually high burnt ratio associated with the quartz from Phase B (10.7%) suggests that some Bronze Age or redeposited Neolithic quartz artefacts were exposed to, and affected by fire, probably in connection with the burial/ritual activities taking place during the Bronze Age.

- The nature of Phase B Bronze Age Building BG24 is uncertain. Crawford suggested that it might be a structure associated with ritual activities, but the fact that the quartz assemblage recovered from this structure is composed very much in the same manner as those recovered from the Neolithic buildings suggests that this may in fact be a Bronze Age domestic structure.

- Most likely the quartz artefacts recovered in connection with the so-called 'libation pit' entered the feature with the backfill.

- The recovery of notable amounts of chips immediately east of the hearth in Neolithic Building 2 suggests that extensive primary production took place within the house, but the composition of the finds from this scatter indicates that the use of any tools produced here may largely have taken place elsewhere in the building or outside it.

- A similar scenario is suggested for Neolithic Building 1, although the assemblage from this structure does not include many chips. It is thought that the near absence of chips despite the recovery of many flakes and some tools, is a result of selective sieving during the excavation, with the soil from Building 2 having been sieved consistently with the use of a fine mesh, whereas the soil from Neolithic Building 1 was probably not sieved, or an unacceptably large mesh size was used, or the residues were lost.

Dating

It is difficult to date the present quartz assemblage, as the upper phases (particularly Phase A and to a lesser extent Phase B) have been heavily affected by erosion, and the collection includes very few typo-technologically diagnostic elements. It is thought that continued aeolian activity, in conjunction with Bronze Age digging and construction work, mixed Neolithic elements into the upper levels, and that most of the lithic finds in those layers may in fact be redeposited Neolithic pieces. However, as mentioned in the distribution section, it is possible that the quartz artefacts in Phase B Bronze Age Building BG24 may date to the Bronze Age, potentially affecting the interpretation of the nature of this structure.

The quartz finds include no strictly diagnostic pieces, but a small number of slightly less diagnostic finds are helpful. It is for example important that a small number of highly regular platform blades with parallel lateral sides and dorsal arrises all derive from Phase D (Table 5.18), and that they were all recovered from Neolithic Buildings 1 and 2. It is generally accepted that 'true' blades were not produced after the Neolithic/Bronze Age transition (Ballin 2002a).

During the characterization and cataloguing of the finds from Barabhas (Elliott Collection; Ballin forthcoming a), two distinct types of single-platform cores were defined, both being fairly flat and with cortical platforms: one has a typical pointed apex, whereas the other one is based on the detachment of medium-sized flakes from the circumference of large hard percussion flakes, and the 'apex' of these cores is defined by the flat ventral face of the parent flake (Figure 5.23). The core shown as Figure 5.23-1 corresponds to cores recovered at RUX6 (Figure 5.28) and may be a Neolithic form (this will need verification from future excavations on the Western Isles), whereas the core shown as Figure 5.23-2 is absent from RUX6, and may be an early Bronze Age form (this also needs verification).

The almost complete absence of small thumbnail-scrapers amongst the quartz artefacts could be seen as confirmation that RUX6 is dominated by Neolithic rather than Bronze Age material (cf. Butler 2005), but as demonstrated by the finds from the Barabhas machair (Elliott Collection), which is dominated by early Bronze Age material, there is a clear trend on the Western Isles of quartz scrapers being large, and flint and mylonite scrapers small. This probably entirely reflects the different sizes of the collected/quarried pebbles, cobbles and blocks of the different types of raw material.

One scale-flaked knife (CAT 23803/007) is only broadly diagnostic, and its semi-invasive retouch indicates a post-Mesolithic date (cf. Butler 2005).

Site	Quartz	Flint	Mylonite	'Greasy' quartz
Barabhas 3, Lewis	10	58	50	
Barabhas 2001, Lewis	2	30	50	
Barabhas (Curtis Collection) [B3]	1	34	29	
Barabhas (Murray Collection)	2	14	16	
Barabhas (Elliott Collection)	2	15	16	
Calanais, Lewis	5	20	27	
Dalmore, Lewis	1	8	5	
Guinnerso, Uig, Lewis	4	49		
Udal RUX6, N Uist	1	15	(Present)	
Rosinish, Benbecula	1	62		
Kilmelfort Cave, Argyll	2	26		
Shieldaig, Wester Ross	1	13		2

TABLE 5.19:
The tool ratios of a selection of sites and raw materials. The sites have been sequenced from north to south, with non-Western Isles sites inserted at the bottom of the sequence (Ballin 2000b; 2001; 2002b; 2002c; 2010a-e; 2016; forthcoming a; forthcoming b; Saville and Ballin 2009). When the ratios for the Guinnerso collection were calculated, the finest quartz debris was disregarded, as this may have been produced as temper for Iron Age pottery.

In contrast to the assemblage from Barabhas (Elliott Collection), which due to its inclusion of numerous (36) diagnostic arrowheads (leaf-shaped, oblique, and – in particular – barbed-and-tanged pieces) was datable on its own terms, the quartz assemblage from RUX6 can only be dated precisely by its association with other finds categories (primarily pottery) and features (e.g. Beaker burials), as well as by radiocarbon dating (see PARTS 3 and 5, this volume). Interestingly, the flint, which would usually have been preferred for tools in preference of quartz (Table 5.19), does not include other diagnostic material than an indeterminate fragment of a bifacial arrowhead (post-Mesolithic). This piece was recovered from Phase B, and may be a redeposited piece from lower levels.

Summary and discussion

In connection with the excavation of RUX6, almost 10,000 quartz artefacts were recovered. It is thought that most of the artefacts from Phase D are Neolithic and relate to domestic settlement (first and foremost Neolithic Buildings 1 and 2, as well as activity areas outside the structures), and that most of the plain artefacts from the pits (such as burials) in Phases A-C may mainly be redeposited Neolithic pieces. Some specimens from the upper levels may possibly relate to domestic early Bronze Age activities, for example in and around temporary building BG24 which might be a domestic structure.

The quartz from the domestic Neolithic settlement is characterized by an unsophisticated, but well-defined and schematic operational schema (Figure 5.22), based on the exploitation of local vein quartz and, probably mostly, quartz pebbles and cobbles from the adjacent beaches. Like most Neolithic/early Bronze Age quartz assemblages from the Western Isles, the quartz was largely reduced following a two pronged approach: hard percussion reduction following the approach described in Figure 5.23, the final stage of which is likely to have been the total exhaustion of the cores by bipolar technique; and the reduction of pebbles, cobbles, and blocks by the application of bipolar technique from start to end of the reduction process. This left numerous waste flakes and flake blanks with cortical platform remnants, and platform cores (predominantly single-platform cores) with cortical platforms, as well as many bipolar flakes and cores. It is also thought that some surprisingly well-executed platform blades were manufactured in and around the two Neolithic buildings.

The only core preparation used at the site appears to be the occasional initial cresting, and crude platform-edge trimming along the way. Most likely, a (large?) proportion of the site's split pebbles are damaged hammerstones, rather than early-stage bipolar cores.

A total of 108 tools were retrieved from the site, and they were divided into 'fine' tools (95 pieces) based on flaked blanks, and 'coarse' tools (13 pieces) which tend *not* to be based on flaked blanks, although there are exceptions. Compared to the mainly early Bronze Age assemblage from Barabhas (Elliott Collection – 260 implements), the basic composition (fine and coarse tools) of the present mainly Neolithic assemblage is exactly the same, at 89% fine tools and 11% coarse tools.

In Table 5.20, the quartz assemblage from RUX6 is compared with other well-known quartz assemblages from the Western Isles. Only published assemblages, or assemblages characterized by this author, with more than 100 pieces are included.

	Approx. date	Total lithics	Quartz ratio	Flint ratio	Other ratio	Comments	Total cores	Platform: bipolar ratio	Total tools	Arrh. ratio	Scraper ratio	Knife	Piercer ratio	Other ratio
Point Braighe, E. Lewis	LBA/EIA	155	98	2			15	87:13	6		50			50
Barabhas 3, Lewis	EBA	294	90	6	4	Includes 2 mylonite	18	56:44	29	28	24	10	3	35
Barabhas 3 (Curtis Collection)	EBA	729	586	50	93	Includes 42 mylonite	14	64:36	34	9	71	9		11
Barabhas 2001, Lewis	EBA	111	67	9	24	Includes 4 mylonite	2	00:100	9		44			66
Barabhas (Murray Collection)	EBA	173	53	30	17	Includes 11 mylonite	5	00:100	19	11	21		5	63
Barabhas (Elliott Collection)	EBA (Neo)	6,856	83	11	6	Includes 208 mylonite	208	30:70	231	16	57	2	2	23
Dalmore, Lewis (Sharples)	EBA	2,564	93	4	3	Includes 63 mylonite	30	40:60	29	17	41		3	39
Dalmore, Lewis (Curtis)	EBA	101	58	28	14	Only cores and tools; incl. 12 mylonite	42	57:43	59	24	49		5	22
Olcote, Lewis	EBA, MBA	15,456	98	1	1	Unclear how many 'other' materials are mylonite, but mylonite definitely present	264	23:79	40	8	73			19
Calanais, Lewis	Neo, EBA	313	74	14	11	Includes 34 mylonite	10	60:40	30	20	53			27
Valtos, Lewis	Neo, BA, IA	n/a				mylonite present	n/a		n/a					
An Dunan, Uig, Lewis	IA	1,087	100						8		38		13	49
Gob Eirer, Uig, Lewis	LBA/EIA	210	67		33	Other: Granite, gneiss, igneous			3		67		33	
Guinnerso, Uig, Lewis	BA/EIA	1,822	98	2		In addition, 4,710 minuscule pieces were collected from sieving (pottery temper?)	5	40:60	82	1	26	1	1	72
Northton, Harris	Neo, EBA					Includes mylonite	n/a		n/a					
Udal RUX6, N Uist	Neo (EBA)	9,969	99	1	trace	Includes 5 mylonite	267	33:66	95		62	2	6	25
Eilean Domhnuill, N Uist	Neo	n/a					n/a		n/a					
Bharpa Langais, N Uist	EN	186	95	4	1	Includes 1 piece of pitchstone	21	67:33	4		75			25
Rosinish, Benbecula	EBA	3,568	99	1			77	21:79	36		81	3	8	8
Allt Chrisal, Barra	EBA	3,621				Includes 4 bloodstone								

TABLE 5.20:
Summary of the main lithic assemblages from the Western Isles.

In the original reports, some quartz assemblages were analysed with other lithic raw materials and coarse stone tools, but in this table, all coarse stone tools have been excluded to make the collections more directly comparable. Sites highlighted in grey Olcote; Valtos; Northton; Eilean Domhnuill (Warren 2005; Lacaille 1937; Nelis 2006a; Armit 1992) are only used in the comparison in a general sense, as they have either not yet been published, or the characterization/quantification of them differed so much from the principles followed by this author that it would have been difficult to compare them directly with the present collection.

Inspection of Tables 5.19 and 5.20 shows that the finds from RUX6 in a general sense fits the picture presented by most other Neolithic/early Bronze Age lithic assemblages from Western Isles domestic sites. The tool ratio for the quartz is very low, and much higher for materials like flint and mylonite, and the quartz assemblage is heavily dominated by scrapers (usually c. 50-80% of the tools) and simple edge-retouched pieces (usually c. 25-65% of the tools), with other tool categories being present in low (usually single-digit) numbers.

In addition, Table 5.20 includes quartz collections from sites dominated by burial and/or ritual activities, such as Calanais (elaborate stone circle, cairn, residual domestic material; Ballin 2016) and Olcote (large cairn, residual domestic material; Warren 2005), as well as a group of assemblages from later Bronze Age/early Iron Age sites (Point Braighe; An Dunan; Gob Eirer; Guinnerso; Ballin 2011a; 2013; 2015a; forthcoming b). The composition of the Calanais collection is clearly influenced by the burial/ritual activities taking place there, with a number of fine barbed-and-tanged arrowheads deposited within the cairn. Olcote is characterized by the use of crushed quartz to cover the cairn. Point Braighe stands out, as the quartz was reduced as competently as any Neolithic/early Bronze Age knapper(s) could have done it, but with most of the quartz having been superficially burnt; this latter phenomenon is presently unexplained. The other three later Bronze Age/early Iron Age assemblages are all characterized by somewhat unschematic knapping (probably suggesting that they generally post-date Point Braighe), and by their inclusion of large volumes of crushed quartz, possibly to be used as temper in pottery.

Most of the assemblages are characterized by an approximately 50:50 distribution of waste material across platform and bipolar flakes and cores, but usually with a dominance of bipolar material. However, it is possible that these specific ratios are defined at least partially by recovery policies (excavation/collection; sieving/no sieving) and by the people collecting/excavating the sites, as exemplified by the two collections from Dalmore – one with a platform:bipolar core ratio of c. 40:60 and the other c. 60:40. The inclusion of arrowheads may partially be a function of site type, with some finer pieces having been deposited in connection with burials, but functional ('everyday') pieces have been found on several domestic settlements (hunting, defence?), and at Dalmore (Ballin 2002b) and Barabhas (Elliott Collection; Ballin forthcoming a) workshops were found for the production of quartz barbed-and-tanged arrowheads, including their associated debris and preforms.

RUX6 is exceptional in the sense that this numerically very large quartz assemblage included no arrowheads at all, neither finished pieces nor preforms. It also stands out as having very low numbers of pounders (two) compared to the 14 pieces from Barabhas (Elliott Collection). The relatively small 333-piece lithic assemblage from Barabhas 3 included as many as 30 pounders (Ballin 2010a). Overall, the above (in conjunction with the presence of numerous scrapers) defines the quartz from the Neolithic levels at RUX6 as having been produced in a domestic setting, but with few other definable activities than knapping, some of which clearly took place within the buildings and some probably outside, and the processing of skin, hides and possibly hard materials like wood, bone and antler by scrapers (Juel Jensen 1994). It is highly likely that RUX6 in prehistory would have included (peripheral?) areas not touched by the excavation, where other activities would have taken place, involving the tool forms commonly recovered from other Neolithic/early Bronze Age sites on the Western Isles but missing from this assemblage.

In terms of assemblage composition, it is equally surprising that no well-executed lithic objects were found in connection with the burials investigated at RUX6. As mentioned above, fine lithic pieces are commonly found in connection with Scottish Neolithic and early Bronze Age burials, and the quartz arrowheads from the Calanais cairn exemplify this tradition. In comparison, the Phase A-C RUX6 burials dating to the early Bronze Age, all appear somewhat depleted in terms of grave goods. This may reflect social differences (the RUX6 burials possibly being low-status graves), as for example any individual buried at the Calanais ritual centre is likely to have been a person of some standing – but it could also simply reflect different burial traditions across space and time.

In summary, the quartz assemblage, supported by other evidence from the site, such as lithic and bone artefacts, pottery, etc., seems to reflect the everyday lives of a domestic, rural late Neolithic society in the Western Isles and, through the apparent absence of quartz objects from graves, the burial traditions of this community at a later stage in the early Bronze Age.

Pumice found at The Udal

By Anthony Newton

A total of 237 pumice pieces were found at The Udal, of which 133 survive from RUX6; 42 from RUX1, 2 and 3 and 26 from UN. This pumice was found in all phases of the site, ranging from the pre-Neolithic Phase E to the insecurely dated Bronze Age to modern Phase A. Pumice has been found around the coasts of the North Atlantic on both natural raised and present day beaches and in archaeological sites (Binns 1972; Newton 1999a). These widely distributed sites stretch from Arctic Canada, Greenland, Iceland, Svalbard, Russia (the Kola Peninsula), the British Isles, Norway, Sweden, Denmark to Germany. In areas with well-developed raised beach sequences, the deposits range in age throughout most of the Holocene (Newton 1999a).

Pumice has been found at over 150 archaeological sites in Scotland, with the most numerous finds being from the Western Isles, Orkney and Shetland (e.g. Newton 1999a; Newton 1999b; Newton 2014). Pumice has been found in archaeological sites throughout Mesolithic, Neolithic, Bronze, and Iron Age contexts right through to modern times in Scotland. The oldest archaeological pumice finds have been from Mesolithic sites, including Staosnaig (c. 7800-8900 cal. BP) on Colonsay (Newton 2001) and Camas Daraich (c. 8400-8500 cal. BP) on Skye (Newton 2004). Sites in the Western Isles include, amongst others Allt Chrisal, Barra (Newton and Dugmore 1995); Kildonan, South Uist (Newton 1999a); Baleshare, North Uist (Newton and Dugmore 2003) and Cnip, Lewis (Newton 2006). All of the pumice so far found in the Outer Hebrides (Figure 5.43) consists of dark brown to black dacitic pumice which can be geochemically correlated with deposits elsewhere in Scotland, Norway, Iceland and Svalbard (Larsen et al. 2001).

Pumice would have provided the local populace with a useful abrasive tool. Many of the pieces found on archaeological sites show obvious signs of wear, with grooves and holes produced by sharpening of antler, bone or wood, and the pumice pieces from RUX6 are no exception (see Ballin Smith, Worked Pumice below). Larger pieces with flattened sides are also found where it would have been used for rubbing of various materials, including skins and wood, for example. Pumice also was, and is still, used as floats for fishing nets and lines. The pumice may have been recovered from a contemporary beach or older eroding shoreline. It is also possible that the pumice may have been taken from an older settlement site, perhaps a midden or some other abandoned building.

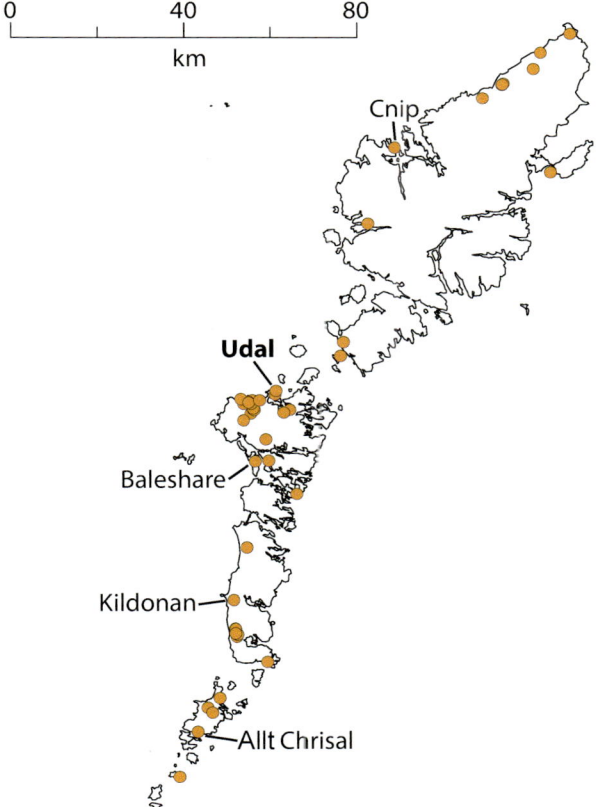

FIGURE 5.43:
Map showing the distribution of archaeological sites in the Western Isles where pumice has been found. Sites mentioned are labelled.

Geochemical analysis of the Udal pumice

The standard technique for analysing pumice is to use an electron microprobe to determine major element composition. This enables comparison with other pumice deposits and with tephra (volcanic ash) layers that may have been produced during the eruption that produced the pumice. Previous experience has shown that pieces of brown pumice of the type found at Udal have indistinguishable major element geochemical compositions. In order to test this five pieces of pumice were selected (Table 5.21) and at least five analyses were carried out on each piece. There was insufficient physical variation between the pumice pieces to warrant a more in depth study. The data collected were then compared with over 600 analyses of pumice from sites in Scotland, Norway and Iceland and with analyses of tephra layers from Iceland.

Pumice sample/ SF No	Colour	Phase	Approximate age
23788	dark brown	A	unclear
23751	brown	D	c. 4200 BP
23814	brown	D	c. 4200 BP
26890	light brown	D	c. 4500 BP
24007	light brown	E	>6500 BP

TABLE 5.21:
Details of pumice analysed

Major element geochemical analysis of the pumice was carried out on a Cambridge Instruments Microscan V electron microprobear the School of GeoSciences, University of Edinburgh. A small sample from each piece of pumice was incorporated into resin on a glass slide. These were then ground and polished to a thickness of 75 µm and carbon coated. Wavelength Dispersive (WDS) analyses were carried out using an accelerating voltage of 20 kV and a beam current of 15 nA.

Results and discussion

Table 5.22 contains the results of the analyses of the pumice from RUX6. All of the pumice analysed is dacitic in composition with SiO_2 abundances between 63% and 67%. Figure 5.44 shows both the comparison between the 5 pieces of pumice and the Udal pumice with other analyses of pumice from sites in the British Isles. It is clear from Figure 5.44 that the pumice from Udal can be correlated with other pumice deposits in Scotland, which are similar to pumice found on raised beaches in Norway and Iceland. Equally, the amount of variation within a single piece of pumice shown in Table 5.22 does not allow the separation of the pumice from Udal into separate events. This is typical of the brown dacitic pumice found on raised beaches in Iceland and Norway and archaeological sites in the British Isles. Even the oldest pieces from Phase E are geochemically indistinguishable from the other pumice pieces. There is no geochemical variation between the lighter and darker coloured pumice.

All of the dacitic pumice found in Scotland appears to have been erupted from the Katla Volcanic System, southern Iceland since about 7000 BP (Newton 1999a, Larsen et al. 2001). Evidence for this is found around the Katla Volcanic System in southern Iceland, where tephra layers (known as SILK layers) are found with similar major element geochemical characteristics to the pumice (Larsen et al. 2001). The precise number of eruptions responsible for the pumice is not known. The Udal pumice is similar to other pumice deposits found in Scotland. Figure 5.45 illustrates that the geochemistry of the analysed Udal pumice correlates with SILK-YN, SILK-MN, SILK-LN and SILK-N4 tephra layers. These tephra layers have been dated to between approximately 1600–3900 cal. BP. The original Udal analyses were undertaken in 1996 and recently (2016) new analyses on a more modern electron microprobe have been undertaken. The SILK tephra data presented here are the new analyses. These new analyses have increased precision and it may be possible to refine the correlation of the pumice to particular eruptions in future.

Conclusions

The pumice found in area RUX6 is dacitic and can be correlated with pumice from other archaeological sites in Scotland and with deposits found on raised beaches in Iceland and Norway. Furthermore the pumice can be correlated with SILK tephra layers found around the Katla Volcanic System in southern Iceland. From this evidence it can be concluded that the pumice was produced by one or more volcanic eruptions from the Katla Volcanic System between approximately 1600 and 3900 cal. BP. Further work is ongoing to try to improve the correlation of pumice to individual dated eruptions, which would be of benefits to archaeological research, as well as volcanology.

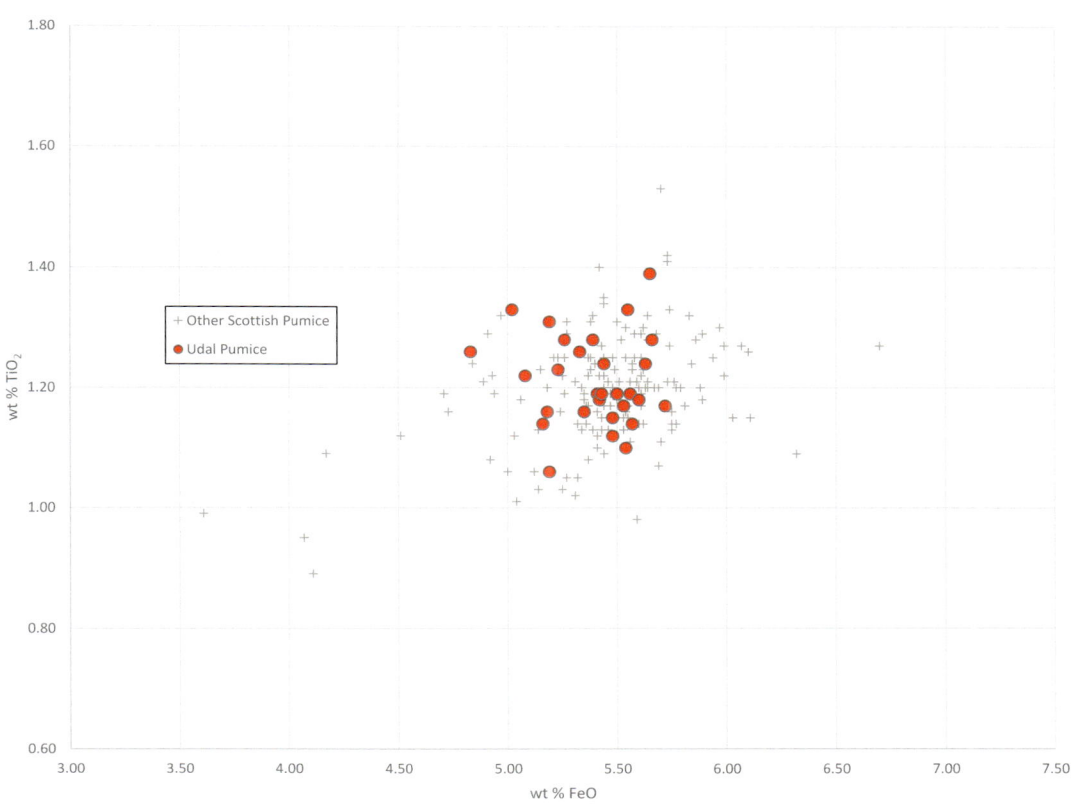

FIGURE 5.44:
Graph showing the correlation of the Udal pumice with the other dacitic pumice from archaeological sites in Scotland (Newton 1999a).

	SiO$_2$	TiO$_2$	Al$_2$O$_3$	FeO	MnO	MgO	CaO	Na$_2$O	K$_2$O	Total
24007	66.78	1.19	13.99	5.41	0.16	0.93	2.63	4.43	2.85	98.42
	66.63	1.33	13.81	5.02	0.2	0.91	3.03	4.54	2.9	98.37
	66.46	1.22	14.43	5.08	0.18	0.89	2.36	4.64	3.02	98.28
	66.24	1.18	13.51	5.42	0.18	0.97	2.59	4.56	2.92	97.57
	65.62	1.14	15.06	5.16	0.17	0.83	3.16	4.99	2.64	98.77
26890	66.62	1.19	14.23	5.56	0.22	1.05	3.01	4.43	2.82	99.13
	66.02	1.16	14.43	5.35	0.21	1.02	2.75	4.97	2.69	98.6
	65.97	1.31	14.33	5.19	0.17	0.96	2.78	4.33	3.03	98.57
	65.58	1.28	14.05	5.26	0.21	1.11	2.93	4.78	2.88	98.08
	65.39	1.24	14.2	5.44	0.28	1.11	2.79	4.72	2.76	97.93
23751	67.01	1.26	14.07	4.83	0.16	0.93	2.61	4.91	3.09	98.87
	66.18	1.16	13.96	5.18	0.15	1.06	3.09	4.89	2.82	98.49
	65.65	1.28	14.4	5.66	0.18	1.16	3.2	4.85	2.87	99.25
	64.97	1.19	14.06	5.5	0.21	1.22	3.29	4.94	2.92	98.3
	64.6	1.39	14.08	5.65	0.26	1.21	3.18	4.78	2.86	98.01
23814	66.37	1.33	14.25	5.55	0.23	1.04	2.98	4.58	2.64	98.97
	66.33	1.19	14.33	5.43	0.22	1.18	3.11	4.8	2.8	99.39
	66.28	1.24	14.45	5.63	0.21	1.18	3.04	4.8	2.76	99.59

TABLE 5.22:
Major element analyses of pumice from RUX6. Total iron is represented as FeO.

	SiO$_2$	TiO$_2$	Al$_2$O$_3$	FeO	MnO	MgO	CaO	Na$_2$O	K$_2$O	Total
	65.79	1.26	14.26	5.33	0.2	1.19	3.27	4.62	2.75	98.67
	65.47	1.23	14.41	5.23	0.29	1.11	3.04	4.87	2.86	98.51
23788	64.94	1.14	13.71	5.57	0.22	0.82	3.04	4.43	2.67	96.54
	64.9	1.17	13.59	5.53	0.23	0.91	2.97	4.36	2.9	96.56
	64.86	1.12	13.76	5.48	0.17	0.89	3.16	4.5	2.62	96.56
	64.81	1.18	13.91	5.6	0.19	1.74	3.1	4.43	2.8	97.76
	64.75	1.28	13.54	5.39	0.16	0.91	3.02	4.37	2.69	96.11
	64.7	1.17	13.8	5.72	0.24	0.77	3.41	4.63	2.8	97.24
	64.7	1.15	13.68	5.48	0.18	1.51	3.14	4.27	2.82	96.93
	64.57	1.1	13.56	5.54	0.2	0.83	3.12	4.83	2.75	96.5
	63.63	1.06	13.79	5.19	0.19	0.81	2.96	4.67	2.78	95.08

TABLE 5.22 continued:
Major element analyses of pumice from RUX6. Total iron is represented as FeO.

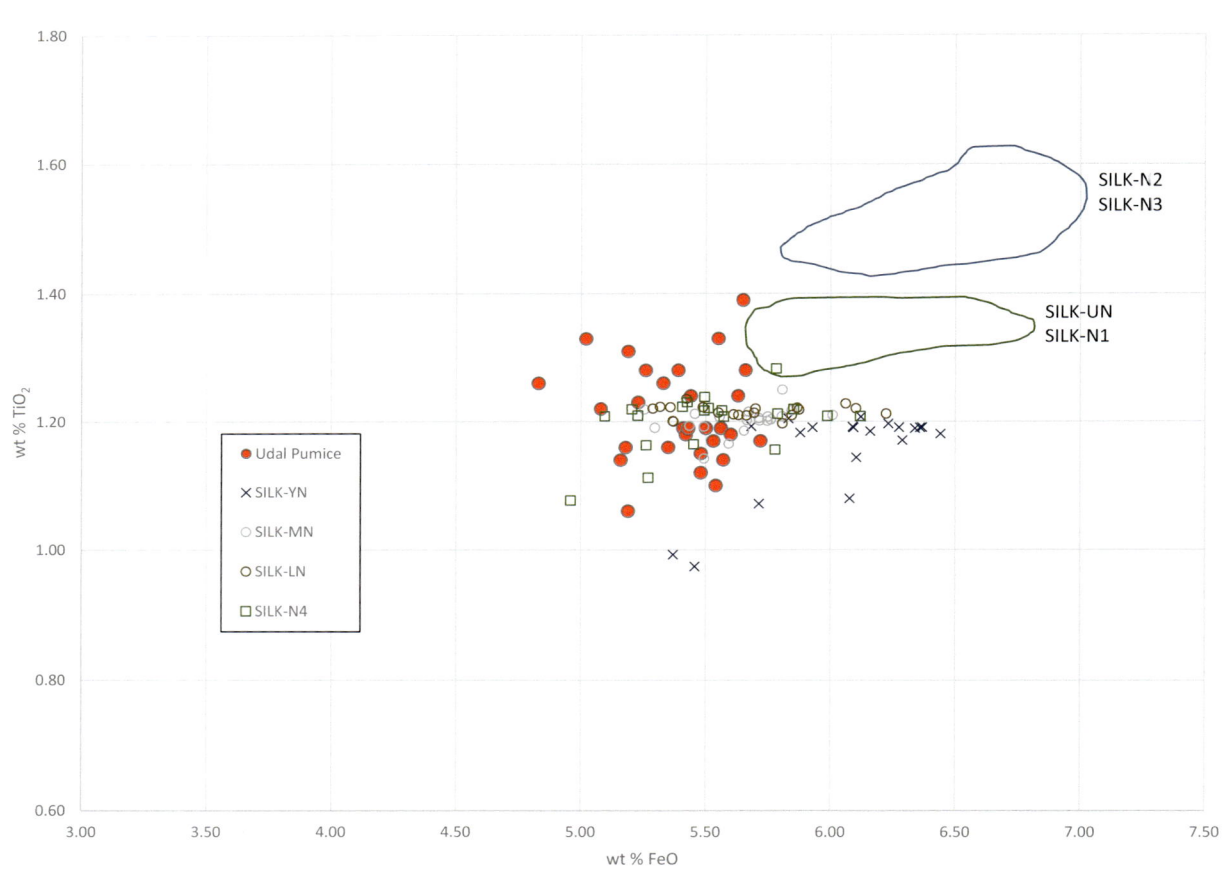

FIGURE 5.45:
Graph showing the correlation of the Udal pumice with the SILK tephra layers erupted from Katla, southern Iceland (Newton 1999a; Larsen *et al.* 2001).

Worked pumice and pumice artefacts

By Beverley Ballin Smith with identification of the raw material by Anthony Newton

Description of the pieces

Of the total number of pumice pebbles analysed by Newton above, 133 are from RUX6. This number does not include six missing pieces, one of which (SF 24007), was the oldest piece identified by him. Table 5.23, displays the numbers and percentage of pumice from each phase, with 63.2% being natural unworked pieces with the remaining 36.8% used as tools. Pumice was, and remains, a lightweight resource brought to land from the sea by a combination of ocean currents and wind. The Udal was no exception and pumice in its waterworn pebble form would have been a useful material for the inhabitants at RUX6 to have collected.

	Worked	Unworked	Totals	%
A	10	2	12	9
B	19	39	58	44
C	4	3	7	5
D	13	22	35	26.3
D/E	3	18	21	15.7
Totals	49	84	133	100
	36.8%	63.2%		

TABLE 5.23:
Comparison of worked and unworked pumice by phase.

Analysis

Each pumice pebble was examined by using a X6 hand-lens for the identification of worked or worn areas. In the raw state, pebbles are rough to the touch in spite of their predominantly rounded shape, and feeling their surfaces with the tips of fingers was important to locate subtle alterations in their appearance from light wear or use. Those pebbles that were unmodified were not analysed further but all were measured and weighed. The total weight of the collection is 1098.2 g. The remaining 49 pebbles were tools used in various ways and in various degrees, and they exhibit marked changes to their appearance. The patterns of wear have been distinguished as eight distinct types or attributes. These are:

Worn - light use, often revealed as less rough surface areas or edges, but not smooth enough to create easily noticeable surface alterations.

Smoothed – where an area of the pebble had been rubbed against something harder to create a flat, smooth surface, often with noticeable edges. The pumice vesicles are often filled in and the surface is noticeably smooth to the touch.

Convex wear was noted on highly smoothed pieces where the ends or sides had been worn more than the middle surface areas of a pebble.

Concave wear was also noted on highly smoothed pieces but only the middle surface areas or edges of a pebble were hollowed by use.

Trimmed – some pieces, such as thin roundels appeared to have been cut by a harder stone tool especially around their perimeters to create sharp edges and exposure of unworn vesicles.

Grooved – a thin, narrow but hard material, such as bone or wood, was rubbed along or across a surface or edge to create narrow V or U-shaped indentations.

Perforated – some pieces had been bored through by a sharp hard stick or bone, to create a hole. Other surface hollows indicate that the piece was not perforated completely.

Facets – some pieces were so worn that facets or sharp edges between surface alterations were very marked.

The results of the analysis were recorded in a database that forms part of the site archive.

Results

The vast majority of worked pumice pebbles displayed more than one attribute of wear. This suggests that when it was needed, pumice was a useful resource for a variety of purposes that included mainly rubbing and smoothing. Table 5.24 displays the number of attributes of worn pebbles and the phase or structure of the site they came from.

The majority (63%) of pebbles have only one or two attributes, 22% have three attributes but only 14% have more than three. This suggests that pieces, especially in Phase B were used for a shorter length of time before they were discarded, than those from Phase D. It may also imply that the raw material was more plentiful but smaller in size in Phase B than in the late Neolithic. Alternatively, as the use of the site had changed from settlement (Phase D) to ritual activities (Phase B), the utilisation and requirement of pumice for tools may have also changed.

Phase	Number of Attributes						Totals
	1	2	3	4	5	6	
A	3	2	3	1	1		10
B	8	4	3				15
BG24			1				1
BM cairn		1	1	1			3
C	2	1	1				4
D			3				3
Building 2 (DH)		2	1	2		1	6
Building 1 (DJ)	2	2					4
D/E		1	1	1			3
Totals	15	16	11	5	1	1	49

TABLE 5.24:
Number of attributes (types of worn areas) on pumice pieces by phase

All the worked and unworked pebbles were plotted to compare differences in size between the two main phases, and to account for the fewer attributes in Phase B. In Phase B the pebbles are smaller, clustering at the lower end of the scale and generally up to 40 mm in length and 30 mm in width (Figure 5.46a). In Phase D there is more variety in size with larger pebbles present. In general, their sizes lie between <10 mm to 70 mm in length and 45 mm in width (Figure 5.46b).

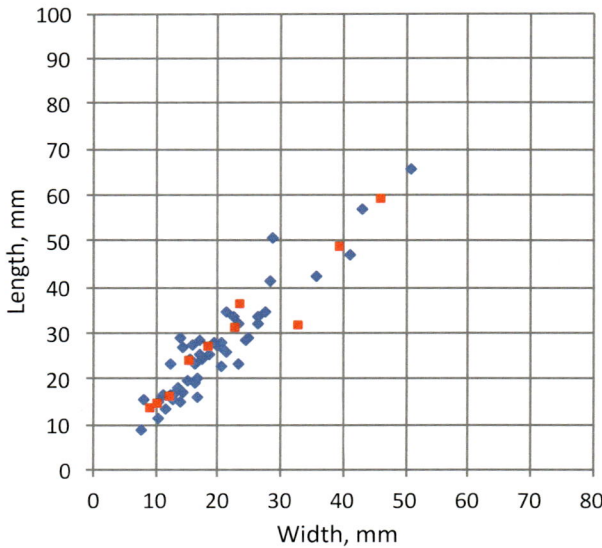

FIGURE 5.46a:
All pebble sizes, a) Phase B, red – structures, blue – all other contexts.

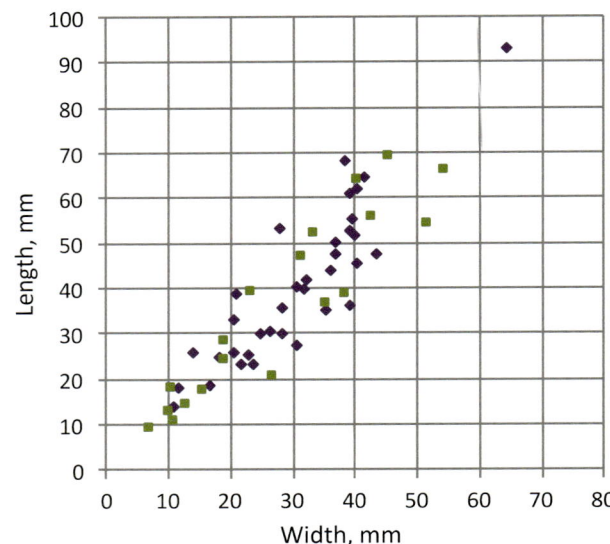

FIGURE 5.46b:
All pebble sizes, b) Phase D, purple – structures, green – all other contexts

An analysis of the size of worked pebbles from all phases (Figure 5.47) shows a similar pattern to Figure 5.46, with Phase B clustering at the bottom of the scale, but with larger pebbles distributed within the size range of pebbles from Phase D. Phase D demonstrates no clustering but a greater range of pebble sizes. The tendency for pumice pieces to become smaller over time, from the Neolithic through to the Iron Age was noted by Parker Pearson (2014: 143-4) in an assessment of pumice weights and numbers from a variety of sites in the Western Isles.

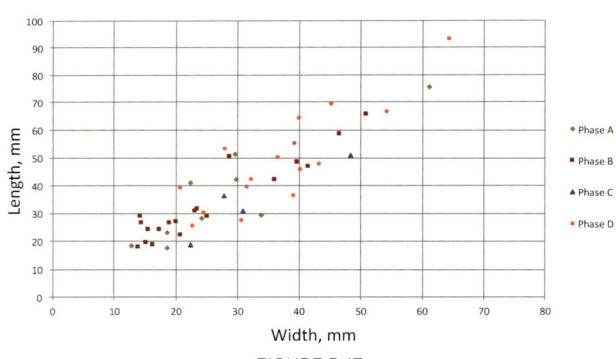

FIGURE 5.47:
Comparison of measurements of worked pumice from all phases.

Description of illustrated pumice

Phase A (Figure 5.48)

SF 23238a is a fragment of a pebble with a relatively flat to concave base, which is worn smooth. The domed top of the piece has one definite groove and one fainter. The pebble measures 29.3 by 33.9 by

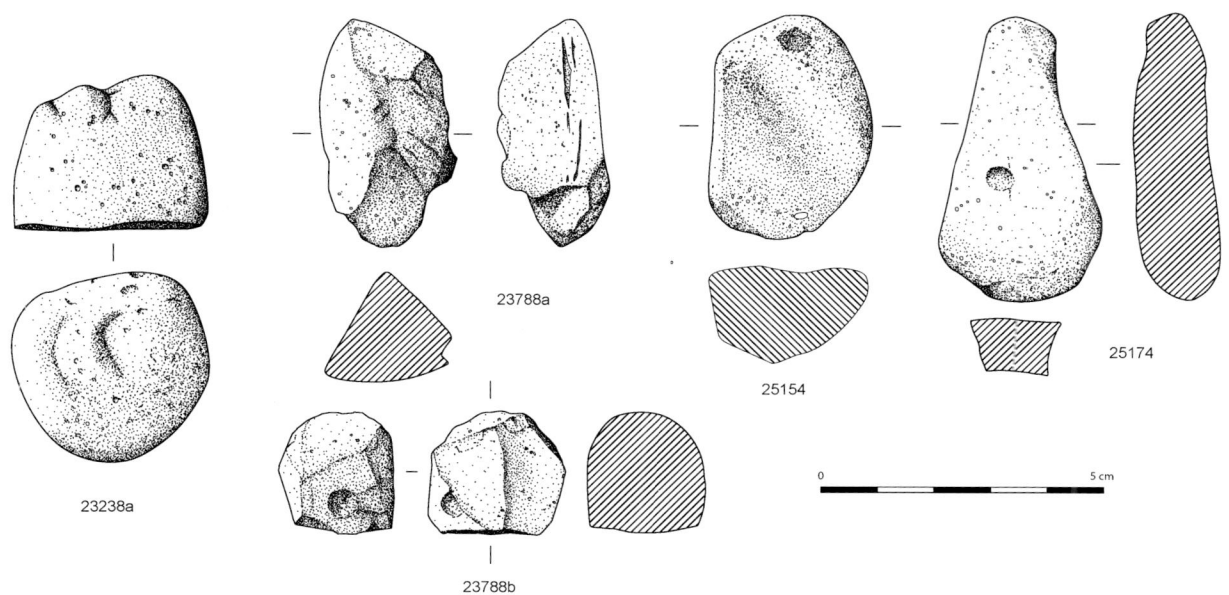

FIGURE 5.48:
Pumice artefacts from Phase A SF 23238a, SF 23788a, SF 23788b, SF 25154 and SF 25174. Drawn by Leeanne Whitelaw.

33.8 mm, and the grooves are 5-6 mm wide and 2 mm deep. Its weight is 11.5 g.

SF 23788a is a three-sided piece with two smooth surfaces and one rough. It is also a slightly lighter shade of grey pumice than the majority. There is a sharp edge between the two worn faces, which are both slightly convex suggesting that the apex between them was used for rubbing. Towards the opposite end is an asymmetrical groove, possibly using a natural fissure in the stone. The pebble measures 41.1 by 22.3 by 23.3 mm. The groove is 4 mm wide, 3 mm deep and c. 8 mm long. The piece weighs 6.6 g.

SF 23788b is a piece that was used as a probe sample, which left two rounded (indentations). The piece has been trimmed from a pebble as it has sharp edges between some of the seven faces. Only one face appears to be slightly worn and hollowed. It measures 28.4 by 24.2 by 21.5 mm and weighs 6.9 g.

SF 25154 is a smoothed and worn pebble. It upper surface is partly smooth, and one side is smoothed, slightly concave and facetted. The lower surface is partly smoothed, and slightly hollowed towards one edge by a shallow, broad groove. The piece measures 42.3 by 29.7 by 24.7 mm with a groove width of c. 12.9 mm. The piece weighs 9.7 g.

SF 25174 is a flattened tapering piece has a broad rounded end. Both its surfaces and sides are smoothed with wear, but one surface is slightly concave with parallel faint grooving, and the other more noticeably convex. This latter surface has a shallow hollow, which may be a natural vesicle in the pumice. Both sides of the tool are concave through use. It measures 51.2 by 29.6 by 16.3 mm and weighs 8.9 g.

Phase B (Figure 5.49)

SF 23907 This shaped and flattened pebble has trimmed by cutting around it sides to produce a tear-drop shape. Both surfaces are flattened and in the pointed apex, one deep depression 5 mm deep has been drilled with a shallower opposed depression on the reverse surface. The piece is unfinished. It measures 28.9 mm by 25.0 mm by 10.2 and weighs 3.9 g.

SF 25319d is a fragment of a pebble with one deep and other two faint grooves. It measures 32.0 by 23.4 by 9.7 mm with the deepest groove 20.7 mm in length, 6.6 mm in width and 4 mm in depth. It weighs 2.8 g.

SF 23375b is the broken end of a pebble, with evidence of trimming by cutting on one side. It has a V-shaped groove on one surface and two faint grooves to one side of the more prominent groove. Overall it measures 29.1 by 14.2 by 18.4 mm with the largest groove 5.5 mm wide and 2-3 mm deep. It weighs 2.7 g.

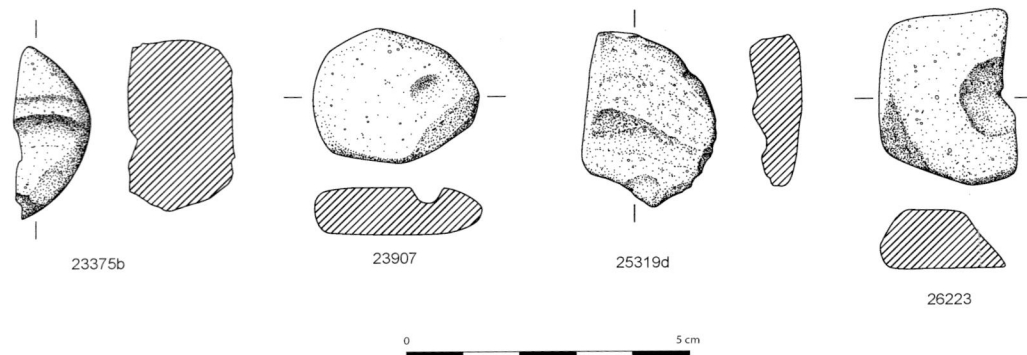

FIGURE 5.49:
Pumice artefacts from Phase B, SF 23375b, SF 23907, SF 25319d and SF 26223. Drawn by Leeanne Whitelaw.

SF 26223 (from BG24) is a rectangular, flat piece with trimmed edges, which split or broke during manufacture at an unfinished perforation. Both faces are smooth but one has a 9.8 mm diameter ground hollow which almost pierced the piece. It measures 31.2 by 23.1 by 10.4 mm and weighs 4.4 g.

Phase D (Figure 5.50)

SF 23853b is a large pebble, irregularly wedge-shaped with a relatively flat base partly worn smooth, and with a wide, deep groove in the middle of the broad end. The other surfaces of the pebble are natural. It measures 69.4 by 45.3 by 33.7 mm. The groove is 9.8 mm wide and c. 5 mm deep. The piece weighs 34.2 g.

SF 26621 is a large pebble with lower surfaces and sides smoothed and worn. The broad end of the tool is chamfered by rubbing to create a facetted surface with a defined edge. The opposite end of the same surface is polished smooth with wear. One side surface is especially smooth, facetted and worn with a sharp edge to the adjoining surfaces. The opposite side is also worn but less so. The upper surface and narrow end is largely natural, except for a groove with an irregular base near the end and edge of the tool. The piece measures 66.6 by 54.3 by 31.8 mm. The groove is 19.8 mm long, 3.6 mm wide and 3 mm deep. The piece weighs 48.8 g.

SF 23952b (DH building) is a fragment of pebble is abraded. Its curved edge is predominantly cut and shaped smooth. One surface is smoothed and possibly worn, as is one formerly broken end. This face is concave through use. It measures 39.8 by 31.6 by 18.9 mm and weighs 7.6 g.

SF 25497 (DH building) is a cushion-shaped pebble slice. Its two sides are broken but the surfaces and the rounded end are smooth. One surface to the narrow edge is very smooth indicating much wear. It measures 45.8 by 40.3 by 28.6 mm and weighs 25.7 g.

SF 25943a is a large, irregularly shaped pebble, partly abraded but with a worn lower surface, which is smoothed and slightly chamfered. Part of it is also slightly concave. The edges of the narrower part of this surface have been altered by cut marks and grooving. One edge has three prominent parallel grooves which continue onto the adjoining side. This tool is well worn. It measures 93.3 by 64.4 by 37.7 mm with the grooves measuring 8.2 - 8.9 mm in width by 2-3 mm in depth and a maximum length of 19 mm. It is one of the heaviest pieces weighing 69.6 g.

SF 26776 is an elongated pebble slice with all three surfaces worn. The upper convex surface is smooth and abraded with sharp edges to the lower surface. This surface is convex with much wear and is very smooth. The third surface is smooth and flat. It measures 53.3 by 27.9 by 19.2 mm and is 9 g in weight.

Discussion

As Newton has remarked (this volume) all the RUX6 pumice derives from one Icelandic source, but not necessarily from the same volcanic eruption. Larger pieces could have been picked up from beaches on the Udal peninsula in the late Neolithic allowing a wider utilisation of that resource. These pieces were most likely used to smooth wood including driftwood, or for the working of hides or leather, while smaller pieces were more likely to have been used for smoothing and polishing smaller wooden or bone items such as arrows, pins, spatulas, or even small

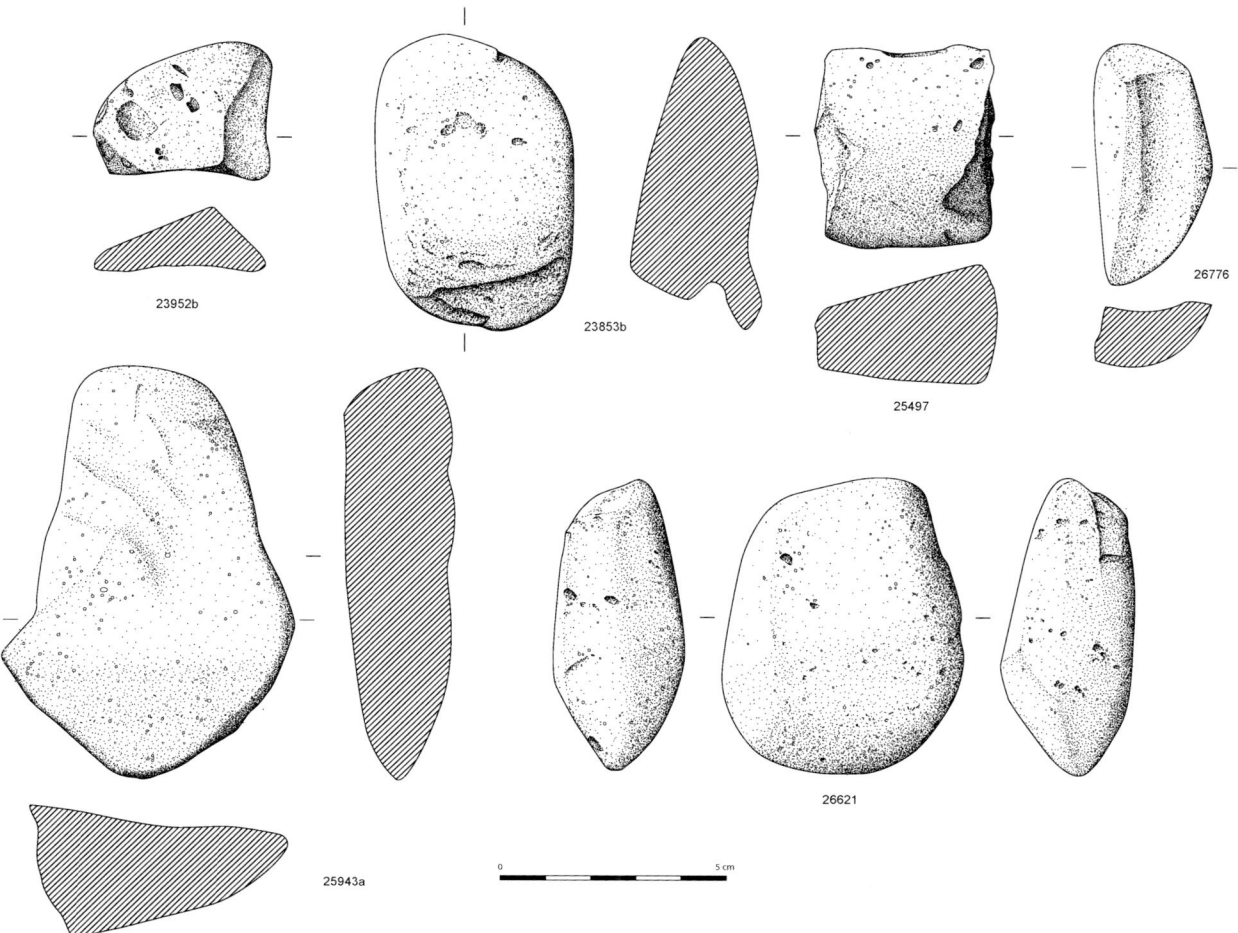

FIGURE 5.50:
Pumice artefacts from Phase D, SF 23952b, SF 23853b, SF 26621, SF 25497, SF 25943a and SF 26776. Drawn by Leeanne Whitelaw.

leather and other organic items. Grooving caused by the abrasion of twine does not seem to have occurred on RUX6 as it did at Allt Chrisal (Branigan, Newton and Dugmore 1995: 145, Figure 4.49), and none of the pieces were considered large enough to have been perforated and used as weights for fishing lines or nets. Small perforated pieces from RUX6 may have been intended as personal items, such as beads or pendants worn round the neck. Highly modified pumice pebbles may have also been used in the manufacture of pottery vessels (see Ballin Smith, this volume).

Interpreting the data is difficult, as worn pumice was found in the anthropogenic layers of Phase E, most likely originating from Phase D disturbance of the subsoil. The pumice recorded from Phases B and C could have derived from Phase D, as ploughing, construction and digging of pits disturbed the earlier settlement contexts. The largest number of pebbles (15) from Phase B are not associated with structures and could have been washed up or blown inland as a result of periods of sea inundation with redeposition of deposits. The pebbles from Phase A are likely to be more recent in date but they too could have originated from any of the earlier phases due to coastal erosion and redeposition. Ballin in his analysis of the quartz (this volume) argues that pieces of that material found in Phase B were likely to have been redeposited from Phase D. It is not inconceivable that the situation with the pumice pebbles is the same.

Almost two thirds of the pumice was unmodified and this suggests several things. It was a resource that was plentiful, possibly being washed up on beaches regularly, but it was not a resource that seemed to have been in great demand. For example, it did not appear to have been horded or cached. It, therefore, may have been an expedient resource that was used and probably discarded afterwards. Some of the larger pebbles from Phase D were kept and reused for much longer periods, probably forming parts of tool kits with specific purposes. The contrast with the occurrence of smaller pebbles in Phases A and B is stark, which indicates the changing patterns of land use, the movement away of settlement from the coast to inland areas and the building of ritual monuments.

Pumice is frequently found on coastally situated sites occupied in prehistory where the ratio of unmodified to worked pieces is consistently around 2:1. At Northton, Harris, most pumice was unmodified from Neolithic levels, but five pieces had grooves and seven had flat surfaces. Several pieces were also found in the Beaker horizons, again with flat sides or a groove (Gregory 2006b, 133). Several slightly modified pieces were found at Rosinish, Benbecula from the Beaker settlement (Shepherd 1976: 210), and three pieces of modified pumice were found in the Neolithic settlement horizons at Loch a' Choire, South Uist (Henley 2012: 192, Figure 9.5). A smaller number of pumice pebbles, but with a larger weight, were found in Neolithic and Bronze Age levels at Allt Chrisal, Barra, where a little over one-third of the collection was modified (Branigan, Newton and Dugmore 1995: 144). Pumice was also frequently found on later settlement sites down through Harris, the Uists to Barra including Cladh Hallan (Parker Pearson et al. forthcoming), but rare pieces have been also been found at Barabhas, Lewis (Torben Ballin pers. comm.).

The occurrence of pumice pebbles on mainly coastal settlement sites of Neolithic, Bronze Age and of later date in Orkney and Shetland is well known (see Clarke 2014: 183), indicating its widespread dispersal and utilisation. The use of pumice at RUX6 is consistent with other contemporary settlements, as it was a useful and easily won resource in a landscape where other raw materials were often scarce, in short supply or were difficult to find.

Stone tools and other items in stone

By Beverley Ballin Smith with additional information by Torben Bjarke Ballin

Introduction

All the stone finds from the site were initially assessed and reported on by Ann Clarke (1997) at the request of Iain Crawford, but over the last twenty years there have been significant changes to the composition of the collection. As with all finds and samples reassessed from all the Udal sites between 2010 and 2012, some artefacts have disappeared from the record, and others that were originally considered missing, or not identified during the compilation of the initial site record, have been found and added to the database. The disappearance of at least 34 samples of stones and ten artefacts can be accounted for by several possibilities. Some stones may have been sent to other specialists and not returned, or put in a 'safe' place from which they have not reappeared, or were simply lost during storage. Significantly large and heavy finds were left on site but for this collection in particular, the geological composition of some finds, combined with storage conditions, has led to their disintegration (see below).

Reconciling Crawford's, and therefore Clarke's original lists, with current finds has not been entirely successful. From a total of 737 [21] recorded finds of stone (not including quartz or flint) Clarke recognised 27 artefacts in her catalogue but only 13 of those present have been re-identified. In addition, the Great Auk Stone as it was named on site has been included on the database with some of its packing stones when it was returned to the collection in 2014. It had been housed in the National Museums Scotland former store in the Custom's House, Leith, after being displayed in the British Museum, London from 3 July 1986 to 15 February 1987 in the exhibition *Archaeology in Britain: new views of the past*. Two of the several pinning stones that helped to secure it in a vertical position are worked or used pieces, and are discussed below. All its worked and unworked pinning stones, like the standing stone itself, have been kept as part of the story of the site, and together they form a piece of composite architecture, that can be redisplayed in the future.

The fragmentary second standing stone SF 26910, associated with the Bronze Age cairn and the BG24 structure, was left on North Uist but two unworked packing stones were retained.

The remaining 91% of unworked stone (Table 5.25) is an issue in terms of reanalysis and storage, and the questions are why was it collected on the site and why was it retained after examination in 1997? An answer to the first part of the question is that it was site policy that if anyone working on the excavation was unsure if stones were worked or not, they were

---

21 The number has varied considerably since the end of the excavation.

kept. They were retained because some looked like 'choppers' or 'bashers' (terms used on site), points, hammers, axes or something else. However, there appeared to be little general understanding of the complexities of the underlying bedrock, how it was formed and how it fractures, and the types of other stone occurring naturally near the site. Many of the unworked pieces of light-coloured gneiss looked like handles to tools, or flat-bladed objects broken off from the bedrock, hence the descriptive terms used in the record. Crawford thought that there may have been an industry on the site, based on his observations of the collected gneiss pieces, that produced specific forms of artefact, like the 'choppers', but there was no evidence for this. Rough and smooth types of gneiss were identified (Graham 1970 in Fettes et al. 1992: 45-46) and both were collected as unworked stone samples or tools. Other examples of local stone, and stone derived from further afield, had found its way onto the site from the coast by natural or by human agency. Cobbles and other smooth stones of basalt/dolerite, diorite, dark-coloured gneiss and the occasional sandstone were used unmodified for the construction of buildings and the erection of at least two standing stones. Most of these stones types are extremely hard and several, described below, were also used as artefacts.

	Artefacts numbers	Missing artefacts	Unworked stone and samples	Missing stones and samples	Total
Number	22	10	671	34+	737
Percent	3%	1.40%	91%	4.60%	100%

TABLE 5.25:
Stone artefact and samples numbers

The retention of the unworked stone, as with many other materials, was done deliberately to maximise the amount of information recorded from the site. Crawford posed questions to one archaeological specialist, and if they were not adequately addressed, he would ask another, and he kept the stone samples for that purpose. He would no doubt have had other thoughts, such as whether the find location of a stone was significant, irrespective of whether it was worked or unworked. This included the many unworked stones found on or within the floors of the buildings. Some larger stone undoubtedly came from the walls of the structures during use and after abandonment, but small pebbles, often in interesting colours or shapes may have been deliberately brought into buildings by their occupants. Crawford presumably wanted to explore the occurrence of stone as well as its possible function. However, as well as addressing those ideas as far as we can, we now have to ask questions such as, is this worked, and if not, why are we keeping it? A balance has to be struck between the original intentions of the excavator and use and future function of a collection.

The raw materials

The need for durable construction resources and specific robust tools was largely met by stone as a raw material. Stone was the only resource identified during the late Neolithic and Bronze Ages that was available, whether as cobbles or boulders from the beach, layers of bedrock, or veins or cobbles of quartz (see Ballin, above), small pebbles of flint (see Wickham Jones above) or pumice pebbles (Ballin Smith above). The importance of stone for the survival of communities in this island setting during prehistory cannot be overstated. Every occupation from hunting, butchery, fishing, harvesting to cooking would have required the input of tools made of stone.

Gneiss (in various colours), dolerite/diorite (in various grades), granite and light weight basalt (usually with vesicles) were the predominant stones used for tools, and all were available locally. Where folded fragments and slabs of light-coloured gneiss had broken away from the bedrock they were used for building purposes. An unusually large flattish slab was chosen for one standing stone and an irregular pillar of gneiss was used for another. The main source for all the other stones came from nearby beaches, where smooth, waterworn cobbles and larger stones were collected for tools. Outcrops of glacial deposits in low coastal cliffs may have also contained suitable tool-sized cobbles or pebbles that were easy to collect.

The inhabitants of RUX6 made good use of their often limited resources, and where possible specific stones were chosen for particular uses. Hard stones such as diorite or dolerites in various grades and basalts were chosen for tools that were used primarily for hammering, cutting and polishing, although occasional amphibolites were also used for these purposes. The blacker varieties of gneiss were much more successful as tools than paler types as they were harder, heavier and had more compact and dense structures.

The use of gneiss, the most common resource, was

dependent on its texture as well as its hardness, as frequently it was more suited to grinding and pounding, than hammering. Other factors in addition to hardness and functionality influenced the choice of rocks. The dense texture and dark blue/black colour of pseudotachylite contrasted with the paler colours and coarse textures of gneiss, and may have been chosen for those qualities as well as its sharp edges when struck. The range of colours of stones, as well as their banding, for example of gneiss and the patterning of other stones, may have appealed to the tool user for aesthetic reasons.

The paler, coarser varieties of some gneiss tends to break down or decay into its constituent parts of quartz and feldspar (degrading to sand and clay), once the inner core of the cobble or stone is exposed through shattering or erosion of its external cortex. Several finds bags contained little more than sand and dust where the original artefact had crumbled away beneath surviving fragments of harder cortex. Although Crawford used some conservation techniques in his *Chrystal Palace* (the on-site finds processing area), they were not sufficient to save some of these stones. The natural properties of some gneiss meant that cobbles of it had to be carefully chosen in prehistory for successful and lasting tool production. But this was not always the case as the presence of degraded gneiss objects demonstrate.

Methodology of analysis

All the stones were identified to geological type, where possible, and were examined for signs of surface alteration and weighed. Stones that had no surface alteration, including naturally split cobbles or pebbles, were classified as unworked. All other stones (classified as worked) were measured and their attributes recorded in detail.

Description of artefacts and possible artefacts

The artefacts are described below in Small Find (SF) number order.

Possible hammerstone SF 23237

This is a rounded, flattened and predominantly smooth cobble of granodiorite/diorite measuring 148.5-150 mm in diameter and 36.2 mm in thickness. The most noticeable characteristic of the stone is the rough scars from one small and one large chip that have flaked away from the tool when it was struck at one point of its circumference. There may have been a natural flaw in the stone at that point as two small scars, now smoothed on the reverse surface could have been the result of natural damage from being in the sea. The scaring at this point may have occurred by the stone being banged rather than it being used as a hammerstone. Both surfaces of the stone are smooth and flat to very slightly concave, but there are no definite indications that this stone was used as a tool. It is a possible artefact that was found in Phase A within the slight cliff caused by coastal erosion. The excavators considered it to have come possibly from a floor of a Neolithic building.

Axehead SF 23762 (Figure 5.51)

This object was found within the intra-mural packing of Neolithic Building 1 (DJ), at coordinates 337.8/733.5. It is a small symmetrical stone axehead, which measures 106 by 65 by 29 mm. Its raw material is likely to be dolerite, or possibly metadolerite which are common rocks on North Uist (cf. Fettes *et al.* 1992: 81). It has been polished all-over, but most finely towards its cutting-edge. The characterization of this piece follows the methodology defined in connection with the classification and description of the North Roe felsite axeheads from Shetland (Ballin 2015b).

The blade is 'waisted' in the sense that its lateral sides curve slightly inwards, rather than following straight or slightly convex lines. At a first glance, the axehead seems to be a four-sided piece, where most British axeheads tend to be two-sided with approximately oval or pointed-oval cross-sections, but the sides of this implement curve gently into the broad surfaces without forming facets. The sides have only been lightly polished, and the original pecking that shaped the piece is still visible. The two sides almost meet at the butt-end, and the piece is approximately drop-shaped, although the butt has been cut off, rather than finished off neatly. The thickest part of the tool is near the cutting-edge. In terms of its general shape, the axehead corresponds well to the piece illustrated by Evans (1897: figure 80), found near Cottenham, Cambridgeshire. The axehead most likely dates, as indicated by its recovery within a Neolithic building, to that period.

Although the object is visually pleasing, it may still be a discarded functional piece. Towards the butt-end, both faces display damage from having been

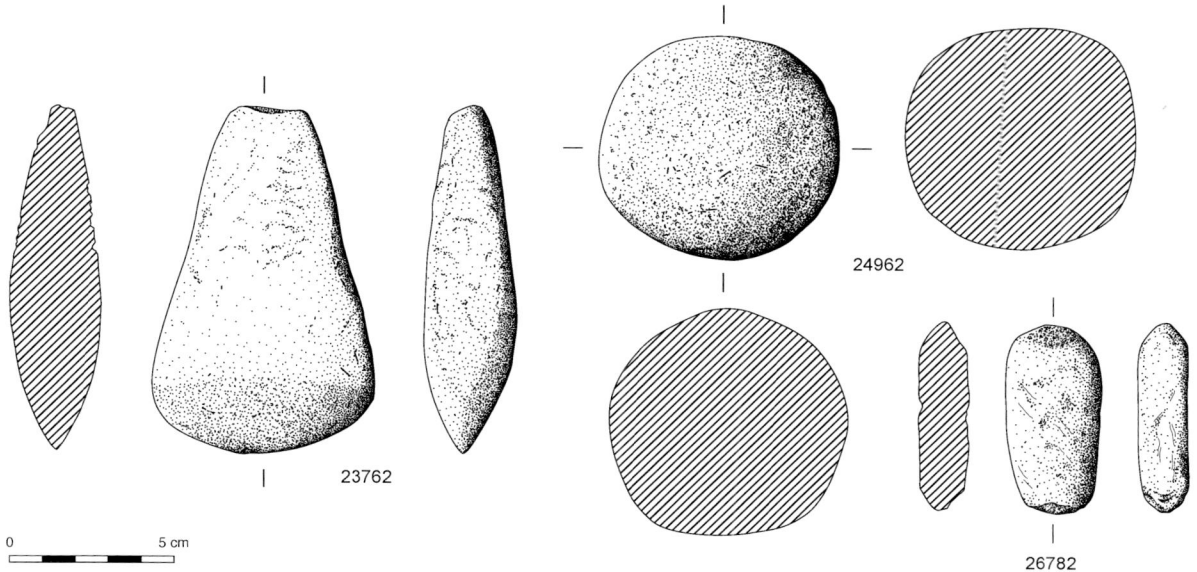

FIGURE 5.51:
Axehead SF 23762, stone ball SF 24962 and quartz percussoir/anvil SF 26782. Drawn by Jo Bacon and Leeanne Whitelaw.

used as an anvil, most likely in connection with bipolar knapping of small flint pebbles or quartz. It is uncertain how this object ended up within the wall core of Neolithic Building 1, but two options are available; the piece was simply waste and it was 'shovelled' into the wall core with other rubble; or it was deliberately and carefully deposited and therefore represents symbolic behaviour. It was either concealed as a charm to protect the building or, even today, was part of the process of putting down a commemorative foundation or corner stone of a new building.

Stone ball SF 24962 (Figure 5.51)

This object was found in the site's upper levels, Phase A, and is basically uncontexted and therefore unstratified. It is an all-over pecked stone ball, in contrast to, for example, polished stone balls (e.g. the piece from the Crantit tomb, Orkney; Ballin 2014c) and carved stone balls (cf. Marshall 1977; Clarke *et al.* 1985, text and catalogue), and with a diameter of 68-72 mm it is almost spherical. Its raw material is probably metadiorite, microdiorite or granodiorite (Pellant 1992: 187), which forms dykes across the southern parts of the Western Isles, although most notably on Barra (Fettes *et al.* 1992: 41).

Although the piece is almost spherical, it has barely visible, rounded facets between a number of domed faces. Seen from one direction it is practically spherical; seen from another angle, it appears to have four domed faces; and from a third angle, five domed faces. The question is, whether this form is a result of the way it was shaped, and possibly reflects the motor habits of the craftsperson responsible for the creation of the piece (i.e. how he rotated the object while pecking it), or whether it reflects intentions (a mental template) regarding a future final product, for example, if the next step of the creative process would have been to carve intricate patterns into the ball. In the first case, the piece would then probably be the final object, in the second case, a preform.

The ball has been pecked all over its surface in a highly regular manner, and it is unlikely to have been used, for example as a hammerstone. In comparison, the originally polished stone ball from the Crantit tomb had been used extensively for knapping or hammering, which had removed almost all its polished areas, and which had left it with a highly scarred and pitted surface.

By association with similar objects this stone ball is likely to be a Neolithic object that through natural wind erosion of the sands capping the Neolithic settlement, by coastal erosion, or most likely through nineteenth century or later disturbance, was removed from its original context.

Worn stone fragment SF 23302 (Figure 5.52)

This piece is the tip, broken off a larger cobble of vesicular basalt which weighs only 45 g and measures

48.8 by 38.8 by 23.3 mm. It is semi-circular in plan, flat on one surface and slightly domed on the other, which has also lost half its cortex or outer layer. One side of the stone is worn to a clear edge from both surfaces through the action of using it as a rubber. The surface areas close to this edge are smooth. Lime, probably from weathered crushed shell sand, is noted in some of the pores on the surface of the stone. The piece was found with pottery in feature BI identified as a pot pit in Phase B.

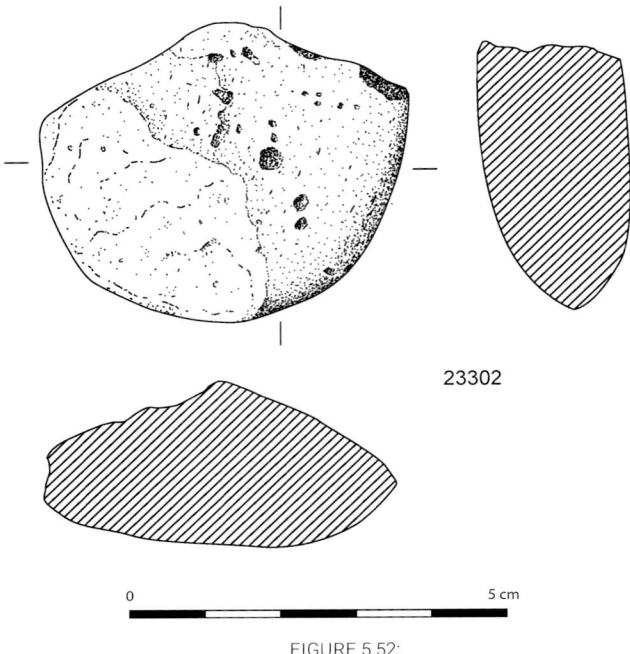

FIGURE 5.52:
Worn piece SF 23302. Drawn by Leeanne Whitelaw.

Pebble polisher SF 23324

This small pebble measuring 32.2 by 26.3 by 41.1 mm is possibly basalt. There is no evidence of changes to the pebble surface except for a worn edge noticed on the flattest surface of the stone. It came from the pot pit (BI) in Phase B with SF 23302 (above).

Cobble fragment SF 24982a

This is a fragment of the tip of a split cobble of basalt which measures 50.5 by 46 by 22.3 mm. Only half the piece has its cortex surviving, the remainder is rough. The shoulder has had a flake removed during the splitting of the cobble but there are no further signs of the piece being worked. It was found in Phase A in the uppermost contexts of the site.

Indeterminate flake SF 24982b

This indeterminate flake is heavily sand-blasted but measures 24 by 14 by 6 mm. Due to the abrasive nature of the machair environment (aeolian activity) affecting the upper levels of the site in which it was found (Phase A), all original surface features have been rounded and obscured. The raw material may be either pseudotachylite or dolerite (Fettes et al. 1992: 136; Pellant 1992: 192). It is uncertain which percussion technique was applied to produce this piece.

Indeterminate flake fragment SF 25036

This distal fragment of a tertiary indeterminate flake measures 26 by 29 by 12 mm and the raw material is possibly granodiorite (Pellant 1992: 187). The piece was found in Phase A in the upper layers (IX) of the site.

Core or indeterminate coarse stone tool SF 25165b

This piece is either a fragment of a core or an indeterminate coarse stone tool that measures 53 by 48 by 39 mm. The parent piece may have been a dolerite cobble, as the object has surviving abraded cortex. A large thick flake was apparently struck off this nodule, and a number of medium-sized flakes were then detached from two opposed edges by using the ventral face of the original thick flake as a striking-platform. One end may have broken-off. It was found in Phase B in layer IX.1/2.

Tertiary flake SF 25495

This tertiary hard percussion flake, measures 48 by 33 by 16 mm. The raw material seems to be feldspar (Pellant 1992: 171), although it was practically never used for flaking by prehistoric people. However, the surfaces and edges of this piece are highly irregular, indicating an attempt to make this into a tool. The interesting information to be gained from this piece is that some of the gneiss or granite from the local area had exceptionally large feldspar (and probably quartz and mica) crystals, in this case exceeding 50 mm, i.e. approaching migmatite or pegmatite in nature (Pellant 1992: 185). It was found in Phase D, within Neolithic House 2 (DH).

Hammerstone/anvil SF 25977

This is a cobble of basalt/dolerite broken across its width that measures 124 by 75.8 by 41.3 mm. The flatter of its two surfaces is lightly chipped at the broken edge on one side and more prominently at the surviving cobble end, where it has been used repeatedly as a hammerstone and where the scars are most noticeable. In the middle of the surface and on the broken edge there are areas of pitting with small surface scars indicating the use of the tool as an anvil. The other surface is more domed and uneven but it too is lightly pitted in one definite and two more diffuse areas, demonstrating that the tool had been turned over and used again as an anvil. Given the number of quartz tools, it is possible that this stone was used for their manufacture. It was found in contexts to the north-east side of Neolithic Building 2 in Phase D.

Small hammerstone SF 26031

This irregular shaped pebble is flattened and slightly elongated. It is of porphoritic and vesicular basalt and measures 70 by 45.2 by 16.5 mm. The narrow end of the pebble is slightly chipped with flaking scars on both faces. The tool was found on the floor (floor 1) but east of the hearth in Neolithic Building 2, Phase D.

Grinder SF 26211 (Figure 5.53)

A large, pale coloured, banded gneiss cobble was used as a grinder. The stone is oval in shape and measures 148.8 by 112.7 by 66.2 mm. Evidence for the use of the cobble has survived at both its ends but the sides of the stone have suffered from heavy abrasion, and there is clear evidence of granulation and disintegration of the core due to loss of the cortex. The broader end of the tool has facetted wear but only a small amount of similar wear survives at the other end. The loss of the stone's sides and the removal of its cortex indicates that the tool was larger and that it was used all around its circumference for grinding, perhaps with a quern. It was found in Phase B, context BM2 where it might have formed part of the cairn capping the cist associated with structure BG24). If this was indeed the case, the stone may have been reused after being discarded during the late Neolithic/early Bonze Age.

Hammerstone/pounder SF 26340

This rounded and irregular but almost wedge-shaped cobble of dark coloured, banded amphibolite was originally used as a hammerstone and pounder. It measures 113 by 103 by 69.6 mm. The widest part of the stone is also the thinnest, and it is on this part of the circumference that there is a circular area c. 33 mm in diameter from it being used as a pounder. The opposed part of the tool has moderate scaring on its edge and both faces from being used as a hammer, and there is evidence that the scaring continued down one side. Like some other tools in this collection, it was reused as a convenient packing or pinning piece for the Phase D Great Auk stone.

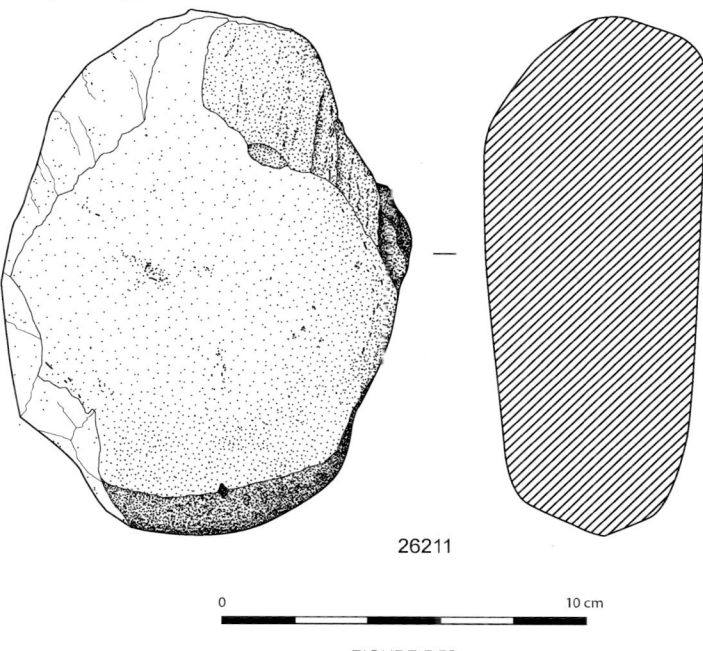

FIGURE 5.53: Grinder SF 26211. Drawn by Leeanne Whitelaw.

Possible hammerstone/grinder SF 26421

This piece is an irregular shaped cobble of pink banded gneiss measuring 113.5 by 68.2 by 40.5 mm. The slightly narrower end has a broad scar possibly from its use as a hammer. The opposed broader end is more complex with some evidence of faceting, with one small and one larger scar from hammering. The tool was found in Phase C, in level IX.3 but to the west of the Phase B structure BG24. Given the high disturbance of the site during Phase C, this tool may have also derived from Neolithic contexts.

Hammerstone/anvil SF 25519

This cobble of basalt/dolerite is sub-rounded and measures 125 by 111.5 by 57.2 mm. The wear on the stone is confined to small discrete areas of fine scaring round its circumference and to one area on one surface. The latter has fine linear incisions suggesting the material struck there had a sharp edge, or that it moved when struck creating linear indentations. It was found within a metre of the saddle quern SF 26508 in structure BG24 in Phase B. Additional information suggests it was used as a chock or packing stone in floor 1 or 2.

Pounder SF 26520

This is a slightly elongated cobble of light coloured gneiss which measures 128.5 by 76.3 by 58.3 mm. Both ends of the stone have areas of light scaring and flaking from the use of the tool as a pounder or light hammer. This tool was one of several that were reused as pinning or chocking stones around the Phase D Great Auk Stone to secure it vertically in the ground.

Indeterminate flake SF 26657

This is a medial-distal fragment of a primary indeterminate flake, measuring 40 by 30 by 8 mm. The raw material is most likely amphibolite (Pellant 1992: 215). It was found in Phase D, in level XI.2 of the Neolithic settlement.

Hammerstone SF 26686

This is a rounded but triangular-shaped cobble of light coloured gneiss that measures 102.6 by c. 78.5 by 73.3 mm and fits well in the hand. The two opposed ends of the triangle have light scaring from its probable use as a hammerstone rather than a pounder. It too was reused as a pinning stone for the Phase D Great Auk Stone.

Pebble polisher SF 26699

A pebble of probably pseudotachylite was used for a small tool is that measures 26 by 21.5 by 13.2 mm. It is black in colour and shiny in appearance, and one surface is flat to very slightly concave where it has been worn smooth. This is due to it being used to smooth a less hard substance, such as wood, leather or pottery etc. It came from Phase E, a mixture of natural and disturbed contexts beneath the Neolithic settlement.

Hammerstone SF 26887

This large cobble is of amphibole and measures 158 by 92.5 by 72.3 mm. One third of the surface has been removed by its heavy use as a hammer at the narrow end of the cobble. Flakes may have also been removed at its other end, but this is less certain. It was found in association with the whale bone vertebra situated over the Phase D shaft (DC), suggesting it was a packing stone that supported the whale bone.

Great Auk Stone SF 26904 (Figure 5.54a and b)

The wedge-shaped projecting head of this dense, pale piece of banded gneiss emerged during the last year of excavation in 1984 followed by the shaft of the stone. Its shape appeared to Iain Crawford to resemble an extinct Great Auk, and it retained that name on site. The 'impressionist zoomorphic' piece remained in situ, pinned by eleven packing stones SF 27027 (seven unworked pieces), SF 26685 (a single unworked stone), SF 26520 and 26340 (hammerstones/pounders) and SF 26686 (hammerstone), until it was excavated.

The maximum measurements of the stone are 0.70 m (length) and 0.43 m (width) with the head projecting c.140 mm beyond the shaft. The profile through the shaft is lozenge-shaped, with the natural angles of the stone representing the front, back and sides (wings) of the auk. The base of the stone is thin and tapers to a width of only 70 mm. Crawford considered the stone to have been worked but its lack of symmetry and its overall shape is most probably the result of natural weathering. For example, the left side of the stone has lost some surface cortex due to spalling of the rock where a natural fault comes to the surface. However, some evidence of working of the stone is demonstrated near its base, which is waisted by flaking. Five flaking scars are also noted on the right side of the front edge from below the head towards the base. There are also areas on the stone where the edges are sharp indicating some trimming, such as on the tip and lower edge of the head on the right-hand side, and on the left side of the bottom edge of the stone.

Missing stones identified as artefacts on site

Saddle quern fragment SF 26508

This piece is now missing from the assemblage but it was recorded as part of a large saddle quern formed from a slab of gneiss with maximum dimensions of 380 by 300 by 80 mm. Its base was highly abraded, its upper surface was concave and it measured c. 30 mm in depth at the broad, flat rim edging the stone. This rim measured 120-220 mm in width. It was found in a wall associated with floor 1 of structure BG24 in Phase B.

Standing stone SF 26910

A slab of foliated gneiss from the exposed bedrock was left in North Uist. It was the lower part of a

FIGURE 5.54a and b:
Great Auk stone SF 26904.

standing stone supported by a plinth of stones and turf closely associated with structure BG24 in Phase B. It measured 100-150 mm in width, 50 mm in thickness, and measured 0.6 m from its base to its surviving height. Given the robust construction of its plinth Crawford considered the stone would have been approximately 3 m in height, but that it had been broken off at ground level during the nineteenth century, when the trench for a saw-pit broke the stone (see PART 2). Two of its unworked packing stones SF 26538 and SF 26569 have been retained.

Discussion

The number of stone artefacts from the Neolithic and Bronze Age phases summarised in Table 5.26, is disappointingly small compared with the vast numbers of quartz objects recovered (Ballin this volume), but the simple explanation for this is that quartz was the preferred stone for tools, because the raw material supply was plentiful. However, that may not be the whole picture. Gneiss cobbles were also exploited, but the reuse of a grinder and a pounder in this material as wedges or packing stones to support the Great Auk stone, suggests the unsuitability of some forms of gneiss for prolonged use. Some of the hardest rocks like basalt, dolerite, diorite and amphibolite were used for a polisher, a hammerstone/anvil, hammerstone/pounder, grinders and the two 'special' objects - an axehead and a stone ball. Most were from Phase D, the late Neolithic, but others tools in these materials found in Phases A and B are most likely derived from disturbed levels of Phase D. This suggests that a greater variety of raw materials were used for tools in the late Neolithic and especially for grinding and hammering, some of which were associated with Building 2 (DH) and one hammerstone with Building 1 (DJ). The occurrence of indeterminate flakes or fragments in Phase A may have resulted from the use of easily available cobbles for hammering in wooden posts (see Botanical remains this volume) and other more recent activities on the machair.

Of particular interest are the stone ball and the axehead, described above. The finding of the axehead in the wall of Building 1 implies a deliberate act of deposition. Both of these objects are typical of other late Neolithic artefacts occurring on settlement sites and in connection with funerary monuments. They are also examples of a common culture found across Scotland and the Scottish island groups. One of the closest sites to RUX6, with similar artefacts is that of Eilean Domhnuill in Loch Olabahat. A number of stone balls, mostly plain, but one incised and others abraded, were found throughout the excavation sequence there. An imported polished greenstone axe was found in the earliest phase (Armit 1988, 22-25), but further work at this site (Armit 1990, 16) produced a large range of tools, including grinders, querns, hammerstones and an imported miniature stone axe, again from the lowest phase but with Unstan ware bowl fragments confirming that the stone tools were late Neolithic in date.

The Neolithic levels at Northton, Harris, produced only two possible cobble artefacts, one being a likely hammerstone (Gregory 2006a, 25-26), while the later Beaker levels produced a flake of a polished serpentine axe and a polished mudstone axe, beach pebble pounders and a mudstone flake (Nelis 2006b, 136). The paucity of larger stone tools from this site in comparison with flint and quartz mirrors that of RUX6.

Tool type	SF Nr	Rock type
Phase A		
Cobble fragment	24982a	basalt
Hammerstone	23237	granodiorite or diorite
Indeterminate flake/fragment	24982b	pseudotachylite or dolerite
Indeterminate flake/fragment	25036	granodiorite
Stone ball	24962	metadiorite, microdiorite or granodiorite
Phase B		
Core or indeterminate tool	25165	indeterminate
Grinder (BM2)	26211	gneiss
Hammerstone/anvil (BG24)	26519	basalt/dolerite
Pebble polisher	23324	basalt?
Worn stone fragment	23302	basalt
Phase C		
Hammerstone/grinder	26421	gneiss
Hammerstone/anvil	25977	dolerite/basalt
Indeterminate flake/fragment	26657	amphibolite
Phase D		
Grinder - Building 2 (DH)	26031	basalt
Grinder - libation pit (DC)	26887	amphibolite
Grinder - Gt Auk packer	26686	gneiss
Hammerstone/pounder - Gt Auk packer	26340	amphibolite
Pounder - Great Auk packer	26520	gneiss
Stone axehead - Building 1 (DJ)	23762	dolerite or metadolerite
Tertiary flake - Building 2 (DH)	25495	feldspar
Phase E		
Pebble polisher	26699	pseudotachylite

TABLE 5.26:
Summary of tool types by stone type, SF number, and phase

Bronze Age excavations in South Uist produced equally few stone tools. At Cill Donnain (Parker Pearson (2014: 35), only three worked stones were recorded, one of which was a hammerstone. Work by Hamilton and Sharples (2014: 205-207, Fig 10.7, 1 and 2) at two early Bronze Age settlements at Machair Mheadhanach and Cill Donnain in South Uist, produced a hollowed stone or anvil and a rare early Bronze Age battle axe, which could have been imported. In contrast, the assemblage from early Bronze Age levels at Kilellan Farm, Ardnave, Islay, produced 60 stone tools (Clarke 2005: 133-141), which included hammerstones, cobbles, flakes and grinders.

It has to be considered that the stone tool assemblage is largely a product of taphonomic conditions (the disturbance of earlier layers by later activities) and erosional and depositional factors caused by coastal and wind erosion. It was evident from the Phase B cairn that stone robbing took place in Phase A, and a similar situation may have occurred during Phase B and C activities, which dug down into the late Neolithic levels and may have removed stone (including stone tools) in the process. Coastal erosion has also played its part, as we do not know how much scouring of deposits removed archaeological evidence from the late Neolithic and early Bronze Age. The site evidence for Phase C included ploughing scars (Figure 2.23), but there is no record of any stone ard point being found at the site. A saddle quern from Phase B was recorded but the piece was left near the excavation and can no longer be found. We know from other samples (bone) and artefacts (quartz) that sand abrasion was suffered by much of the collection. The question remains as to whether sand blasting was so severe that during the excavation other stone tools may not have been recognised because of significant abrasion.

The prehistoric pottery

by Robert Squair and Beverley Ballin Smith

This is a reworked, updated and expanded version of the report produced by Robert Squair in 1998 that formed part of his PhD.

Introduction

The excavations at RUX6 produced a substantial quantity of prehistoric pottery, but much of this comprises very small fragments. The processes of wind, sea and sand, as well as human activities on the site, combined to break the pottery down into smaller and smaller fragments. Nevertheless, through the analysis of the collection, distinctive vessel types have been identified from the late Neolithic and early to middle Bronze Age (Phases B, C and D). This assemblage adds further to our knowledge and distribution of typical Hebridean pottery styles of the time.

Methodologies

All pottery from the site was recovered by hand or through on-site sieving. Taphonomic conditions affected the survival of sherds (see Post-depositional changes, below) but the deeper the pottery was recovered from the excavation the more friable and distorted it became. Many of the pieces were conserved with PVA either on site or in the finds hut (Crawford's *Chrystal Palace*), in order to preserve them. Unfortunately, some of this pottery could not be analysed in its entirety because of the fusion of sherds, sand and PVA.

The pottery was generally bagged on site after its recovery, and each sample was given a unique small finds number (SF no.) and recorded with its stratigraphy, date of retrieval and and sometimes its grid coordinates (see General site methodologies). Where it was possible to remove adhering sand grains, sherds were gently brushed as they were too friable to be washed. Prior analysis of the assemblage was also attempted in the finds hut, with an initial sorting of sherds from the same contexts into possible vessel groupings, i.e. those with a similar appearance.

In spite of the care taken with the packaging of the pottery, many sherds broke in storage accounting for a substantial amount of small pottery fragments and dust. During specialist analysis the majority of sherds larger than 10 by 10 mm were individually bagged to prevent further abrasion. The remainder, the much abraded, undiagnostic small sherds or fragments were bagged together. All sherds are identified by small find number and boxed by phase.

The information on the pottery was entered initially onto record cards, and later into a Microsoft Access database. Vessel numbers (with a letter for the Phase followed by a number) were allocated during the post-excavation analysis in order to identify similar sherds from within contexts and also phases. Analyse and identification of the sherds was undertaken using standard diagnostic criteria. Similarities in production, style and context were also used to ascribe sherds to vessels. Refitting was almost impossible to achieve because of the generally small size of sherds and the effects of post-depositional abrasion on them. Some sherds demonstrated there was more variation of manufacture within the same vessel than between individual vessels. However, the identification of vessels represents an interpretation of the pots recognisable in the assemblage and in phases (see Catalogue, Appendix 2).

The small size and high abrasion of many sherds prevented measurement of rim diameters and percentages, and identification of the profile of vessels on all except for the larger fragments. In general, the shape and function of many vessels remains unknown or uncertain. Since the analysis of this assemblage by Squair in the 1990s, the Phase A pottery has been lost, and discussion of it is only possible because of Squair's report.

Analysis and description of the pieces

The assemblage comprises 6244 sherds, small fragments and dust, weighing a total of c.15.6 kg. It contains generally small, severely weathered or abraded, friable sherds. The average sherd size is less than 20 by 20 mm, and the average sherd weight is less than 3 g. Details of sherd thicknesses and weights can be found in the Catalogue (Appendix 2).

Sherd type	Number	Percentage	Weight (g)	Percentage
Rim	158	2.5	1009	6.4
Neck	18	0.3	177	1.1
Shoulder/carination	50	0.8	425	2.7
Body/indeterminate	5235	83.8	10166	65
Base/base edge	44	0.7	1384	8.8
Fragments and dust	c.737	11.8	c. 2487	15.9
Total	6244	99.9	15650	99.9

TABLE 5.27: Composition of the pottery assemblage.

The composition of the assemblage is tabulated in Table 5.27. The majority of it comprises body sherds with a large amount of indeterminate fragments with dust. The occurrence of other sherd types (rims, bases, shoulders and necks) account for 19% by weight, but only 4.3% of the assemblage by number, which is low for an assemblage of this size.

The raw materials

The raw materials necessary for pottery production are all available in the immediate locality of RUX6. The glacial till on which the site was located was a readily available source of clay suitable for pottery manufacture. Within the Phase D Building 2 (DH), a vertical shaft provided access to the glacial till lying beneath the building's floor (See PART 2; Crawford 1981: 4). It is not unreasonable to speculate that some of the vessels represented in the assemblage, particularly those from the late Neolithic, were manufactured with clay from this source. The various minerals and rocks forming the temper in the pottery derive from the local geology and from the weathering of gneiss that outcrops in the vicinity (cf. Brown nd.).

In this assemblage, it is generally impossible to distinguish between inclusions occurring naturally in the clays and temper added deliberately to it to make it more workable, except for shell fragments. The uneven distribution and size of rock temper within SF 26192 (Vessel B65), for example, demonstrates the inconsistency of composition of the coils. However, the analysis suggests that some temper was added deliberately (cf. Gibson 1995: 100). Indeed, there is some evidence to suggest that the density of temper within a fabric was altered to facilitate the manufacture of different parts of the vessel. For example, SF 25784 formed the base of Vessel B06, and contains a higher proportion of temper than the pot's corresponding body sherds.

The analysis of the assemblage suggested there was little variation in clay and temper composition. Although Squair subsequently identified nine fabrics at a macroscopic level, there was little hard evidence to indicate this was more than different mixes of the same or similar raw materials. Temper added to the clay included shell, burnt shell, organic matter, quartz, quartzite, feldspar, mica and various other unidentified minerals or rock including large fragments of stone.

The evidence for the addition of vegetable temper to the clay and stone raw materials was noted in a number of sherds from Phase D, including those comprising Vessels D50, D85 and D88, but due to the condition of the pottery, and the large numbers of sherds with carbonised food deposits, this could not be explored further. Organic matter in the form of chopped grasses, possibly straw or even hay, was generally added to clay to improve its plasticity from the early Neolithic in Scotland. In places such as Iona, this practice continued into the Iron Age (Ballin Smith forthcoming). The elongated negative impressions or voids on the surfaces of some sherds are the result of the organic material burning away during the firing process.

The most significant change in the raw material composition of the assemblage lies at the boundary between the late Neolithic (Phase D) and the early Bronze Age (Phases B and C). Pottery containing larger pieces of shell temper dominates Phase D, whereas pottery with rock inclusions is more prevalent in Phase B, although shell as a temper was also present. This difference could represent a change in the availability of shell sand, or the potters' preference for rock temper. It is also evidence of a degree of experimentation in the manufacturing process, where rock-tempered thicker pot bases were found to last longer and perhaps have better thermal qualities.

Post-depositional changes

The physical properties of the assemblage are difficult to assess when the vast majority of sherds, 76.5% by number and 59% by weight, are small, abraded, possibly concreted, and have been subject to taphonomic changes. Only about 1470 sherds (c. 23.5%), of the assemblage, weighing some 6424 g, (41%), escaped from some degree of abrasion, concretion, or both. Abrasion of the exterior surfaces of vessels was noted on a large number of vessels, which left rock temper protruding from the clay. This was possibly the result of taphonomic processes and the burial of the sherd in sand.

The fragmentary condition of the assemblage is not entirely due to the disturbance of occupational material, by agricultural (spade digging or mattocking and ploughing) or building activities, into which pottery was discarded. Its condition is largely due to abrasion by mechanical movement from wind and sand (sand blasting), by water incursion, and the redeposition of deposits. Although only a few sherds were found in Phase C, most of these are likely to have derived from Phase D due to the digging of pits that disturbed earlier layers.

Some pottery in Phase D, including conjoinable sherds that broke in situ, was compressed and embedded within larger concretions of machair sand and degraded gneiss. This was likely to have been the result of changes to the high water table causing the waterlogging of occupation and floor deposits, or to a more acidic pH of the pre-machair soils. The poor condition and compression of some pottery sherds was probably also due to the build-up of sand deposits, and the weight of the construction of the Phase B cairns.

Manufacture of the pottery

There is considerable evidence concerning the manufacturing techniques used in the making of the pottery. The prevalence of partially exposed or even entirely detached coil joins, and the frequency of fracture along sherds, confirms the use of coils as the main technique employed to manufacture the vessels. Approximately one third of all sherds in the assemblage and approximately half the identified vessels display fracturing along the coil join. This suggests that that building vessels by the repeated addition of coils to form their shape was not entirely successful. The predominance of this technique in the Western Isles had been noted by Stevenson (1953).

The base of the pot was made first, with the coils forming the body of the vessel added to it. Sloping coil joins (N-shaped), which appeared to maximised the area of the clay surface available for bonding, were invariably weakly joined together (see Ballin Smith 2014: 41). Moulding marks surviving on some sherds indicate the shaping and forming of the coils during manufacture. Rims were usually formed by folding the last coil back upon itself and then it being moulded to create the desired shape at the mouth of the vessel.

One problem Squair had in analysing the assemblage was the problem distinguishing between rim sherds with simple or everted (convex) shapes and broken coil joins. For example, the breakage of the coils in SF 26259 (Vessel B10), from the BG24 cist, superficially looked very much like everted rim sherds, even though this is a heavy vessel with a surviving folded over flattened rim with a slight internal flange (Figure 5.55). Table 5.28 has been produced to emphasise the number of vessels (a total of 173) identified to structures and features, which have diagnostic sherds such as rims, bases, carinations/cordons and also decoration.

Part 5

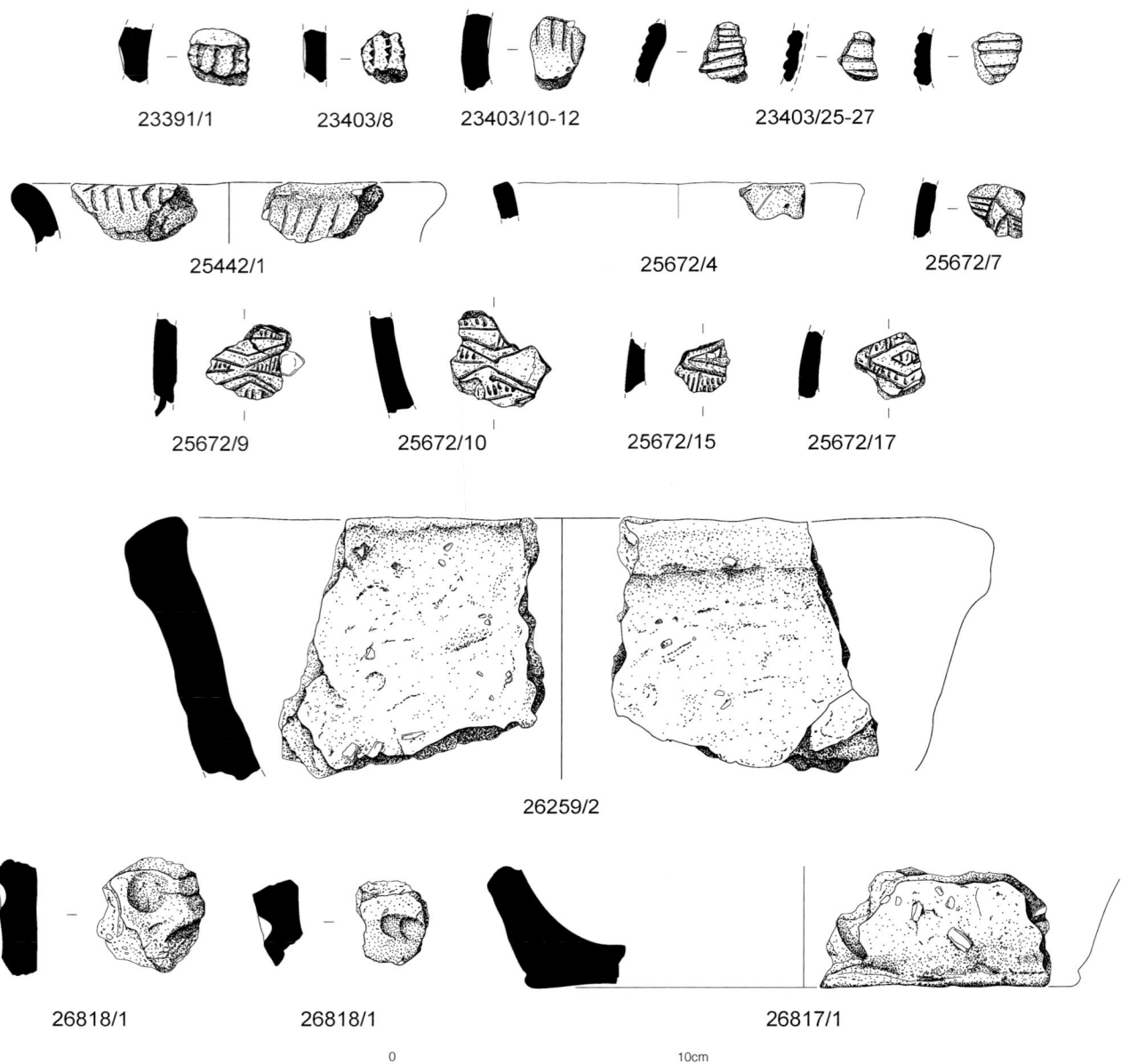

FIGURE 5.55:
Beaker sherds SFs 23391/2, 23403/8, 23403/10, and 23403/25-27, Grooved Ware SFs 25442/1, 25672/4, 25672/7, 25672/9-10. 25672/15 and 25672/17, early Bronze Age pot SF 26259/2, finger dimple sherds SF 26259x2, and late Neolithic base SF 26817/1. Drawn by Jo Bacon and Leeanne Whitelaw.

Rims

Rims are the commonest diagnostic forms of most prehistoric assemblages and they were identified throughout all the phases of this site (Table 5.28). Present are everted rims, flat rims (Figure 5.55 and 5.56), rounded rims, irregular rims, inturned rims, rims with an internal bevel, straight rims with flat tops, and one with an external bevel.

In fact, such a wide variety of forms suggests that no specific design was dominant, or that the rim shape changed over time, with vessels with internal bevels being noticeably more prominent in later periods of use (the Bronze Age of Phase B). The evidence indicates that the forming of the vessel by adding the rim coil and shaping the rim was the most difficult task of the manufacturing process to achieve. It was noted that the rim shape of individual vessels was not consistent. This could be due to the skill of the potter, but also the finished pot could have been inverted during the drying process, with the weight of the pot distorting the shape of the rim. SF 26259 (Vessel B10), is an example of this as its heavy rim coil was turned over the pot's exterior surface, in order to thicken it and to prevent distortion before firing.

	Rims	Bases	Cordon/carination	Decoration
Phase A	6	0	0	4
Phase B	23	4	11	19
Phase C	9	3	3	1
Phase D	41	12	11	19
Phase E	3	0	1	2
Totals	82	19	26	45

TABLE 5.28:
Number of vessels with diagnostic sherds.

The remains of SF 26142 (Vessel B12) from the eastern side of the site produced a rim with an uneven profile, but it is the only one that could be measured in the assemblage. The vessel's rim diameter was estimated to be c. 160 mm, but the percentage rim surviving could not be calculated. A rim SF 24000 (Vessel E05), which was thought by Squair to be a beaded rim, and therefore evidence of a residual early Neolithic pot, was re-identified as a later Neolithic straight rim with thick carbonised food residues adhering to its surface.

Bases

All the base sherds that have been recognised in the assemblage are flat. They are usually heavy and thick, but they are also the rarest sherds surviving in the assemblage (Table 5.28). Only one base was measurable and that was SF 23294 (Vessel C01), with a 180 mm diameter. The pottery of this particular base was 15 mm thick, but the base of SF 26817 (Vessel D20) (Figure 5.55) was almost equally robust at 14 mm thickness. Due to taphonomic conditions and the application of PVA, it was not possible to examine the surviving base sherds in more detail to see if some had had prolonged use on the hearth and whether their exterior surfaces had been burnt away to expose the inner core of the fabric.

Carinations and cordons

Carinations were created at the junction of the neck and body of the vessel, where a coil of clay was added during manufacture to create a shoulder, and from where the pot began to narrow to form the neck and rim. In most cases the carination is slight and is formed by the potter pinching the clay once the coil had been fixed to the body of the vessel. By this action a slight ridge was made, which was subsequently smoothed over during the finishing of the vessel. In SF 23608 (Vessel D17) and SF 26820 (Vessel D29), the carination is acute and marks a distinct change of angle between body and the neck of the pot. With SF 26186 (Vessel C06) and SF 17611 (Vessel B55) the carination also marks a change in the wall thickness, from 14 mm for the bellies of the vessels to only 8 mm for their necks. This suggests that the manufacture of these vessels made their lower halves robust enough to sit on the hearth, while the thinning of their upper parts lessened the overall weight of the pots, but also allowed better control of the shaping of their necks and rims.

In the examples of SF 26142, SF 23665 and SF 27214 (Vessels B71, D26 and D37), a separate strip of clay (a cordon) was applied horizontally around the vessel after forming to emphasise the shoulder. The clay was smoothed over to complete the surface finish of the pots.

Surface treatments

The surface treatments applied to vessels during their manufacture was largely lost due to the severity of post-depositional mechanical abrasion. However, sand could also affect exposed sherd surfaces to produce smoothed or polished areas. During the final stages of manufacture the majority of vessels were most likely wiped (see also moulding marks in Decoration, below) smoothed or burnished, but evidence for this and other surface treatments is rare. The evidence of the deliberate roughening of the exterior surface of SF 23942 (Vessel B03), remains dubious. The burnishing of the exterior of vessels was presumably carried out for both functional and decorative reasons, for example, SF 23403, SF 26689, SF 25976 and SF 23841 (Vessels B04, D13, D23 and D59).

Decoration

Surface treatments or the finishing of vessels includes their decoration, and although decoration appeared on the surfaces and rims of 46 vessels (Table 5.28), most of it was recognised on relatively small sherds of less than 20 by 20 mm (Figure 5.55). Most of the decoration is in the form of incised or impressed motifs.

Incised lines

The majority of the decorative motifs noted in the assemblage come from Phases B and D and are in the form of one or more parallel lines incised into the vessel surface, probably by a small pointed bone or stone, to create narrow grooves. The parallel lines were positioned horizontally, vertically and obliquely in bands or motifs around the vessel. Most of the decorated vessels in Phase A have horizontal incised lines, but SF 16234 (Vessel A16, not illustrated) has also incised decoration on both surfaces of the rim, with horizontal parallel lines on its exterior and with diagonal parallel lines forming a herringbone motif on its internal surface.

Sherds of a late Neolithic Grooved ware vessel SF 25672 (Figure 5.55, 5.57 and 5,58) and SF 25578 (Figure 5.58), all ascribed to Vessel D11, are decorated with all-over incised geometric motifs, including lozenges divided into quadrants that are either a quarter, half or fully infilled with parallel incised lines. Some lozenges are enclosed by a larger lozenge with the intervening space infilled with parallel lines creating an elaborate border, and including a smaller infilled lozenge at its centre - the focus of the elaborate design. Some sherds, also attributable to this vessel, have stabbed and impressed designs infilling larger motifs. Concentric curvilinear lines, also forming indeterminate motifs, are also present but due to post-depositional abrasion and the generally small size of the sherds, the overall design is unclear. One sherd from SF 25672 has a large fragment of shell protruding from it (Figure 5.57), indicating the use of shell as a tempering medium. The rim sherds from this pot were generally plain (Figure 5.56) but SF 25672/4 was faintly incised (Figure 5.55).

Other decorated Phase D sherds include SF 25644 (Vessel D16), which has faint cross-hatching below the rim, and SF 25442/1 (Vessel D53), which has horizontal parallel linear grooves on the lower, surviving portion of the vessel near the base, while its rim is decorated on its top, and externally, with incised parallel incised lines (Figure 5.55).

Decorated Phase B sherds include SF 25928 etc. (Vessel B20) and SF 23403 (Figure 5.59), which have an incised herringbone motif aligned horizontally, but SF 23187 (Vessel B55, Figure 5.60) is profusely decorated with incised motifs separated by its carination. The neck of this vessel has parallel lines in a herringbone or lattice design, and below the carination the decoration comprises diagonal lines. Also incised, but with a different motif is SF 23372 (Vessel B61), which has a horizontal dashed line immediately below the rim. The rim of SF 25712 (Vessel D07) has an external bevel which carries diagonal, parallel line incised decoration with horizontal and diagonal lines on its interior surface similar to SF 25442/1 (Figure 5.55). Other decorated sherds with incised lines from this phase include SF 23182, SF 23187, SF 23391 and SF 26143 (Figure 5.60).

Stab and drag

Although the evidence is slight, another technique used was the action of stabbing a tool into the clay and dragging it rather than gently lifting it from the surface once the incision had been made. The mark on SF 25303 (Vessel B67) could be incidental rather than deliberate, as it comprises a vertical groove with a circular expansion at one end where a tool was stabbed into the clay. A single incised horizontal line beneath the groove may form part of the design. The flattened straight rim of SF 25588 (Vessel D33) is decorated with a single horizontal row of stabbed dots (not illustrated), formed by jabbing a tool, possibly toothed, into the clay and then pulling it out again.

Shell edge impressions

Apart from the late Neolithic dated Grooved Ware sherds of Phase D, Phase B produced a number of examples of typical late Neolithic/early Bronze Age Beaker pottery. The form of decoration noted in was created possibly using the edge of a cockle shell (*Cerastium sp.*) or a limpet (*Patella vulgate*). Both sides of the rim of SF 23403 (Vessel B04) are decorated using the edge of a shell. The rim edge and the exterior surface of the rim are incised with diagonal, parallel lines, and below it internally are vertical, parallel lines. The neck of the vessel had diagonal parallel lines forming a herringbone motif in repetitive horizontal bands. Immediately above and below the carination were both vertical, parallel lines and diagonal lines, presumably in bands running down the body of the vessel (all Figure 5.59).

Rim SF 25442 and SF 25588 (Vessel D56) was similar to SF 23403 (Vessel B04), as below it are shell-impressed diagonal and parallel lines. The rim top, and the interior of the rim are also decorated with diagonal and vertical shell impressed parallel lines. The motifs are arranged in horizontal bands.

FIGURE 5.56:
Phase D Grooved Ware sherds from Building 2 (DH) all SF 25672.

FIGURE 5.57:
Phase D Grooved Ware sherds from Building 2 (DH) all SF 25578.

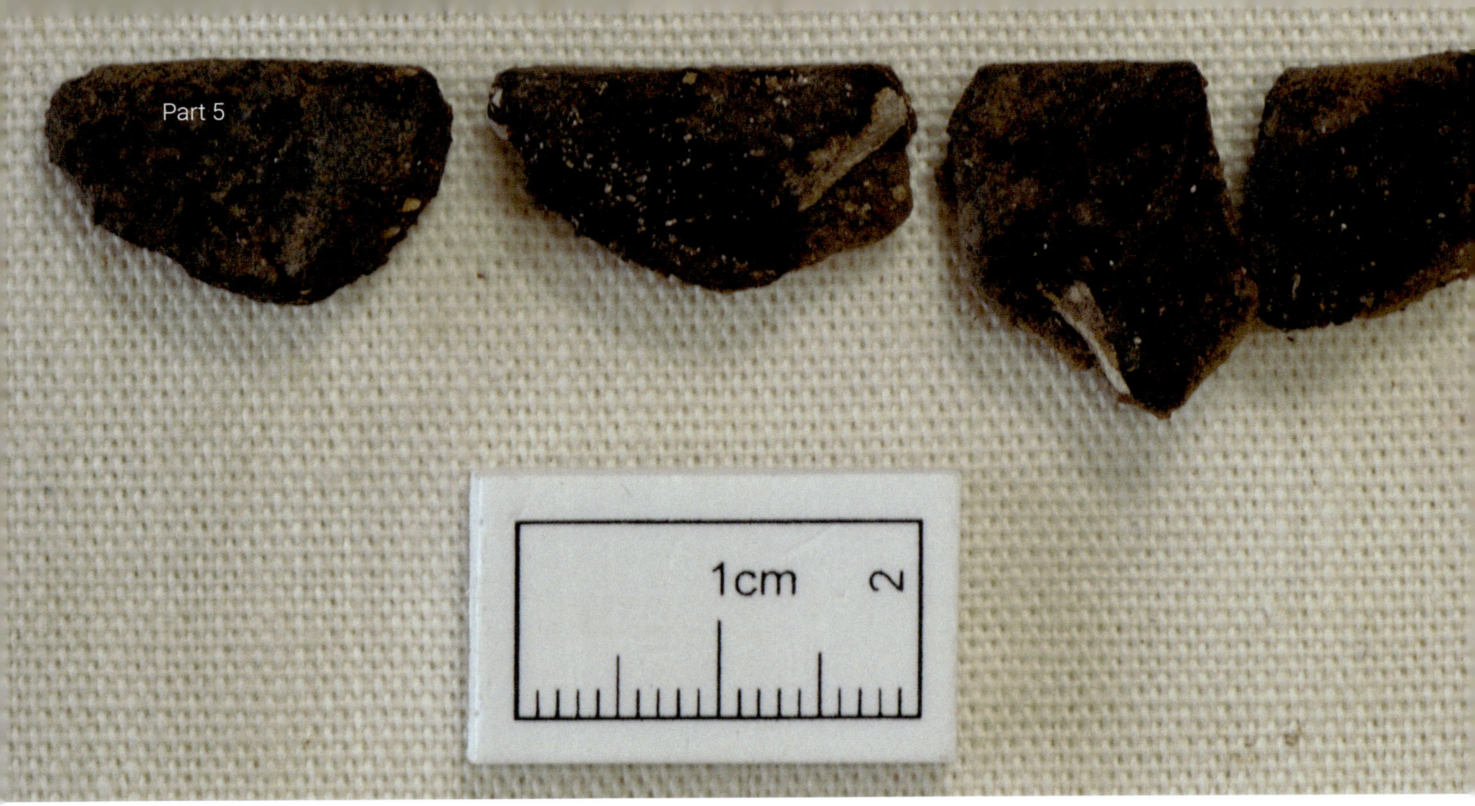

Part 5

FIGURE 5.58:
Phase D rims from Building 2 (DH) SF 25744 (left), rest SF 25588.

FIGURE 5.59:
Phase B shell edge decorated sherds from the east half of the site, all SF 23403.

FIGURE 5.60:
Phase B decorated sherds: SF 23391 (top left) and SF 26143 (bottom left) from the east half of the site, SF 23187 (centre) from the west cist, and SF 23182 (right) from the west half of the site.

Fluting and moulding marks

These marks could be a product of the moulding of vessels and also of the finishing. SF 25411 (Vessel C22) appears to have an external fluted decoration in the form of curvilinear lines below the rim, which may be a product using the hands and fingers to finish the vessel. SF 27214 (Vessel D94) has fingertip impressions on its exterior, which were considered to be decoration. A finger depression, most likely from moulding, was noted on the exterior of sherds from SF 26830 (Vessel D01) and and a finger nail depression on SF 26818 (Vessel D09) (Figure 5.55).

Firing conditions

As the colour of most sherds has been affected by their use on the hearth, the carbonised food residues and by taphonomic conditions, it is difficult to assess their conditions of firing. Oxidised and reduced sherds appear to be in the majority, but there are a number of sherds, mainly from Phase D that are either mainly oxidised or fully reduced, which perhaps reflects the lack of control of the processes needed to gain consistent firing. The colour of most sherds fall into the Munsell Hue 10YR 6/2 light brownish grey, 5/2 greyish brown to 3/2 very dark greyish brown. Where carbonised food residues are present the colours are predominantly Hue 10YR 3/1 very dark grey to 2/1 black.

Surface residues

The most notable residue on the exterior but also on some interior surfaces of vessels was the remains of carbonised (burnt) food. These remains, sometimes 1 mm more in thickness, are present on approximately half the assemblage, indicating that the vast majority of vessels were used on the hearth as cooking pots. Some of the residues were confined to rims, shoulders or above the carination, on others they were found near the base, or internally near the rim. SF 23294 (Vessel C01) and SF 25588 and SF 25672 (Vessel D11) exhibit a glossy residue probably acquired through prolonged use over an open fire, while SF 26249 (Vessel B10) (Figure 5.55) has thick encrustations around its rim and on its surfaces, indication its use for cooking.

Vessel form and function

The fragmentary condition of the assemblage and the small size of the majority of sherds, as mentioned previously, prevented any reconstruction of vessels.

The paucity of larger sherds with the potential to interpret vessel shape, combined with a low percentage of diagnostic sherds with measurable dimensions, inhibited the interpretation of the form and function of most vessels. A total of 216 vessels were recognised across all phases by Squair (Table 5.29) but only 80% derived from structures (Table 5.28).

Phase	Number of vessels
A	16
B	74
C	23
D	98
E	5
Total	216

TABLE 5.29:
Number of vessels in each phase.

The majority of pots were used for cooking and probably storage, due to their evidence of carbonised food residues and the narrowing of the mouth of the vessel from the shoulder or carnation. However, the occasional identification of what appears to be open bowls and small pots suggests the assemblage may have contained a limited variety of different shaped vessels for different functions. Abrasion through use was tentatively identified on the interior surfaces of 23 vessels, including on their rims, as a result of stirring or scraping the vessels contents, for example SF 26258 (Vessel B10). Some of the exterior abrasion (see Post-depositional changes, above) may have been brought about through handling and moving vessels on the hearth.

Pottery forms

Only c. 36 pots or c.16.7% of the assemblage have an identifiable shape (Table 5.30) and they are discussed below by phase. They are also described in more detail in Appendix 2.

Phase	Number of vessels
A	2
B	8
C	4
D	20
E	2
Total	36

TABLE 5.30:
Number of vessels in each phase with a discernible vessel profile.

Phase A

It must be noted that all the prehistoric pottery of Phase A derived from earlier phases, most notably from Phase B.

SF 24976 (Vessel A03) is a plain rounded pot with an everted and irregularly profiled rim sherd which has a flattened top. The heavy sooting and carbonised food residues on both surfaces of the rim indicates it was a cooking pot. Other sherds with everted rims include SF 23025 (Vessel A14) and SF 16234 (Vessel A16). The latter is, however, decorated with horizontal incised parallel lines on the exterior, and diagonal parallel lines forming a herringbone motif decorate the internal surface of the rim. The decoration and its location suggest this was a Beaker vessel. SF 23598 (Vessel A05), SF 23534 (Vessel A06), SF 23312 (Vessel A09) and SF 17274 (Vessel A15) are decorated with horizontal, parallel, linear grooves on their exterior surfaces, are also interpreted as fragments of Beaker pots.

SF 23598 (Vessel A04) has a possible carination indicating the pot had a narrower neck and mouth.

Phase B

The majority of vessels from this phase have straight or slightly rounded profiles and are interpreted as jars. Their thick, coarse walls suggest large, flat-based vessels. SF 25324 etc. (Vessel B01) has a possible burnished shoulder sherd and several body sherds with amorphous depressions which are interpreted as the possibly vestiges of impressed decoration. This coarse jar has external carbonised food residues indicating its use as a cooking vessel. SF 25961/3 (Vessel B25) has finger moulding.

A number of pots have everted rims, or possibly everted rims that also carry internal bevels. These include SF 26259 (Vessel B10), SF 26176 (Vessel B23), SF 25324 (Vessel B38), SF 23182 (Vessel 56) and SF 26504 (Vessel 70). This rim form is typical of late Neolithic and later wares from elsewhere in the Western Isles and the Hebrides in general. The internal bevel may have been formed in order to take a lid, but the size and shape of these pots is not known.

Heavy bowls are suggested by SF 25315 (Vessel 39) and SF 23391 etc. (Vessel B48) which are coarse pots with a carination, indicating a bipartite shape. SF 25438 etc. (Vessels B2), SF 25819 and SF 26232 (Vessel B18), SF 25578 (Vessel B28), and SF 25396

(Vessel B32) are possibly heavy open bowls with everted rims. The rim of SF 25515 (Vessel B31) is inturned suggestive of a round or barrel-shaped pot.

Several vessels, of indeterminate form and type, have decoration typical of late Neolithic / early Bronze Age pottery and are considered to be Beakers: SF SFs 23391 (Vessel 51), and sherds SF 23403 (Vessel B04) (Figure 5.55 and Figure 5.59). The latter has diagonal, parallel lines as does SF 23317 and SF 23395 (Vessel B52) which are both decorated with horizontal, parallel grooves. Other decorated sherds include SF 26143 (Vessel B20), SF 23391 (Vessel B51) and SF 23187 (Vessel B55) (Figure 5.60). The decoration of SF 23182 (Vessel B56) (also Figure 5.60) comprises short, vertical, parallel lines, arranged in horizontal bands immediately below the rim on the vessel exterior, and is also a Beaker. It is paralleled by two vessels from Site T26/T26A at Allt Chrisal on Barra (Gibson 1995: Figure 4.38, nos. 203, 216).

SF 23372 (Vessel B59) is a very small body sherd which has lost most of its external surface except for a solitary deeply incised line that would probably have been part of a wider motif of parallel lines.

Phase C

This pottery is most likely derived from Phase D contexts, due to the activities and conditions present on the site at the end of the late Neolithic.

Vessels, frequently flat-based, with indeterminate profiles, and interpreted as jars, include vessels SF 23294 (Vessel C01), SF 25967 (Vessel C15), SF 26205 (Vessel C19) and SF 26285 (Vessel C21). SF 23464 (Vessel C02) and SF 23492 (Vessel C05), with tightly curved body sherds, were probably cylindrical shaped cooking pots. SF 25411 (Vessel C22) is a small rim sherd representing fine pot, possibly a beaker, with fluting below the rim, and SF 26026 (Vessel C11) with its everted flat-topped rim was probably an open bowl. SF 26206 (Vessel C10), with a round topped rim was probably a jar and SF 26186 (Vessel C6), is interpreted as a heavy bowl, as it has a carination.

Phase D

This phase is characterised by a largely undecorated late Neolithic assemblage. It includes vessels which were carinated, and others with have flat bases. Their shapes can be described as generally indeterminate, although some are cylindrical and others open bowl shaped. The vessels from this phase were predominantly used on or near the hearth as cooking pots and probably storage vessels, due to the amount of carbonised food remains on their surfaces. Some smaller vessels are also possibly present, but due to the small size of rim sherds and and their uneven manufacture and finish, it is difficult to be precise about the actual size of vessels as none of the rims from this phase could be measured.

Some of the finer pots (with wall thicknesses of less than 6 mm) had generally simple straight, thin rims with flattened or gently rounded tops. They included, for example, SF 27214 (Vessel D35) and SF 25547 (Vessel D52). SF 23889 (Vessel D62) has a straight rim with a rounded top, as has SF 26452 (Vessel D88), which is irregular in profile. Other rims, such as SF 25994 (Vessels D41 and D42) are entirely irregular in form and others are slightly everted, for example, SF 25588 (Vessel D79) and SF 26424 (Vessel D82).

There are eight examples of rim sherds with bevels. One SF 25712 (Vessel D07) came from Building 2. The remaining examples all have internal bevels, of these SF 26164 (Vessel 019) is from Building 1, but SF 25712 (Vessel 49), SF 25443 (Vessel D72) and SF 26820 (Vessel D96) are all from Building 2.

Vessels with heavier rims and wall thicknesses of over 6 mm are few, but two examples SF 25442 (Vessel D51) and SF 25443 (Vessel D95) have rounded rims, although the latter is slightly inturned. The variation in rim thickness and shape was not standardised around the individual vessels, and it is quite likely that had reconstruction been possible, that examples would have been found with straight, everted or inturned rims on the same vessel.

Grooved Ware sherds (see Appendix 2 for details) identified with geometric decoration are confined to SF 25578 (Figure 5.58) and SF 25672 (Vessel D11) (Figure 5.57) and SF 25712 (Vessel D48). All these sherds are probably part of a single open mouthed bowl-shaped vessel (Alison Sheridan, NMS, pers. comm.). Associated with these sherds from the same building (Building 2 – DH) were two carinated pots SF 25620 (Vessel D02) and SF 26819 (Vessel D09). Both have fingertip/nail impressions either from the moulding of the pot or as a deliberate decoration.

Sherds of what are interpreted as late Neolithic/early

Bronze Age Beaker sherds with incised lines, both vertical and oblique, and lattice decoration, have mainly come from the east side of the site. They include SF 26689 (Vessel D13), SF 25575 (Vessel D16) a carinated bowl, and SF 25574 (Vessel D21) also a carinated bowl, SF 25588 (Vessel D33) is a dot or possible comb impressed pot. Two sherds of a shell edge impressed Beaker pot (Vessel D56) were also found in the infill of the late Neolithic Building 2.

Phase E

The evidence from this phase includes sherds of a carinated cooking pot SF 23891 (Vessel E01), SF 24000 (Vessel E02) is a shouldered cooking pot, and SF 24000 (Vessel E03) is slightly finer and possibly smaller pot that may have been decorated.

Distribution of vessels and depositional practices

The majority of the assemblage can be interpreted as comprising broken vessels that were discarded into deposits that were disturbed by later activities, causing the abrasion and further degradation of sherds into smaller and smaller fragments. Only a very few vessels from Phases B, C and D were interpreted as being in-situ pieces or deliberate deposits. With the loss of sherds from Phase A, what survives is mainly derived from Phases B and Phase D, indicating the disturbance of earlier contexts.

Phase B

East side of the site

SF 26259 (Vessel B10 with fragments B09) was deliberately placed in cist (BM) with a highly flexed inhumation, a calf and three worked bone points (Crawford 1984: 2). The vessel was deposited whole, but the cist was disturbed by recent coastal erosion causing it to fragment. In addition to this vessel was another, which was associated with the cairn over the cist and the area immediately around it. This comprised sherds of a single Beaker identified subsequently to the analysis of the assemblage. The pot, which was decorated with incised parallel lines comprised SF 26285, SF 23391 and SF 23395 (Vessels B17, B51 and B52). A second Beaker vessel, SF 23403, SF 23391 and SF 25112 (Vessel B04), decorated with vertical and diagonal parallel lines of shell edge impressions was associated with the cist and the floors of the temporary structure. Tiny decorated sherds may imply that two other decorated Beaker vessels had been in use on the east side of the site. The most convincing is the single sherd SF 26143 (Vessel B13) with a deep incised chevron executed in parallel lines and the less definite SF 23372 (Vessels B59-62) with faint grooves and a dashed line.

This structure (BG24) contained SF 26495 and SF 26504 (Vessel B15) in a pit beneath its uppermost floor. The presence of large sherds indicates that it was probably intact when it was buried or deposited. A second pot comprising SF 26584 and SF 26566 (Vessel B14), came from earlier floors of the same structure, as did SF 26526 (Vessel B69).

West side of the site

Sherds comprising a decorated Beaker, SF 23187 (Vessel B55), with incised chevrons were associated with the cist (BB3) and the construction of the cairn complex. Another Beaker pot is represented by sherds SF 23182 (Vessel B56), with incised vertical dashed grooves. The single incised line on sherd SF 25303 (Vessel B67) is less convincing as decoration.

These Beaker vessels were most likely the result of ritual or feasting activities that took place at the time of the burials. Two Beakers were clearly identified and associated with activities on the west side of the site, with two or three Beaker vessels on the east side, with a tentative fourth. These vessels are represented by small sherds, often poorly preserved, which may have been deliberately broken as part of the burial rituals.

Phase C

The majority of vessels identifiable from this phase are very fragmentary and incomplete, often represented by single sherds. Only SF 23294 (Vessel C01), from a pit (CE – labelled Judith's pot pit during the excavation), was considered a deliberate deposit or placement, but even it is difficult to interpret. The absence of its rim suggests that it was removed by later disturbance. Other vessels recognised from this phase are likely to have derived from late Neolithic levels.

Phase D

Less than half the vessels recognised in this phase were associated with structures. The rest came from external deposits.

The shaft, platform, stone settings and alignments

Some pottery, SF 17623 and SF 26722 (Vessel D14) comprising two rim sherds, and SF 26737 (Vessel D24) a base sherd, were found in midden deposits connected with the ritual complex of the stone platform and its whale bone capped shaft. Vessel D24 was found within the shaft and could have been a deliberate deposition or discarded piece.

Burnt area DK

In the external angle between Buildings 1 and 2 was a discrete burnt area (DK) which included SF 25984 and SF 26818 (Vessel D57). It is suggested that fragments of this vessel are the same as SF 26818 and SF 26819 (Vessel D09) found in Building 2. The possible relationship of these sherds suggests that debris, including broken pottery, was cleaned or removed from the floor and the hearth and dumped outside the building.

Building 1 (DJ)

Approximately 90 sherds (1%), weighing nearly 300 g (4%), of the total number and weight of sherds from this phase, derive from contexts associated with this building (see PART 2). This amount of pottery is small in comparison with that excavated from the adjacent structure (Building 2). However, there are some similarities in depositional practices.

SF 26830 (Vessel D01), was found beneath the wall of the building and could have been deliberately placed there as a foundation deposit. SF 23685, SF 23664 and SF 23665 (identified as Vessels D25, D26 and D28), were all found together in the fill of one pit within the floor of the structure. It is quite possible that these sherds, which include base and body sherds and a possible cordon, were all part of the same vessel, which was either deliberately deposited or placed in the pit for a specific use. Carbonised food remains on its external surfaces indicate it was a cooking pot. However, the fragility of the sherds may indicate that it broke during use and was simply buried in the floor.

Building 2 (DH)

This structure contained c. 1100 sherds or c. 25% which weigh c. 3700 g (40%) of the total number and weight of sherds from this phase. Like Building 1, there is some evidence to indicate deliberate disposal or placing in specific features within the structure. However, the number of sherds representing vessels is small and they are very fragmentary.

SF 26027 (Vessel D05) derives primarily from the hearth and the floor area to its immediate east with SF 25994 (Vessel D43) located in the floor area to the west. A number of pots were found in shallow pits or hollows within the building. They include a smashed pot SF 25951 (Vessel D31 with fragments of D33, D11 and D66) in one pit, SF 25442 (Vessel D06) and SF 25442, 25443 and 25447 (Vessel D51, D53 and D95) in another, and SF 25491 (Vessels D03, D39 D40) in a third. Their presence may not indicate broken pots that were discarded, but pots deliberately positioned in the floor for everyday use, for example, as containers for limpets or fresh water.

Distributed in the uppermost floor level of this building was SF 25578 and SF 25672 (Vessel D11) with SF 25712 (Vessel D48), comprising the Grooved Ware pot. Other vessels, represented by small numbers of small sherds were found in the deposits infilling the building. The most significant sherd of the infill was a rim of a Beaker vessel SF 25442 (Vessel D 56) decorated with shell-edged impressions (Figure 5.60), not unlike sherds of another decorated Beaker SF 23403, SF 23391 and SF 25112 (Vessel B04) found in deposits in Phase B on the east side of the site.

Phase E

All of the vessels identified in the anthropogenic and natural contexts are not associated with any features. Their fragmentary condition suggests that they are most likely intrusive sherds derived from Phase D activities.

Chronology of the assemblage and comparison with other sites

From the pottery analysis, two or three typological distinct vessels can be identified that provides a rough chronological framework for the assemblage. These include the Grooved Ware vessel from Building 2 of the late Neolithic (Phase D), and Beaker and other vessels from the early to middle Bronze Age burial and cairn activities in Phase B.

The use of the Grooved Ware vessel in final floor Building 2 indicates it appears late in the currency of this vessel, which for other Hebridean and Orcadian dated sites was in the range of 3000–2900 BC (Alison Sheridan pers. comm.). Her detailed analysis of the pottery, and especially that of the single Grooved Ware pot from the ritual complex at Calanais, Lewis, highlights the rarity of this style of vessel in the Western Isles, with the only other reported examples being from the passage tomb at Unival, North Uist (Sheridan et al. 2016: 594) and sherds found recently at An Doirlinn, South Uist (Garrow and Sturt 2016; Copper 2016: 18-21). At this latter site, some of the Grooved Ware is finely made and indicative of an open bowl(s) similar to Vessel D11 at RUX6. Interestingly, the An Doirlinn examples are also considered to be later than the accepted date range lying between 2780–2480 cal BC and 2480–2330 cal BC (ibid, 26). These dates compare reasonably well with the radiocarbon dates from pottery residues and a sheep bone from the uppermost floor of Building 2 (see Table 3.1), which indicated a date range around the early to middle part of the third millennium BC (c. 2618–2464 cal BC) for the last use of this structure and therefore, also the use of the Grooved Ware bowl (Vessel D11) within it. The similarly shaped and decorated Grooved Ware vessels from Barnhouse, Orkney are earlier (see Ashmore 2005: 387-388) as that settlement appeared to end c. 2900 BC.

Unlike the vessels from Barnhouse (Jones 2005: Figure 11.8) and Calanais (Sheridan et al. 2016: Illus 18.11) both the An Doirlinn and the RUX6 samples were not sufficiently well preserved to provide the actual size of the vessel. The plain rim sherds, and a decorative band or two of elaborate geometric designs, based on variations of plain and infilled lozenges and their borders, together with a few plain sherds are all that remain of Vessel D11. The motifs of the pot are finely executed, and from the description of the Unival vessel by Sheridan et al. (2016: 594) and its illustration in Young's (1966: 46-47 Plate 2a) paper, it may have been very similar. In the An Doirlinn examples, the vessels were decorated from their rims down (Mike Copper, pers. comm) and the same is the case for Calanais and also the Grooved Ware found at Machrie Moor (Haggarty 1991: Illus 6). It is quite possible that the heavy sooting and carbonised food residues have obscured other parts of Vessel D11 but from the evidence that survives its decoration seems to be limited.

Sheridan (2016: 594-595), suggests strong links between timber and stone circles and the use of Grooved Ware vessels, with the ideas of both moving south and west from Orkney down the Atlantic seaboard. This places the pottery from both RUX6 and An Doirlinn in a different situation as both appear to derive from non-ritual domestic contexts (Copper 2016: 26). The later dated occurrence of the RUX6 example may explain its survival to the end of the late Neolithic and its use in a domestic building. It is possible that this is a local copy of a relic from the past or a vessel that may have had a special purpose, but even so, it remains the only Grooved Ware example from the site. If Sheridan is right in making the links between Grooved Ware and stone circles, the question is what happened at RUX6 where this one special pot survived? The RUX6 settlement certainly had a ritual element at its core – the bedrock shaft and its plinth, with its radiating lines of stones with wooden posts and the Great Auk stone. Was this perhaps the local expression and interpretation of the ideas imbued within the stone circles elsewhere?

In addition to the Grooved Ware vessel, plain, coarse pots with carinations and flat bases, often identified as cooking vessels were the ubiquitous domestic pottery vessels in use at RUX6, and throughout the Uists and the Western Isles in general, during the late Neolithic (see Parker Pearson 2012: 403, Figure 20.1).

Sherds of Beaker vessels were found in contexts at RUX6 associated, or probably associated, with the ritual activities on both the west (earlier) and eastern (later) sides of the site. The human remains buried in a large cist beneath the cairn complex on the west were dated to c. 2119–1892 cal BC (Table 3.1). The remains of two Beaker pots SF 23187 (Vessel B55) and SF 23182 (Vessel B56) (Figure 5.60), also found on the western half of the site were represented by single sherds only. Vessel B55 was found in the backfilling material between the pit and its cist during its construction, and appears to be closely linked to the burial itself. This sherd with its pronounced carination was made locally but in the international style of Beakers, and is similar to ASH 39 found at Calanais, but with a more expansive, possibly overall, decoration (Sheridan et al. 2016: 598, Illus 18.15). If the stylistic identification of this piece is accepted, its date range appears to be later than the currency for these early Beakers, ending roughly c. 2200 BC (Ibid, 603). However, it can also be argued that it is a residual piece, but given the late date range of the Grooved Ware vessel from the late Neolithic, an equally late date for Vessel B55 does not seem unusual in this context.

The other vessel, SF 23182 (Vessel B56) (Figure 5.60), is even more problematical as its precise location on the site is unknown. Due to its internally bevelled rim, it would seem to stylistically belong to Clarke's (1970, Vol 2) N3 style of Beakers commonly found in Scotland.

The occurrence of sherds of Beaker pottery on the eastern part of the site is also difficult to interpret and date. Both SF 25442 (Figure 5.55) and 25588 (Vessel D56) found in the infilling of the late Neolithic Building 2 on the west, and SF 23403 (Vessel B04) (Figure 5.59) on the east are probably sherds of the same shell-edge decorated vessel with an out-turned rim. It is has the largest number of sherds (36) of any of the Beaker vessels, and was obviously a whole pot (27 sherds from SF 23403), with SF 23391 accounting for two sherds, SF 25112 for three and SF 25385 for four, which are now missing. Although this pot became dispersed across the site, the greater part of it was associated with 'floory', interpreted as material over(?) the floor of the temporary building into which the eastern cist was constructed and over which its cairn was built. The association of this vessel with the cist and cairn, or even the temporary structure is, however, based on poor evidence. The shape of the vessel may have been similar to N4 Beakers, such as the one from Auchrynie, Aberdeen (722) (Clarke 1970: 368). Its decoration is also comparable to some of the shell-impressed sherds from the Beaker II levels at Northton, Harris (see Gibson 2006: 119, Figure 3.34,5-6), which are dated to a range between 1940 and 1680 cal BC (Gregory 2006c, 90), perhaps indicating that the Beaker vessel the east side of the site is most likely later than Vessel B55 on the west.

The third pot which is evidence of the chronology of vessels and activities on the site is SF 26259 (Vessel B10) placed in the cist inserted into the temporary structure, along with human remains dated towards the end of the early Bronze Age 1877–1658 cal BC and the bones of a calf dated to the middle Bronze Age c.1518–1409 cal BC (see Table 3.1). It is plausible that this heavy, plain cooking pot with its tapering sides was contemporary with the calf and is therefore the latest vessel from RUX6. It was not possible to date the carbonised food residues from it due to treatment with PVA in the field, but its shape is not inconsistent with the middle Bronze Age bucket-shaped pots found at Cladh Hallan (Parker Pearson 2012: 403-404, Figure 20.3).

Conclusions

The RUX6 assemblage, in contrast to the larger assemblages from sites such as Northton and Kilellan Farm, Islay, has only a few rare diagnostic pots, but it is possible to demonstrate some chronological development or changes from Grooved Ware to Beaker vessels and to the heavy middle Bronze Age cooking pot from the evidence that has survived. The development of a typology of the majority of plain pots used throughout the site has been impossible to achieve as Crawford wished, because of the assemblage's very fragmentary state and the disturbance of stratigraphy throughout the phases. In contast, the larger assemblage from Kilellan Farm with its plain and decorated vessels demonstrates the continuity of pottery traditions found elsewhere in western Scotland during the early Bronze Age in particular (see Cowie 2005: 63).

In spite of obstacles, such as the lack of vessel shapes, it is possible to demonstrate that the inhabitants of RUX6 made pots that were in common currency throughout the late Neolithic and early Bronze Age of Scotland, demonstrating contact beyond their immediate environment. The survival of pottery from RUX6, and the assemblage from An Doirlinn in recent years, highlights the fact that even relatively small assemblages can provide evidence that broadens our knowledge of the prehistoric pottery traditions of the Hebrides.

Worked bone artefacts

By Beverley Ballin Smith with identification of the faunal material by Judith Finlay Aird and Catherine Smith

Composition of the assemblage

The category of 'worked bone' totals 47 database entries, of which 13 are antler, one is whale bone and two are cartilaginous fish, but not all of them are considered to be modified. Several of the unworked pieces are discussed below as they may have had meaning during the late Neolithic and were used, or could have been used, without alteration.

The worked pieces were separated during the excavation and post-excavation processes from the bulk mammal bone samples. Preliminary reports on the worked bone and antler artefacts were written by Imogen Crawford in 1997 but further analysis of the material took place during the assessment of the collection from 2012, as well as the detailed analysis of the faunal remains by Finlay Aird PART 4. Although Finlay Aird comments on the generally good condition of animal bone in the collection, much of the material identified as worked bone is not well preserved, and there is some uncertainty as to whether pieces are worked or not (Table 5.31). Modified pieces are definitely worked, uncertain pieces could be worked, but 12 pieces considered worked by the excavator, are unlikely to be modified due to their poor preservation. Of the 47 pieces only 22 pieces are definite artefacts.

	Modified	Uncertain	Unknown	Totals
Phase A	2	3	2	7
Phase B	8	1	3	12
Phase C	2	1	0	3
Phase D	10	8	7	25
Totals	22	13	12	47

TABLE 5.31:
Number of bone artefacts considered modified, uncertain or unknown.

Mammalian bone used for artefactual material includes that derived from both wild and domesticated species as well as from a stranded or beached whale, and a large fish (shark or ray), possibly also beached by a high tide or storm waves (Table 5.32). The largest amount of antler came from phases A and B, and although not designated to species, is likely to be from red deer. The pieces from Phase A may be of recent origin and have relatively fresh breaks, while those from Phase B have abraded surfaces. Antler from Phase D includes a cast antler and an antler beam. Only one artefact, from Phase D is considered to be deer bone, and indicates that deer meat on occasion might have been eaten, even though it had little status in the late Neolithic diet of the inhabitants at the site (Finlay Aird, PART 3), and could have been the result of opportunistic wild food gathering.

The main food animals, whose bones were used for making artefacts, are sheep/goat and cattle. Sheep/goat was the dominant food animal during Phase D (Finlay Aird, PART 3), but more tools or possible tools, survived from cattle bones than sheep/goat. This is probably a simple matter of the robustness of cattle bones surviving better in the archaeological record of the site than the more fragile ones of sheep/goat. In Phase B there were very few of either and in Phase C none.

The largest category of bone artefacts are those which cannot be identified to species because of the condition of the pieces. In Phase D most of the unidentified pieces are from large ungulates (cattle or deer most likely). Two of the worked bones from Phase B are from small ungulates (sheep/goat?), while Phase A produced artefacts from both large and small ungulates.

Description of the artefacts

Antler

None of the antler pieces from any of the phases is clearly recognisable as an artefact. An antler beam and other pieces such as SF 24044 (Phase A), SF 23181 and SF 23600 (Phase B) and SF 24959 (Phase D) have been split longitudinally and the core removed. A small antler tine tip fragment (SF 27229b), was found in the latest floor of Neolithic Building 2 (DH), while a more intact fragment of cast antler including the burr, beam, brow tine and base of bez tine was found in a pit in Neolithic Building 1 (DJ). The trez tine was absent but the brow tine has small scratches near its polished end, suggesting its possible use as a lever or pick. It is the largest most intact piece of antler from the site and has a length of 360 mm. With the condition of antler being poor, it is very difficult to determine if polish is due to taphonomic conditions or actual use wear.

Sheep/goat bones

Four worked bone points were made from the shaft of a tibia, two distal metapodials, and a proximal metacarpal. Three were split sagittaly, with part of either the proximal or distal parts of the shafts forming a point that appeared to be worked in all cases. SF 26347 was associated with the BG24 cist

	Antler - possible antler	Sheep/goat - cf sheep/goat	Cattle	Deer	Unidentified	Cetacean - whale	Cartilaginous fish - shark/ray?	Totals
Phase A	4				3			7
Phase B	5	2	2		3			12
Phase C					3			3
Phase D	4	3	6	1	8	1	2	25
Totals	13	5	8	1	17	1	2	47

TABLE 5.32:
Species numbers by phase.

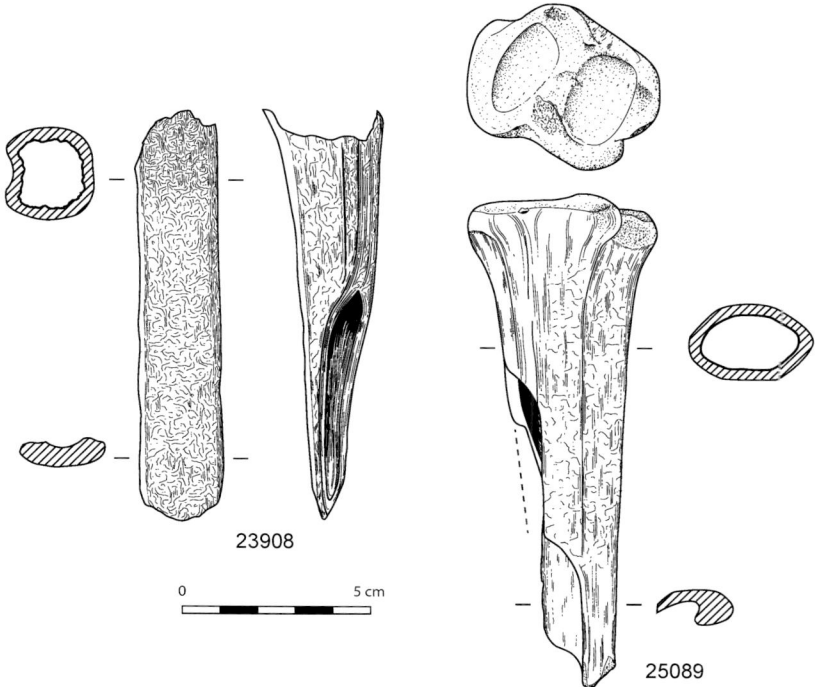

FIGURE 5.61:
Scoop SF 23908 and pounder SF 25089. Drawn by Leeanne Whitelaw.

in Phase B, and the three other artefacts, SF 25997, 25998 and 26041 were found on the uppermost floor near the hearth in Neolithic Building 2 in Phase D (all Figure 5.62). Their lengths varied between 62 and 75 mm. A segment of a possible bead, SF 25226 (Figure 5.62) was found in Phase B. It was cut from a section of metatarsal shaft and measures 39 mm in width.

Cattle or cattle/large ungulate bones

Two samples from the top layers of Phase B were identified as artefacts. SF 23908 is an abraded metatarsal shaft fragment whose medial/lateral aspect has been modified to form a scoop 114 mm in length (Figure 5.61). SF 25089 is a distal (fused) tibia 130 mm long. Its distal end is flattened and worn smooth indicating its possible use as a pounder (Figure 5.61).

Three artefacts were located on the uppermost floor of Phase D building 2 and predominantly in the northern part of the structure. SF 25692 is a femur shaft split sagittally, with the distal end modified by trimming to form a scoop 193 mm long. SF 25614 and 25694 are two possible scoops made from the distal ends of tibia shafts, one was split sagittally. The former measures 205 mm in length and the latter 120 mm.

Two other pieces from the uppermost floor of Building 2 could be butchering debris rather than artefacts. SF 25693 is a femur shaft 128 mm long whose distal end may have been modified to form a scoop. However, sand abrasion of what appears to be a butchered break could have modified the piece. SF 26056 is another femur shaft 130 mm long which is split sagittally.

Found in or close to the libation pit (DC) is SF 26791, a 100 mm long tibia shaft whose posterior part is split dorso-ventrally and its proximal end is cut medio-laterally. The angle of the cut surface indicates possible use-wear. It could be a rubbing tool or a broad chisel (Figure 5.62).

Deer bone

SF 26787 is a possible spatula made from a tibia shaft fragment, 155 mm in length. It was found in association with the libation pit (DC).

Unidentified species

The three pieces unidentified to species from Phase A are very abraded and in poor condition. Three pieces from Phase B include two worn points (SF 25100 (Figure 5.62) and 26346) and a scoop-like piece SF 25960, which is likely to be a product of abrasion. SF 23430, 23462 and 23463 (both Figure

5.62), from Phase C, are fragments of trimmed long bone shafts, which could have been used as points or scoop-like tools. They vary in length from 41 to 78 mm.

The uppermost floor of Building 2 (in Phase D produced seven artefacts, SF 25616 - a scoop/plaque, SF 25618, 25619 (both Figure 5.62) and 26057 bone points, SF 25738 a plaque or polisher (Figure 5.62) and SF 25616 is probably a piece of butchery waste rather than a scraper. Associated with the same building was SF 26240 a broken, but entirely abraded rough point, which may be unmodified. A trimmed and polished piece SF 23991

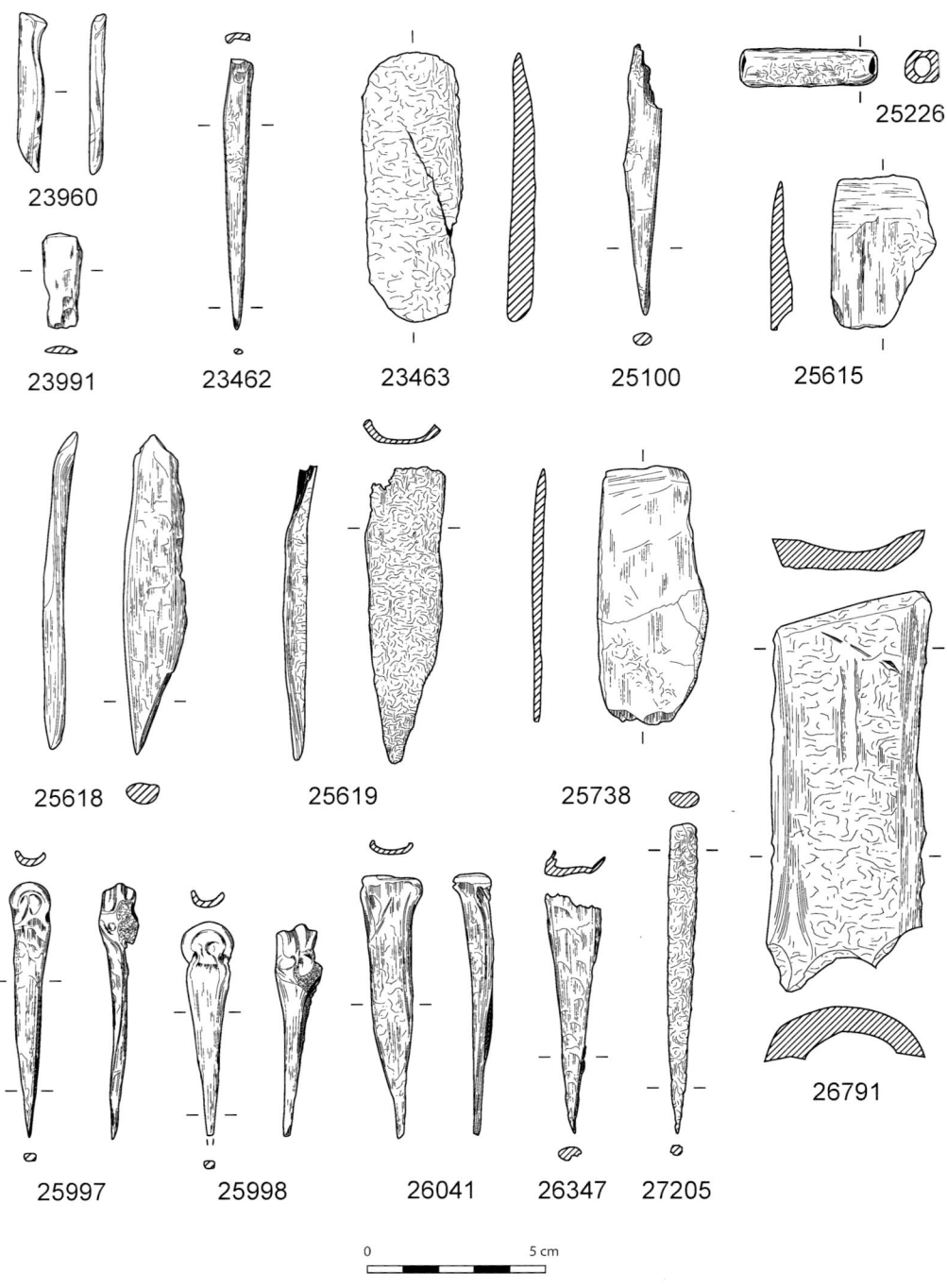

FIGURE 5.62:
Modified piece SF 23960, polished piece SF 23991, trimmed pieces SFs 23462 and 23463, point SF 25100, bead segment SF 25226, scoop/plaque SF 25618, points SFs 25618 and 25619, polisher plaque SF 25738, points SFs 25997, 25998, 26041, 26347 and 27205 and chisel/polisher SF 27205. Drawn by Jo Bacon and Leeanne Whitelaw.

FIGURE 5.63:
Unmodified vertebral discs from a possible shark or ray. © Udal Project Archive

is also from phase D (Figure 5.62). The bones were predominately large ungulate long bone shafts, of which one was split sagittally. They varied in length from 27 to 92 mm.

Description of unmodified pieces

The unmodified pieces include 15 vetebral discs from one or more cartilaginous fish, possibly shark or ray that were found on the uppermost floor of Building 2 in Phase D. These discs (SF 26032 and 27228) were found on two separate excavation days but both samples were close to the hearth area and may indicate they were from the same fish. The pieces were photographed prior their unexplained loss from the collection. Examination of the image (Figure 5.63) indicated that they had probably not been modified (Catherine Smith pers. comm.).

A vertebrum, SF 26888, of a large species of whale is formed of three conjoining fragments, the centrum, lateral process and neural arch. This object was placed as a structural or ritual item above a natural shaft in the bedrock (DC). There was no evidence of butchery, but the piece is stained or bleached from its exposure to the weather during excavation. Its caudal hole was described by Crawford as a libation orifice. A fragment of the bone was radiocarbon dated SUERC-71143 to 2432–2195 cal BC (see PART 3). It is the only whalebone to have been excavated and recorded on the site.

Discussion

Taphonomic conditions have played a considerable part in the survival, or not, of organic materials on the site. Bone artefacts in particular have suffered due to sand accumulation and abrasion, to the extent that there is much uncertainty about what the objects are and what species of animal they were made from. The collection contains points, roughly shaped scoops, plaques and a possible pounder, most of which were found in connection with the last floor in Building 2 in the late Neolithic. The poor condition of other objects in other phases of the site may, as with other cultural artefacts, suggest that they may have been disturbed from their origins in Phase D.

Finding published comparative material in the same geographic area and from the same periods is difficult. However, the small collection (20 pieces) of worked bone from Neolithic levels at Northton (Murphy and Simpson 2006a: 72-73) was well preserved with points, spatulate objects and others indicating intensive use. Longitudinally split long bone shafts were used from both cattle and sheep/goat, as at RUX6, but Northton exhibited a more extensive range of tool types, and well-manufactured antler objects (ibid: figure 2.34). The Beaker levels at the same site produced more examples of spatulate objects/rubbers, points, but also beads, combs and a plaque (Murphy and Simpson 2006b: 140-

41) indicating that more favourable taphonomic conditions allowed the preservation of a more informative collection of personal items and tools missing from RUX6. This is certainly also true of Northern Isles sites, where for example, the Neolithic Knap of Howar on Papa Westray, (Henshall 1983: 75-78), and Skara Brae (Child 1931), Orkney produced an extensive collection of points, pins and awls from sheep long bones, a possible pin or spoon, and larger implements made from cetacean bone.

A useful study of bone and antler use from the Iron Age at Foshigarry and Bac Mhic Connain, west and south of the Udal peninsular (Hallén 1994: 189-231), provides a good overview of a range of artefacts that were found at these later sites. There are examples of antler picks (ibid 203-204), awls, pins and spatulate objects, which shows the continuation of objects and tools types from earlier periods but also their development and refinement. Examination of these few sites emphasises not just how many bone objects have been lost from the archaeological record at RUX6 but also the information they contained.

PART 6 Discussion

By Beverley Ballin Smith

The origins of settlement on the site

During the course of the project it became apparent from the excavated evidence that the two late Neolithic domestic buildings DH and DJ were not the earliest features on the site. The anthropogenically mixed upper levels of the subsoil (Phase E) contained some material cultural samples and artefacts. It is difficult to view this material as deriving entirely from Phase D activities, in spite of the disturbance of the deposits by the digging of postholes and pits. That view is reinforced by the 100 mm thick build-up of turf with some shell sand on the top of Phase E, which contained charcoal concretions with some occupation-type material, indicating human activities prior to the construction of the two late Neolithic buildings. The large mound of ash (DG) situated over a fire pit stratigraphically predated all the recorded late Neolithic activity and the buildings, and led to questions about its date and function. It may have been similar to a slightly later pit and burnt area (DK) which was used, most likely, for firing pottery. It was succeeded by the shaft and platform monument to the east and the very scant remains of an earlier structure (DA), lying partly under Building 1 (DJ) and its levelling. This combined evidence, suggests at the very least, an earlier arrangement of settlement with the remnants of a building being replaced on a slightly different plan by the construction of the two domestic structures.

Sometime after the major sand accumulation at the end of the late Neolithic (level X) the first marine incursion (levels IX.31 and IX3.2) occurred, which included material that was not water-rolled in its matrix of predominantly cobbles and shingle. Crawford suggested that some of this angular stone could have derived from the stone foundations of earlier buildings, situated to the north that were also of Neolithic date. He thought that they had been exposed by strong winds from their covering of sand and then removed by the flooding. This is not conclusive evidence of their being earlier or contemporary structures related to the current settlement, as slabs of rock may have broken off the coastal exposures of bedrock during the event. However, Crawford's hypothesis is worth bearing in mind and there might be an element of accuracy in his view. The evidence also implies that settlement extended northwards beyond the limit of the excavation.

The residues from two pottery sherds found on the uppermost floors of the two domestic Neolithic structures appeared to be earlier than other samples from the same floors (Hamilton, PART 3), possibly indicating that they were residual. Further investigation indicated that the pottery vessels may have been used for cooking marine foods such as fish, which gave too early a date until their ^{13}C had been recalibrated. This acceptable explanation and the recalculation of their dates fits better with other samples dated from the structures. Irrespective of whether the samples were early, their dates remained in the accepted range of c. 3000 - c. 2500 BC for the late Neolithic occupation of the site (Brophy and Sheridan 2012). The data available leads to the conclusion that there is not a single piece of material cultural evidence that suggests an early or middle Neolithic date for the origin of the settlement. If changes to the buildings took place or there was a reorganisation of settlement, it all took place in the time period of the late Neolithic

FIGURE 6.1:
Excavation of the bowl-pits above the plough marks. The coastline is immediately beyond the trench with *A' Croig Bheag* inlet between it and *Rubha Huilis* – the headland beyond. Looking north-west. © Udal project archive.

There is, however, another piece of evidence that indicates that the late Neolithic landscape, and therefore also the settlement, may have continued further west and north. The surviving plough furrows in Phase C (level IX.35) (see PART 2) drawn across the stabilised blown sand of level X, probably date before *c.* 2200 BC during the transition of the late Neolithic into the early Bronze Age (see PART 1). Crawford described the furrows as continuing beyond the shore-face erosion and running towards the present day inlet of *A' Croig Bheag* with some heading north-west towards its mouth (Figure 6.1). He implied that the land was much further out to sea and at that time there was no sea inlet. This is explored further below (in Further landscape changes and new structures).

The above evidence indicates that other buildings probably existed further north, and that some of the anthropogenic material could have originated from them before the construction of the two domestic structures (DH and DJ). The orientation of these two surviving buildings and their location is not unlike other late Neolithic structures in the north of Britain such as the Skara Brae village, Orkney (Child 1931: 4) that was covered in sand, only to be revealed in a storm in middle of the nineteenth century, and the two interlinked buildings on Papa Westray at Knap of Howar that were exposed in a similar manner early in the twentieth century (Ritchie 1983: 41). More recent excavations at An Doirlinn, an eroding islet in South Uist (Garrow and Sturt (eds.) 2016 and 2017) (Figure 6.2), indicated a similar scenario of the survival of partial structures from as early as the early Neolithic through into the early Bronze Age, which indicated a larger settlement with a long time line. In these cases, coastal erosion is responsible for having taken parts of these settlements and has left us with only a very restricted and truncated picture of what was there originally, as in the case of RUX6.

The uses of ritual and domestic structures

Ritual

The difficulty in deciphering the sequence of the construction of some of the earliest of the late Neolithic structures is that the stratigraphy was not entirely clear to Crawford. This was largely due to the Phase A saw-pit which cut vertically

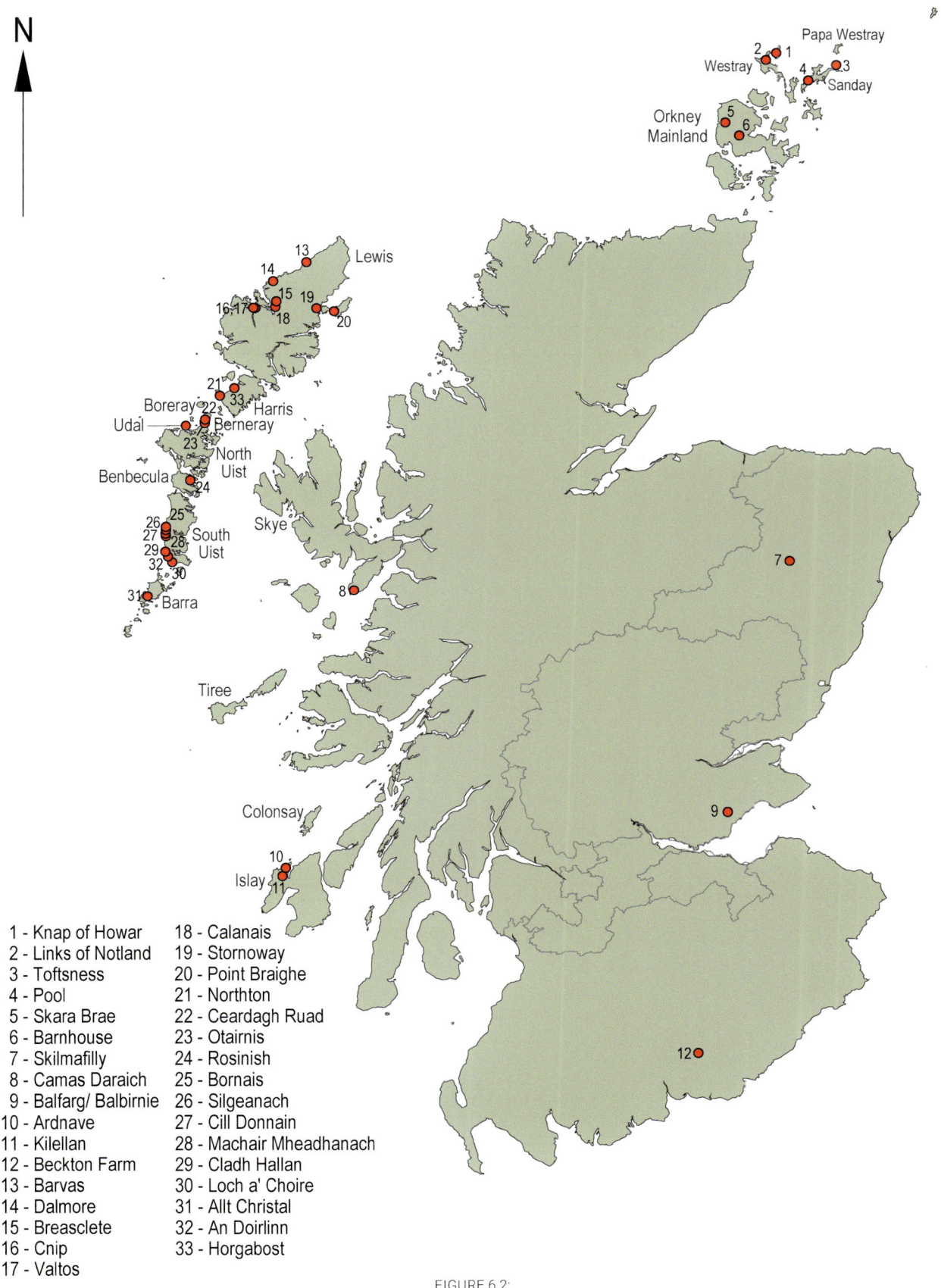

FIGURE 6.2:
Map of sites mentioned in the text.

down through the area, and to the disturbances caused by later activities including marine erosion, especially to the east of the saw-pit. We are left with a number of seemingly related structures such as the stone platform with shaft (DA), a pit (DL), a whale bone vertebra (DC), and a covering mound (DB), which formed one structural unit, but there are other aspects to it (Figure 6.3). Two thin, radiating and curved stone alignments (DF) with wooden posts are considered part of that unit, together with the Great Auk standing stone (DD) situated to one side and half way between the platform and the domestic structures to the west. The pattern of what survives is not easy to interpret especially as Crawford considered that other stone alignments were removed due to later disturbances.

The dominant structure is the platform, which surrounded and formalised a shaft created by a protuberant piece of bedrock, and which had a whale bone positioned to one side of it. Although there is no record of what the shaft actually looked like, or a section drawn through it, one interpretation is that it could have been a freshwater well or water basin, with ground water pooling at its base. The reason why a pit was dug around it may have been to enlarge the catchment area. With landscape and ground water level changes it probably dried up and its function changed to one imbued with a different meaning and associated with different rituals, and where the shaft emphasised the void to a place underground or to the underworld. The identification of votive offerings, suggesting a ritual deposit (Merrifield 1987: 41), is difficult although some artefacts were found in the shaft. These included fragments of butchered bone, a bone chisel and spatula, fragments of pottery vessels, a quartz anvil and scraper, and a hammerstone. Some can be discounted as accidental entries, but some of the artefacts may have been deposited deliberately as a form of offering or for safe keeping. The whale bone may have been a later addition to the design and appeared to be an expression of functional change. Its radiocarbon date suggests it was added towards the end of the late Neolithic (Table 3.1).

The excavations at Beckton Farm in Dumfries and Galloway, produced a number of deep pits arranged in a square, which could have received a wooden platform, and to which Grooved Ware pottery was intimately associated (Pollard 1997: 113-117). The pits were considered to be for ritual rather than domestic purposes, perhaps as shafts for the reception of an excarnation platform with nearby cremation pyres, as this interpretation had been made for similar remains at Balfarg/Balbirnie in Fife (Barclay and Russell-White 1993, and Figure 6.2). RUX6 certainly had a stone-built platform as well as the shaft, and although Crawford's interpretation of part of the feature was vivid and suggestive of the pouring of a liquid though the whale bone orifice as a ritual offering to the gods, or the spirit world that lay below ground, he did not take his interpretation further. He did not explain the remaining features in any way, and certainly had not entertained a connection with the possibilities of excarnation or the laying out of bodies before burial.

At some stage the platform was transformed into a low elongated mound of stone that encompassed the whale bone but left its orifice open at ground level, and from which the stone and wooden alignments curved and radiated away. These changes indicate that there was more than one phase to this structure: the shaft and platform being among the first, followed by the mound and possibly the stone alignments combined as secondary alterations. Together they imply that the structure was important, even though a transformation occurred between its original function and purpose and that of another. The construction of a mound over the feature to partly hide the orifice implies that its burial in stone was deliberate, and that the mound and platform were of significance to the settlement. Importantly, the structure was not slighted or taken apart, inferring it still had meaning, and that the meaning was perpetuated beneath the stone mound. With the onset of the accumulation of blown sand at the end of Phase D, the structure may have been deliberately covered with stone not only to protect it but also to ensure its survival into the future.

The low stone and wooden alignments lay within a few metres of the platform and may have marked off specific areas for individual purposes, especially with small posts at their ends. Alternatively their purpose could have been to simply enhance the platform by leading the eye or the person directly to it. Crawford thought the stones of the alignments were chosen for their colour and shape, adding to the radial arrangement and ritual purpose of this small complex of features.

The Great Auk stone was part of this monument. Its 'beak' did not point directly at the platform or shaft but to the south, away from buildings and towards the sun at its zenith, perhaps acting as a

FIGURE 6.3:
...ale vertebra indicating the position of the shaft and platform
...covering mound of stone with a stone alignment leading away
... The Great Auk stone can be seen on the right opposite the
...tone alignment. Looking south. © Udal project archive.

form of time keeper or primitive sundial. This stone was deliberately placed, positioned and packed with a combination of used tools and unworked stones. It was located on the edge of the complex of stone and post alignments, between domestic and ritual activities, and may have formed a simple demarcation or barrier between the two, with posts behind it and to the south-west reinforcing this idea.

Speculation is not fact, and the facts we have are limited by the amount of evidence that survives. Proposing an understanding of structures is always problematical especially when we do not have a more complete picture, but Crawford's eccentric naming of the stone alignments as 'garden gnomes', possibly indicates his line of thinking of some of this ritual unit. We know that this complex of structures was older than the two domestic buildings and that it survived until they went out of use and the settlement was abandoned. At that point these structures became submerged below the sand and remained largely covered until the excavation in 1980. However, their location may have been marked in the landscape as their survival seemed to influence the siting of cist burial and cairn during the early part of the middle Bronze Age.

A domestic pit (DK)

This feature, which may have been contemporary with the two buildings (discussed below), was located near their junction to the south, and was identified as a burnt area. It contained sherds of discarded pottery and ash from the burning of heather (see Ramsay this volume). It is quite possible that this may have been the site of a pit for the firing of pottery made from clay from Building 2 and possibly manufactured in either of the two buildings. If it was a pit for firing pottery it would have been easy to achieve by the digging of a scoop in the sandy earth and lining it with combustible material such as heather roots, turf and waste from driftwood. The newly made pots could have been layered on top of the prepared pit and then covered over with more combustible material. A final covering of turf would have sealed the pit ready for firing. The process would have left a burnt area after use, with sherds of pottery indicating pots that broke during firing.

This feature is unlike the stone-built kilns with flues found at Eilean an Tighe by Scott (1951) of early Neolithic date (Figure 6.4), but nevertheless burnt pits of this type would have been the commonest mode of producing pottery locally.

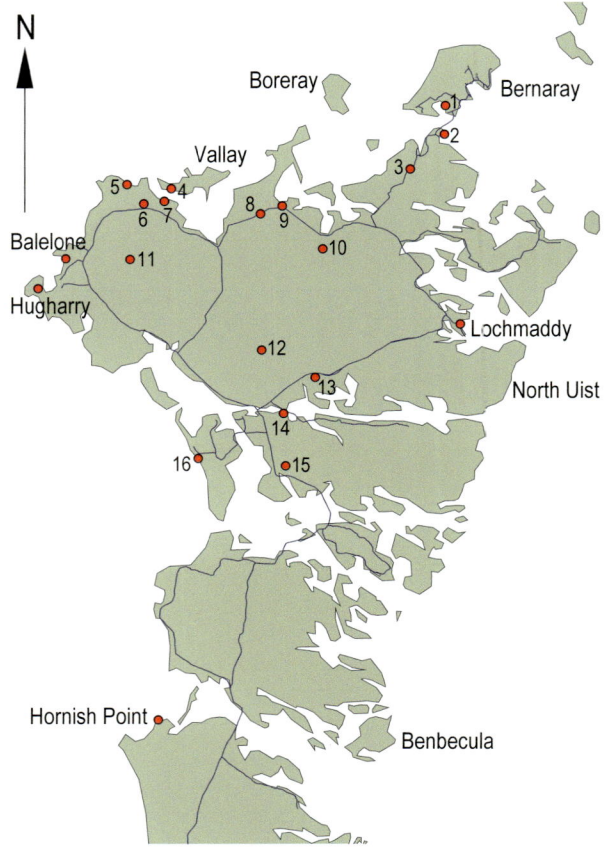

1 - Ceardagh Ruad
2 - Otairnis
3 - Port nan Long
4 - Bac Mhic Connain
5 - Foshigarry
6 - Eilean Olabhat
7 - Griminish
8 - Sollas
9 - Grenotote
10 - Eilean an Tighe
11 - Clettraval
12 - Unival
13 - Barpha Langais
14 - Craonabhal
15 - Bharpha Carinish
16 - Baile Sear

FIGURE 6.4:
Map of North Uist sites mentioned in the text.

The buildings

Two buildings do not necessarily make a settlement but they can be viewed as evidence that there was one, with the remainder of it removed by coastal erosion. Building 2 (DH) to the east was the earlier and the smaller of the two. It was partly built over by Building 1 (DJ) to the west, which was the better preserved.

Both the RUX6 structures were designed as domestic dwellings with circular or sub-circular wall outlines, built to a similar pattern with an inner and outer face of stone infilled with sand and turf. The outer face of stone where it survives of Building 1 is slight, with the inner face being built of heavier stone. The evidence suggests the walls of the structure may not have been high but probably mostly built of turf. Their doorways were positioned in the northern portion of each building, with a fire or wind screen between them and the central hearth. Both hearths were centrally placed within the buildings and were demarcated by stones placed on edge as a kerb. Building 1 had four substantial posts positioned in a square around the hearth to support the roof, and although the arrangement of Building 2 may originally have been similar, the pattern had been altered indicating changes to the roof and the structure over time. Building 1 provided evidence that it had a sleeping or working platform supported on two slight partitions or rows of stones in the southern part of the building with an opposed structure across the centre of the room indicated by two parallel stones set in the floor. The arrangement in Building 2 is less clear but a more substantial partition is indicated in the southern part of the building with a stone arrangement in the north, just east of the door. The evidence of drains in the northern part of these buildings suggests there was an issue with ground water flowing off the bedrock.

The alteration of Building 2 seemed to include the addition of an extension to the east. Although the evidence is not entirely clear, it appears that the building wall was partly taken down to allow for the construction of an area (enclosure or cell) that may have led outside to a yard. The extensive levelling that was needed for this construction abutted the earlier platform and mound and suggests that some of the posts aligned with the Great Auk stone may have edged this yard, but there was too much disturbance to be certain of the sequence of events and what those events actually were.

The evidence indicates that Building 1 replaced Building 2 but with a slightly larger ground plan. It showed no signs of modification, whereas Building 2 was altered, reroofed and possibly extended in response to the construction of Building 1. There may have been an interconnecting door between them, but this is uncertain. With the construction of Building 1 the function of Building 2 may have changed into that of a workshop and this is noted in the material culture evidence and the events that took place there.

During the excavation of these structures and the processing of flotation samples from Building 1, some of the residues were lost or accidentally thrown away. This has affected the amount of evidence, including small bones or shells, and the fine knapping debris from the floors that were expected to be found, and also the interpretation of the activities that occurred there. It has also not been possible to say with certainty where activities took place, although west and north of the hearths seemed to have been the most favoured areas.

Building 1 (DJ)

The most unusual find associated with this structure was the axehead, SF 23762, that was located in the fill of the building wall on its west side. It could have been accidentally incorporated in the infilling material of sand and turf, but its weight and size suggests otherwise. The placing of this functional object in the context, in which it was found, indicates that it was an intentional offering that expressed symbolic or commemorative behaviour. It may have been a curated object from the past or one imbued with communal memories or symbolism, and was most likely considered a protective talisman. Stone axes or 'thunder stones' have been known about since classical times with reference to their magical purposes and protection against lightning, and were often placed in the roofs of buildings (Merrifield 1987: 10-13).

This building produced some heather charcoal from the uppermost floor but no cereal grains, and this limited evidence tells us little about the fuel burnt on the hearth or what covered the roof. However, as driftwood was used in Building 2, it is more than likely that the posts and the roof timbers were derived from tree trunks washed up on beaches. The building generally contained marine shells and molluscs, some of which were found in postholes and some in the drain(s) possibly from abandonment of the building, but shells of limpet and whelks were probably deliberately brought into the building to be processed on or near the hearth. Mammalian food evidence was more informative, with bones representing a minimum of eight sheep, six cattle and the occasional red deer embedded or scattered through the floors of the building. Part of a cast red deer antler was found in a pit possibly indicating it was used as a pick or as a resource for flint and quartz knapping. Finlay Aird Part 4, from her study of the faunal remains, suggested that metapodials from sheep and cattle were used for tools and that there had most likely been an area in the building for their production. The presence of worn, grooved and smoothed pumice pebbles reinforces that idea, even though pumice could also have been used in the manufacture of other materials such as pottery or skins.

It is also likely that other tools found in the building such as quartz scrapers, edge touched pieces and a scale-flaked knife also indicate the production of bone tools and perhaps the skinning of hides. The fine detail of the preparation of quartz and flint, i.e. the knapping debris, was unfortunately lost with the flotation samples. Pieces of flint and a core, as well as additional prepared pieces of quartz including regular platform blades, indicate that the knapping of raw materials took place in the building, most probably around the hearth (see Ballin PART 5). There was also evidence of the presence of the occasional bird bone, and fish – cod and ling in particular, with the rare vertebral disc from a cartilaginous fish such as a shark or ray.

A fewer pottery vessels were found in this structure than in Building 2, but three in particular, SF 23664 a vessel body, and two bases SF 23665 and SF 23685, were found in pits in the floor, indicating the vessels were deliberately positioned there for daily use. Two of these vessels had thick carbonised food remains indicating their function as cooking pots.

The limited material cultural remains and their associated information from this building suggests three scenarios. Firstly the floors of the building may have been regularly cleaned. Secondly, that disturbances from later periods had removed a lot of evidence from the southern and western areas of the building, and thirdly that prior to the abandonment of the building, during the accumulation of blown sand, tools and pots may have been deliberately removed and relocated elsewhere.

Building 2

This building provided a wider range of evidence for its use than Building 1, although the activities taking place in both structures seemed similar. Small amounts of carbonised heather and an even smaller amount of larch charcoal were found on the last floor of the building in and around the hearth and to its east and north-west. The presence of larch indicated the use of driftwood for structural purposes, possibly for its roof supports. The heather charcoal found on the hearth implies that it was used for fuel. A pit in the floor contained limpets and both limpets and whelks were generally found in the floors of the building. Interestingly, the land snails found in a pit in the floor indicated that the conditions inside the building were damp, perhaps due to seeping groundwater.

The floor(s) were littered with the fragmentary bones of domestic species. The partial remains of a minimum of 18 sheep were counted as well as three cattle, and the bones of dog were also present. Red deer had also been brought back to the building as well as their antler. Fish were identified from the bones of ballen wrasse and there were many more discs from vertebrae of cartilaginous fish such as shark or ray than in Building 1. Finlay Aird (PART 4) considered that the animals' carcasses may indicate processing (butchery) for meat as well as bone. Butchery debris was noted in the floor deposits as well as bone tools, confirming that tool making from the carcasses was undertaken in this building. The bone tools analysis (Ballin Smith, PART 5) identified points of various types, scoops, a plaque, and a polisher. Pumice pebbles, which were worn and grooved, indicated the smoothing and finishing of tools, and perhaps the manufacture of pottery vessels, also took place there.

The presence of a pit besides the hearth that was dug down to subsoil to collect clay suggested the possibility that pottery was also manufactured as well as used in the building, even if it was fired outside. Clay had been piled up on the floor in preparation for use, and this indicates that someone living in the buildings was skilled at pottery making. It is possible the clay was collected for other purposes, but there was no structural evidence for its use. More undecorated pottery sherds were found scattered in negative features, by the hearth and on the floor than in Building 1, suggesting pots were regularly used for a variety of activities in addition to that of cooking. One of the most interesting pots SF 25672 was represented by a number of small but highly decorated sherds indicating that this was a late Neolithic Grooved Ware vessel (Figures 5.55-5.58). Located close to this vessel but stratigraphically above it in the infill of the building was the remains of a decorated Beaker SF 25442. The floor of the building and the infilling deposits over it also produced the remains of a possible finger dimpled pot SF 26818, the base of another vessel SF 26817 (Figure 5.55), several other pots represented by SF 25712, and SF 25588 (Figure 5.56). The latter sherds with incised decoration suggest it may have been part of the same late Neolithic Grooved Ware vessel as SF 25672. This collection of sherds and pots indicated that a range of vessels had probably been left there when the building was abandoned. The sherds were often very small, indicating the trampling, reworking and disturbance of floor deposits that must have occurred, as well as that of the material lying above them that infilled the building.

The evidence for stone tool manufacture was more clearly defined in this building than in Building 1 because of the presence of a knapping floor east of the hearth (Ballin and Wickham-Jones, PART 5). Both flint and quartz tools were made there and pieces were retouched and also re-sharpened. Quartz scrapers for skins and flesh were noted with pounders and edge touched pieces, but other tools were not present, implying that they had been removed for outdoor activities. Larger stone tools included an occasional flake, a hammerstone and a small hammerstone/anvil. Together they suggested that the building in its altered last state was a workshop used for a variety of purposes including butchery of animals, the making of stone as well as bone tools, and for the storage of raw clay and perhaps for pottery manufacture.

In contrast to Building 1, the evidence from Building 2 survived to give a much fuller impression of everyday life in the late Neolithic. The evidence of use of its eastern cell or annex is missing, but it is not impossible that animals were stalled there, or it was used as a drying area for skins or meat.

Comparison

Late Neolithic buildings in the Western Isles other than those at RUX6 include Eilean Domhnuill, (Eilean Olabhat) (Figure 6.4), which displayed two conjoined structures of sub-rectangular shape but otherwise of similar wall construction and internal layout to those of RUX6. Structural evidence from other contemporary sites such as Bharpa Carinish (Figure 6.2), Northton, Eilean an Tighe (Figure 6.4) is slight and lacking good preservation (Armit 1992: 311-18). The more comparable examples in size and shape are with the simpler round or sub-round building shapes at Barnhouse, Orkney (Figure 6.2) such as House 6 and to some extent House 1 (Downes and Richards 2005: 109). Their walls comprised small stones with a silty clay infill, a central hearth, a drain, partitions denoting work areas and an interconnecting doorway.

Part of the problem in finding comparable contemporary buildings in machair to those from RUX6 is that they have most probably been lost to rising sea levels, tidal scour and erosion, as well as by sand covering and later disturbance. The Udal buildings, although sharing a similar pattern with other structures of the same period, would have displayed regional and resource differences, which to a large extent, dictated the size and shape of what was built in a given environment, for example, the availability of stone and timber for construction. What is important to consider is the communal aspect of dwellings and the activities that took place there. Due to coastal erosion only two buildings survived at RUX6, and it is impossible for us at this distance in time to calculate how many more buildings there could have been distributed to the north, west and perhaps to the east, but on slightly lower land that was later submerged and removed by the sea in the early Bronze Age. We can gauge from the plans of buildings at Barnhouse and Skara Brae on Orkney that a small village of buildings could have been achieved at RUX6, and it is tantalising to consider, for example, how many other structures could have once congregated close to those that survived at the latter. It is to be hoped that ongoing research in the Western Isles will provide more information on structural complexes to add to that provided by RUX6.

Resource use and identity

It is clear from the surviving environmental and material cultural evidence from the two buildings that a range of resources were used for daily life in the late Neolithic, and as far as can be ascertained there were no imported objects. Although there is little carbonised evidence from the botanical remains from the buildings (Ramsay, PART 4) it is evident that turf/peat/soil was in demand. It was an important part of the construction of buildings, the walls in particular, and possibly for covering the roofs. It was most likely used on the hearths with heather for cooking and providing light and heat. It was also a resource that would have been food for animals and it could have also been required for animal and human bedding. Removing turf from the soil would have had consequences for the fertility of the land near the settlement, which required its replacement with midden-like material and possibly dried or rotted seaweed. We have no direct evidence during Phase D for the tilling of soil and the growing of crops, or even their processing, as querns, pestles or carbonised seeds were not present in the record. However, the lack of evidence is not sufficient to say that no crops were grown in the vicinity. This relatively small segment of what was probably a larger settlement may not reflect accurately the bigger picture of a community needing a much more diverse economy in order to survive, including the cultivation of cereals. The pastoral nature of the agrarian economy at RUX6 was dominated by the presence of sheep and cattle and reflects the situation at Neolithic Northton on Harris, where animal husbandry and herding was the main agricultural activity, although some barley was also grown there (Gregory and Simpson 2006: 83-4).

The general impression from Northton was that people survived by combining various economic strategies such as fishing, fowling, hunting and gathering (ibid), and that was probably the case at RUX6. The number of marine crustacea at the site, predominantly limpets and whelks, could have been eaten or used as bait for fishing. Some fish were eaten but how much was caught by fishing or trapping versus what was brought to land on storm tides is not known, but marine species would have provided some variation to a largely meat and dairy diet. Although there is little evidence of the practice of fowling, some bird bones were identified from the floors of the buildings, but they could have been caught for their feathers or their bones, as well as their meat. Collecting eggs from nesting birds would have been a seasonal occupation, but the gathering of other food stuffs such as seeds and fruits in the natural environment of North Uist may have been an unpredictable and difficult occupation. The botanical record does not include nuts or seeds, and softer parts of plants like tubers are not present either. Although hazel is a native woodland species on North Uist (Ramsay, PART 4) there is little evidence of the collection of its wood and none for the collection of its nutshells during the late Neolithic at RUX6.

Of interest, given the dominance of sheep and then cattle in the diet, is the surprisingly poor representation of deer, only between 6-14% of the total number of bones of individual animals excavated from the site during the late Neolithic (see Finlay Aird, PART 4). Shed antler was brought back to the site, as well as the occasional deer carcass, but whereas antler is generally considered to be a useful resource, especially in historical times, it did not seem to be a preferred item for tools at the site, with very few pieces identified or preserved. The hunting of deer might not have been a lucrative occupation; herds could have been small or difficult to access; or hunting deer may have been restricted by territorial rights. The total absence of quartz or flint arrowheads from the two buildings suggests that they could have been made or stored elsewhere, or that hunting deer was not an economic priority.

The collection of driftwood, mainly larch and spruce, from suitable coastal locations could have been an essential factor in the location of this settlement and in its survival. Tree trunks, a finite resource from forested land in North America were brought over on ocean currents to land on west-coast beaches throughout the Scottish islands. Judging from the number of postholes in the buildings, tree trunks of different diameters were used to form roof supports, as it was probably unlikely that any native timber, at that time, would have been of sufficient size or girth to be able to fulfil this purpose. In fact, carbonised native timber was not present in the botanical record for this phase at all. Timber was a precious resource and it would have been reused. It could have been taken from an abandoned building to be used in a newer, or made into tools or even firewood. Recycling of limited resources may have been an important aspect of life at the site.

The exploitation of stones as a raw material for tool making, such as flint, quartz, pumice and other rocks would have been a localised part of the economy with stony and sometimes sandy beaches providing many of the raw materials needed for essential tools such as scrapers, blades and knives. The flint resource was small but essential, and the maximisation of it would have depended on the skill of the knapper to provide the required tools.

Other items of material culture found at the site indicated that the settlement was part of a regional or wider late Neolithic society. Manufactured pottery vessels, decorated with geometric designs formed by infilled triangles within incised parallel lines and found on the floor of Building 2, exhibit the elements of the late Neolithic Grooved Ware pottery tradition found throughout the British Isles. Most of the applied decoration and the incised spiral motifs are missing but the other decorative elements are unmistakable (Figure 5.55). Only three other sites with Grooved Ware have been confirmed in the Western Isles. These are Uineabhal (Unival), North Uist (one vessel) (Figure 6.4), Calanais, Lewis (Figure 6.2) where small decorated sheds from approximately 10 vessels were retrieved (Cowie and MacSween 1999: 50), and from An Doirlinn (Copper 2016: 26).

Two other Neolithic artefacts, most likely produced in the Western Isles from local stone, are the axehead SF 23762, found in situ in the wall of Building 1 and a stone ball SF 24962, an unstratified find from Phase A. Both of these express Neolithic identity and both probably had symbolic meaning. The axehead was a tool that had a secondary use as an anvil but could have had other associations and significance when it was placed in the wall (see above). Several similar axeheads were found at Skara Brae, Orkney, some on the floor of one building, others in occupation layers or on wall head deposits, indicating some similarities in depositional context with the example from RUX6 (Childe 1931: 98-99). A few axe fragments and flakes have been found in South Uist (1) including An Doirlinn (3), North Uist at Eilean an Tighe (5), and three axes in Lewis (Clough and Cummins 1988: 238 and 241), but where their association with domestic settlement is doubtful.

Stone balls are identified as prestige ceremonial objects belonging not to individuals but to a clan or community (see Ballin 2014c: 53) and although a few have been found in association with burials, by far the largest number is linked to settlements. Carved, incised and decorated stone examples as well as plain stone balls were also found in the buildings at Skara Brae (Childe 1931: 100-101) emphasising the relationship of these items to activities in a domestic setting that could have been either mundane or ritual. It is likely that the RUX6 stone ball came from one of the two buildings, or other contemporary nearby structures that were removed by coastal erosion. A carved stone ball was found as a stray find on Benbecula (Badcock 2008:15) suggesting the use of such artefacts was equally as common as elsewhere in Scotland.

Elements of a wider late Neolithic society, and therefore the identity of the inhabitants, were expressed in the two buildings at RUX6 and in the use and manufacture of certain artefacts. The truncation of what must have been a larger settlement has unfortunately left us with only a narrow glimpse into the life of a community that from the evidence was mainly concerned with survival, husbandry, the butchery of animals and tool making. There may have been much more that would have widened our understanding, but given the negative effects of site formation processes we are lucky to have what survived.

Changes to society after the accumulation of blown sand

The late Neolithic settlement was brought to an end by what can be described as a dramatic and destructive event. The thick accumulation of sand that was blown inland from probably developing dune systems further out along the coast, covered not only buildings, but grazing land and fields and created an inland sand plain (see Hansom 2003: figure 9.3). It is likely, but not certain that the dune system was in the process of forming and therefore at the beginning of the cycle proposed by Hansom. However, the effects of the event were so severe that the landscape would have been largely unrecognisable once it had ceased and conditions ameliorated. The coastline would have altered, houses buried or blown away, and possibly people and animals killed. Those people that survived would have had to move away altogether from their homes taking with them their herds and their possessions, probably to camp further inland initially. It is likely that they had time to develop mechanisms to cope with social stress and the threat, if it started gradually, to move within their own territory but further away from the effects of the sea and wind. They would have also had to build new dwellings perhaps by salvaging anything reusable from their buried buildings.

Crawford excavated evidence (unpublished) of the location of contemporary and later human occupation lying between 100 and 300 m inland to the south-west through test trenching at RUX1-3. However, it is not possible to determine whether the inhabitants of RUX6 moved to this slight elevated ridge to establish a settlement there or whether they moved elsewhere. The accumulation of sand would have caused enormous physical stress as well as the potential loss of harvests and of grazing for animals, and probably necessitating a change in diet for the population (see PART 3). Throughout this period, the transition between the late Neolithic and the early Bronze Age (Phase C), there would have been more accumulation of sand across the area with periods of stabilisation and the establishment of vegetation. There was probably a good passage of time, perhaps several generations, before the sand had consolidated sufficiently for turf to become established. It is, however, without doubt that the former settlement area was overseen and was still considered by the people that moved from it to be part of the community's territorial assets. This can be demonstrated by efforts that were made to cultivate it. The first attempts, seen as colour changes over the filled in buildings were likely to have been made by a mattock to break up the ground, or remove tussocky grass. This was followed, perhaps after a further passage of time and better topsoil

FIGURE 6.5:
Modern ploughing of the machair plain at Udal. Looking north-east. © B. Ballin Smith.

development, by what Crawford termed intensive ploughing that criss-crossed the land leaving plough scars in the sand filled with darker topsoil (Figure 6.5). Although significant landscape changes were still continuing to take place, and the coastline may have moved inland, there was still cultivatable land to the north in what is now the *A' Croig Bheag*. At the time of the ploughing there may have been a marshy area but perhaps not a sea inlet.

The landscape at this time was not entirely flat but hummocky, and one such hummock covered the early ritual monument at the eastern part of the site where ploughing scars reached almost up to it, indicating how extensive bringing the land into cultivation was, once ploughing commenced. Into this transformed terrain a boundary marker was established of three or more large posts set into the ground in large pits, which were heavily packed with stone. The posts were likely to have been driftwood and they were most likely tall. We do not fully understand what this boundary was, why it was marked in this way or why it was oriented NE/SW, but it largely seemed to demarcate activities to the west from those to the east. The activities that took place beyond the barrier to the west were the digging of circular, sub-circular or oval bowl-pits *c.* 1.5 m in diameter and 0.5 m deep, which Crawford puzzled long and hard about. He knew there were both early and later pits and their location seemed to be marked in some way as very few overlapped. What he perhaps did not realise was that they seemed to be dug side by side in rough parallel lines covering the area up to the 1974 eroded coastal edge and therefore originally beyond it. There are aberrations to the overall pattern and one pit was partly stone-lined where it came down on top of the walls of the Neolithic buildings buried below. It seemed likely that pits were dug one after the other, and when they were needed. Crawford thought they had been used and then quickly backfilled with their own spoil and turf. Very little cultural material was found within them, and what there was, most likely came from the late Neolithic building remains below that had been disturbed by their digging. They contained little in the way of recognisable organic material, apart from some carbonised heather, which again could have derived from the remains of the earlier phase.

At the time they were dug, samples were not kept of their infills but Crawford thought they had ritual connotations. Today, scientific testing of their organic fills for lipids, nitrogen and phosphates may have determined their purpose. This author's view is that they were most likely standardised pits for the disposal of human or animal organic waste, which left little evidence in the sandy soils. Not only would the use of these bowl-pits, if they were for waste disposal, have kept buildings more sanitary but their contents would have improved the fertilisation of the area and aided the development of the machair soils on top of the blown sand. Manuring was considered an important activity from earliest times on light soils for crop growing (Fenton 1981). Pits were dug across the site during this phase and their use continued into Phase B indicating a widespread and ubiquitous activity. In a paper by Lauer et al (2014), the fills of enigmatic slot pits (Neolithic) and pit alignments (Bronze Age and later) in the loess soils of Central and Western Germany have been examined scientifically. The pits were located outside settlement sites on areas considered to be arable fields with the Bronze Age pits aligned in rows. The pits contained no cultural artefacts and they were backfilled quickly with humic soil. The analysis of the soils confirmed that the pits contained some charcoal and other organic materials including manure or faeces, and this would have increased the fertility of the light sandy soils. The comparison of the German pits with those at RUX6 is striking, even if the latter cannot be absolutely confirmed as being manure pits. However, one factor that may indicate this interpretation is correct was the widespread cutting of turf for building purposes[22]. Once removed, something had to replace it in order to replenish the soil, and manuring may have been the solution. No matter what their purpose was, pits and the activities associated with them indicated the intensive use of the landscape by people living close by, with perhaps the intention to maintain arable fields.

During the time of pit digging there might have been some habitation to the north-west. A small stone box set into the floor and a pit some metres distant containing the base of a pot with a posthole besides it are all that remain of what could have been two possibly ephemeral structures located close to the north-western limit of the pits as the evidence survived. The occurrence of these features could suggest an attempt at re-occupation of the area for buildings, or that other undefined activities were occurring. Evidence associated with them was washed away in the next major natural event to affect the area.

Ploughing and territory

Discussed above was the evidence that the area covered by the excavation, in spite of the sequence of natural disasters, had been managed or controlled, and was still considered to be part of the territory of a community. Ploughing evidence confirms this, and so might the digging of bowl-pits that may have improved the fertility of the land. With the deposition of a shingle beach by a massive sea incursion, the land was no longer productive in terms of agriculture, and to some extent it had become marginal or even waste land, but it still had great value to the community that possessed it.

Ploughing has been noted at a number of Neolithic and Bonze Age sites in the Western Isles. Ashmore (1975: 36) in his work on the Calanais standing stones in Lewis indicated that the ploughing that was present close to the ritual monuments was a form of ritual cleansing. At Crip, Uig, Isle of Lewis (Figure 6.2), (Close-Brooks 1995: 254-6) marks created by an ard plough in a sand environment, similar to those at RUX6, were found beneath the cairn demonstrating an extensive cultivated area that had been truncated by recent erosion. Mound 18 at the early Bronze Age settlement in the machair at Sligeanach, Cill Donnain in South Uist (Figure 6.2), also had ard marks beneath it but the landscape had been ploughed after the sand had consolidated and turf had begun to grow. Hulled barley was grown but cultivation seemed to be short-lived (Sharples 2012: 219, 221 and Smith in Sharples 2012: 244). Evidence of Bronze Age ploughing with field boundaries was found at Rosinish, Benbecula (Figure 6.2), in connection with a Beaker context, where there the plough marks, including a spade mark, indicated an extensive area was under cultivation with naked barley as the main crop grown (Shepherd and Tuckwell 1977). The Beaker period at Northton, Harris, seemed devoid of ard marks, and cereals did not appear to be grown in the vicinity, as the settlement of the site seemed to be based more on a domestic and wild meat economy (Gregory and Simpson 2006: 153).

The evidence for Neolithic cultivation in the Western Isles is much more limited than that in the Northern Isles, and especially on Orkney where cereals were grown at Skara Brae, Knap of Howar (naked barley), Pool and Toftsness (hulled barley) (Dickson and

---

22 Wet not dry turf was likely to have been used for building as its regeneration and building qualities were superior (Daniël Postma, traditional building specialist, Netherlands, pers. comm.).

FIGURE 6.6:
Modern stooks of barley cut from the machair plain at Udal. © B. Ballin Smith.

Dickson 2000: 48-51 and 62) (Figure 6.2), indicating that different soils and other environmental conditions affected the type and extent of farming carried out. There was no evidence of cereal cultivation at RUX6, although from the slender evidence presented from Northton, barley was probably cultivated in the late Neolithic in Harris (Church 2006: 35). On machair based sites in the Western Isles at the end of the late Neolithic/early Bronze Age, once blown sand had been consolidated by vegetation it was brought into the communities' territorial systems by spading or mattocking and by ploughing. It was a different type of land to what was there before, but it was easy to till and the growing of hulled barley seems to have been the most common crop (Figure 6.6). Although hulled barley was found in the RUX6 Phase B temporary structure, it was not grown in the immediate area around the building at that time, indicating fields were elsewhere in the territory as was the main area of settlement. Where ard marks have survived we can only assume they were for crop cultivation at RUX6, but there is no direct evidence that this was the case. It would seem that ard marks have, under the right conditions, survived under or around later monuments or mounds in machair environments and certainly they stretched out into areas of landscape that have since been lost to marine erosion. It is quite clear that beyond the limited areas known of ard mark survival at that time, more extensive areas of the machair plain were under cultivation. This presupposes that there had been a change in economy in the early Bronze Age of the Western Isles from one that was more pastorally-based in the late Neolithic to a mixed farming system, with corresponding changes in society.

Due to the natural events that were happening in the early Bronze Age, it did not appear that much of the ard cultivation was long-lived. At RUX6, the earlier mattocking had preceded a much more extensive ploughing, but even that might have been short-lived as soils were generally thin. Continuing accumulations of windblown sand with strong winds, followed by marine incursions may have made some of the ploughed areas marginal for cultivation. This idea is reinforced by the construction of burial monuments on formerly ploughed land such as at RUX6 and Cnip, where ploughing could have been part of the preparation or 'ritual cleansing' of the land before a change of use took place, i.e. burial.

Further landscape changes and new structures

The covering of the land with sand was the beginning of further natural landscape changes taking place during the early Bronze Age. What probably initiated the deposition of sand was a noticeable rise in sea level that later caused a major marine inundation. As people from RUX6 had already moved inland, they were largely unaffected by the direct effects of the flooding that occurred and by the subsequent deposition of a widespread shingle beach. However, landscape changes caused by this event would have been extensive and spectacular, as the land to the west and north must have altered again beyond all recognition. Organic soils were removed from the western area of the site close to the top of the now consolidated sand, and any temporary structures were washed away. The deposition of a beach of large cobbles, shingle and other stone indicated the flooding was extensive and forceful. Crawford thought it had even washed away buildings buried in the sand from early phases, but that is unproven. The deposition of a significant and thick beach deposit inland must have been the result of immense tidal power that was likely to have been assisted by strong winds. During this episode the shoreline of the day probably retreated inland to such an extent that the *A' Croig Bheag* was formed by the scouring out of the probably marshy deposits lying there, and a rocky platform was possibly exposed where sand hills and fields had once been (Figure 6.1). Once landscape stabilisation had occurred and, extreme wind and sea episodes had dissipated, and when presumably the threat of further damaging natural events had receded in people's memories, the area of the excavation became occupied again.

The first burial and the first cairn

The first event, which led eventually to the construction of a cairn complex, was the burial of an individual in a pit in the sand below the shingle beach. This burial had been so badly scoured by the 1974 erosion that it was remarkable that it and its contents survived. Any indication that a cist had been built there was washed away, but the pit seemed too small for a cist suggesting the body had been simply buried in the ground. The human remains are too poorly preserved to indicate more than the skeleton was male and a middle adult in age (see Bohling and Buckberry, PART 3). The pit for this burial had been dug down from above the top of the shingle beach and its location, in what later came to be determined as close to the centre of the second of the cairns, could have been accidental rather than deliberate. There was no evidence that an earthen or sand barrow had been heaped on top of it, but its location seemed to be important.

Crawford had mentioned in the annual interim reports that all stone had to be brought to the site, and therefore the presence of it denoted human activities. That was never truer than for the construction of the first part of the cairn complex that survived only as an arc of masonry (BA) that appeared to run across the top of the burial in the pit. Its relationship with this burial is ambiguous and was never made clear by Crawford. However there is an implied passage of time between the two events.

The cairn may have been intended as a single kerbed structure of circular or sub-circular shape and was well-built externally with good stone. Due to the lack of evidence we can only assume it contained at least one inhumation burial, and possibly one that was buried in a cist. There was no indication anywhere on the site of cremation being practised. The resource input to the cairn construction also required the use of turves and sand to fill in behind the masonry, even though a good quantity of stone was used to construct the cairn kerb. What is difficult to ascertain is the overall height the cairn attained and how prominent it was in the landscape. It might not have been built significantly taller than the four courses that survived, but it could have risen to a slight dome at its centre through the piling of turf, sand and smaller stone into a heap there.

The movement of sand and dunes inland, plus the effects of rising sea levels, tidal scour and flooding, probably meant that expanses of bedrock were exposed for the first time that became a relatively easily resource to exploit for the construction of enduring commemorative monuments (Figure 6.7). The positioning of the cairn in the landscape may have been due to several factors (discussed further below) but the availability of good building stone during the early Bronze Age may have been a primary consideration.

The second cist and cairn

The next event to initiate further building on the site was the death of young adult male individual. The preparations for the burial were elaborate and required not just the digging of a large pit but the search for stones of the right size and weight for

Part 6

FIGURE 6.7
Looking south from Hallis across the *A' Craig Bheag* inlet to the site of the excavation, with the exposed gneiss bedrock. © B. Ballin Smith.

lining a rectangular burial chamber. Good quality large slabs of gneiss were used for the construction of the cist that was built below ground level, and a large and heavy gneiss boulder was used to cap it. It is discussed in PART 2 that the capping stone could have dictated the size of the cist, if the former had been located first. An alternative explanation is that it took time to find a single stone big enough to cover the burial chamber. During the interval between cist construction and finding the right sized capping stone, and for it to be dragged to the burial place (Figure 6.8), the body had already lain in the base of the cist some time under a layer of sand. The cist had been left open or only partly covered and it gradually filled in with sand, turf and other debris. The lack of a capstone explains why the cist was left open, instead of it being capped straight away. If indeed this was the cause of the natural filling in of the cist, the situation may have also caused a delay in the construction of the cairn that was to complete the burial.

In the interval between the burial and the construction of the cairn over the cist, one small body sherd of a late Neolithic/early Bronze Age Beaker, vessel SF 23187, was found in the backfilling material around the cist. It had been highly decorated with herringbone and vertical parallel motifs, and it is likely that the vessel may have been involved in the rites accompanying the burial of the individual. Another smaller sherd, a Beaker rim SF 23182, although unstratified was also found in western half of the site associated with the cairn complex. It had short vertical lines in horizontal bands on its rim and represents a second vessel associated with the ritual of burial and the cairn construction.

The position of the cist in relation to the first cairn kerb is curious as the former seems to have been built very close to the older kerb edge, and that part of the kerbed cairn (BA) was removed for the construction of the cist burial. The damaged edge to the original cairn is not easily explained away by modern stone robbing or coastal erosion, and suggests that it was removed by the new construction. The core of the new cairn abutted the old and then it entirely subsumed within its circumference. If, as argued, the early cairn was subject to some disturbance, it implies there was an interlude of time before the later cairn was built. The new construction seemed to have taken some notice that the earlier cairn was a 'special' place as it was not entirely removed for building stone. It survived some stone robbing but also it served a purpose that enabled ritual activities to be centred round it.

FIGURE 6.8: Removal of the cist capping was achieved using the Landrover winch plus rails from the light railway. Crawford bought rails and a bogie to remove sand from other areas of the Udal project. © Udal project archive

The construction of the core of the new cairn over the cist would have required its lid to have been in place, but the necessity for a heavy lid was made apparent by the type of construction that was built next over it. A kerb of waterworn stones (BB) was constructed around the cist to abut the surviving arc of the earlier cairn (BA). Its stones were positioned vertically side by side in a small trench that allowed them to stand upright. Due to subsequent stone robbing the circuit of the kerb could not be traced beyond the south-east corner of the cist, but the overall design must have been devised before the burial took place. Flat stones were laid in the middle of the area enclosed by the kerb and over the cist, with angled stones placed around the circumference at the back of the kerb. By continuing the construction in this way and placing more vertical boulders over the kerb and angled stones behind, the structure was gradually built into a mound of stones that tapered towards its summit at the required height. The central element of the new construction was a dome of stone that was solid, stable and required no infilling of turf or sand. The number of boulders needed for its assembly required considerable manpower to bring them from the coast, but also necessitated a degree of forethought and planning. Its construction was somewhat unusual but it fulfilled two functions. The position of the solid dome meant that the cist could not be opened or robbed, but it also meant that the cairn constructed over it could achieve height, probably between 2 and 3 m, or as high as the core of stones and its diameter would allow.

The cairn was completed by the addition of a c. 2 m wide kerb (BC) of facing stones backed by boulders, sand and turf. It narrowed to the north-east to encompass the earlier cairn and swung round to the south and east to encompass the central dome of stones and the buried cist. Whether it continued its course in a complete circle is not known due to its removal by coastal erosion. The kerb would have been constructed at least a few courses high, with loose stone and turf filling in behind it to slope up the construction to join the height of the central dome, and thereby creating an artificial mound (Figure 6.9). This newly built monument reinforced the idea that the buried individual, as well as the structure above and around him, was important, but it also emphasised the status of the community that built it, and it made a visual statement. The cairn would have been the most prominent man-made edifice on the peninsula and perhaps in the local area, and as such it would have been a land mark. Over time it would have slumped and consolidated and been partly masked by a covering of blown sand and turf to merge with the landscape's natural landforms around it. At the time of construction, the structure that was monumental in appearance.

Land and monuments

Keeping a hold on territorial seemed to have been an issue at the heart of the community and reflected the changes to the land and its use, and perhaps also changes in society. If land could no longer support agriculture, or was not needed for it, land could be put to other uses. One mode of reinforcing a 'ownership' claim on land by a small-scale segmented society was by erecting a burial cairn on it, as at RUX6, which would have been a 'symbolic expression of territory' (Renfew 1976: 205). Renfrew also remarked (ibid, 206) that ritual monuments were a central focus of territory and also for the events that took place there. The erection of the cairn complex at RUX6 was a territorial expression of monumental proportions, perhaps needed at a time of dramatic natural landscape changes, or as Renfrew has argued (ibid 218), possibly because of social stress due to the effects of the natural environment on farming and on the community.

Kerbed cairns for inhumation burials were constructed at a number of places throughout the Western Isles, with either upright stones or coursed masonry formed of larger stone at their perimeters. The most comparable examples to the cairn complex at RUX6 include a kerbed cairn at Cnip that was erected over the centre of an earlier D-shaped cairn (Close-Brooks 1995: Illus 6 and 7), with a slightly elevated central portion over a corbelled cist. A cist burial of Bronze Age date was located close by the kerbed cairn and this too may have also had the remains of a kerb encircling it (Dunwell et al. 1995). Another example in Lewis, the Breasclete kerbed cairn, was found at Olcote near Calanais, which had two kerbs and a central cist (Curtis and Curtis 1999; Neighbour 2005). A corbelled kerbed cairn was excavated by Crawford in 1977 at Rosinish, Benbecula and a possible corbelled cairn was recently exposed in a low cliff by coastal erosion of the machair near Horgabost, Isle of Harris (Roper and Murray 2017) (Figure 6.2), but it did not appear to have a kerb. Two kerbed cairns are also located inland from Udal at Craonabhal but in a complex of other monuments including chambered cairns, and one nearby at Otairnis on Boreray with a large kerb (Badcock 2008: 61 and 52) (Figure 6.4). The remains of these monuments indicated they were

FIGURE 6.9:
The cairn complex during excavation in 1974. Looking north-east. © Udal project archive.

constructions with probably little height but possibly located in prominent positions and therefore seen from a distance, but also their relationships to other monuments were probably of great importance. The seemingly low Horgabost cairn was situated c. 100 m from a standing stone known as the McLeod's Stone (Roper and Murray 2017: 6). The height attained by the central core of the RUX6 kerbed cairn enabled it to be a highly visible territorial marker. It shares characteristics with the cairn at Cnip in its enveloping of an earlier monument and with the cairn at Breasclete because of its double kerb, but it stands apart through its design and the statement it made in the landscape.

Final activities on the east side of the site

After the construction of the cist and the cairn complex in the early Bronze Age, probably no earlier than 2270–1920 cal BC (see Hamilton, PART 3) there is no further recorded human activity on the western part of the excavated area until the nineteenth century AD. Evidence from both the stratigraphic record and the radiocarbon dates suggest that activities on the eastern side of the excavated area took prominence after the cairn complex had been built. Although the eastern area had been affected by sand accumulation it had escaped the sea inundation and the deposition of the pebble beach by some metres. During the building of the cairn complex it is unlikely that the landscape to the east had been abandoned, but there was no record of contemporary activities taking place there.

Part of the Phase C linear barrier of three or more large posts set up at the edge of the main area of bowl-pit digging was still visible. The Phase D shaft and its mound of stone may have been covered by sand and uncovered by wind several times and its position may have been noted. One item recovered from the shaft during the excavation was a late addition, perhaps the result of burrowing rodents that took a bone in there after the monument had been partially sealed and sand had accumulated round it (see Table 3.1).

The standing stone

Crawford considered that the excavated evidence of the early Bronze Age activities in the east was ritual.

He was probably accurate in his interpretation about some of them but others have been reassessed as being more domestic in character. One of the features concerned the erection of a standing stone (BD) in the southern part of the area. The southernmost posthole of the Phase C linear barrier was dug out and the post, if it was still present was removed and replaced by a tall stone of rectangular, dense gneiss aligned almost exactly north/south. When erected it may have stood approximately 1.5 m above the ground but it would have appeared elevated through the creation of a plinth of turf and stone that was packed around its base and capped by two flat stones. The relationship of this standing stone to the cairn complex is unknown, but it seemed to re-emphasise the linear barrier that existed previously, and in the Uists standing stones often seem to be located near other prehistoric monuments (Badcock 2008: 22). Perhaps it was a territorial marker denoting the zone between ritual and domestic areas of the landscape, or perhaps it was a link to the past, marking the Phase D shaft, platform and mound, which lay only 5 to 6 m to its north. Due to coastal erosion and the limits of the excavation we do not know whether it formed a line to the north with additional standing stones. Seen in isolation much of the understanding of the stone and its position has been lost.

The temporary dwelling and workshop

However, the standing stone was not isolated for long. Constructed in the slight sandy hollow formed between its mound to the south and the mound of the Phase D covered shaft and platform (DB) to the north was a structure (BG24) accessed from the east. It was sheltered from the prevailing south-westerly winds by its location and it was probably used as a temporary dwelling or workshop. The structure did not have stone walls, but it was demarcated around its perimeter with stones on edge or laid flat, indicating its flimsy almost tent-like construction. The seven or more large postholes in its floor housed supports for its roof and reflected the various times the structure had either been rebuilt or altered during its long use. Several radiocarbon dates (Table 3.1) indicate its intermittent occupation over a period of at least 200-250 years towards the end of the early Bronze Age and into the middle Bronze Age, ending somewhere around the beginning of the fifteenth century BC.

During its period of use the remains of four or five fire pits were dug into the centre of its floor, each one replacing an earlier one, and generating large amounts of ash. The floors spread out beyond the edges of the building and the botanical record indicated that local wood such as birch and hazel were present, either used as fuel in the fire pits, or in roof construction, along with driftwood. The occurrence of local timber is important as it suggests native woodland was present somewhere in the vicinity but not necessarily on the peninsula. Heather was also used in the building along with turf, peat or soil indicating the different resources available nearby for construction and bedding, as well as for heat, light and cooking. The structure was also informative from the point of view of food as it produced the only evidence of cultivated cereals - that of barley or hulled barley, from the excavated area. Carbonised grain was trampled into the ashy floors but it was likely that some of it been processed (dehusked and ground into flour) in the structure. A saddle quern, possibly originating in the building, was reused as packing for the cist that was inserted into its floor after the structure was abandoned. Associated with the quern, and seen as part of a food preparation tool kit, was a stone grinder also found on one of the floors.

Food prepared or eaten in the building included that from cattle and sheep, crustacea and limpets, with two pits in the floor containing marine shell. The bones of a greater black backed gull were found in the floors as well as a deer tooth, but as their remains did not appear in the record it is unlikely that fish formed part of the diet. Interestingly, a single dog tooth was found. Whether the dog was a working animal or one for the pot, is open to speculation (see also Finlay Aird and Smith, PART 4).

One of the most interesting aspects of the use of this structure was the number of flint artefacts (17) including flakes, and the high volume of quartz with a raw material cache put away for safe keeping in a box (BG3) set into the floor (see Ballin, PART 5). A hammerstone/anvil in basalt or dolerite found in the structure was also part of the tool kit used in the knapping process. The structure was used as a lithic working area or workshop, as copious evidence of flint and quartz tools and debris found there indicated a knapping floor. Small numbers of sherds implied that pottery vessels had been used, although some sherds are burnt suggesting they had sat in the ashes of a fire pit, or when discarded had fallen into the fire. Sherds of a vessel SF 26495, found in a pit beneath a stone in the western part of the floor, were part of a plain cooking vessel, and other random sherds were also from plain but broken cooking pots. A single bone point was also found on the floor.

The appearance of this structure contrasted greatly from the enduring stone-built constructions. The lack of solid walls and the number of postholes indicate it was most likely a tent-like structure with animal skins covering driftwood poles and held in place at ground level by ropes tied round its upright edging stones especially in the west. The trampling of ashes to the north, south and east indicated that either there were no walls there or that they were flimsily constructed with woven heather branches, for example, or that slightly more durable turf or peat was used. Its east side could have been closed with a skin flap.

This single temporary structure, which underwent many alterations is important as it was the only domestic dwelling since the late Neolithic on the site that demonstrates conclusively that people continued to live in the area, used its resources, made tools and ate there (Figure 6.10). In a sense, it is remarkable that evidence survived of it, as it contrasts strongly with the ritual and monumental cairn complex in the west and the permanence of the nearby standing stone, but the information it has yielded about the daily life of people, rather than their care in the afterlife, is important.

A shallow linear trench (BH) had been located during the excavation that seemed to course from the west to the temporary structure but its relationship to it or the cairn and cist were unconfirmed. It was suggested in PART 2 that this may have been the remains of a more recent saw-pit. Sherds of one or two early Bronze Age Beaker vessels from possibly late in the period were found on the east side of the site. The largest number of them (30), SF 23403, seemed to be related to a floor from their recorded description. However, the information is not specific as to whether they came from the temporary structure or from ash outside it. The finds record for the other two samples SF 23391 and SF 23395 only indicates that they came from the east side of the site. These sherds are all small and heavily abraded but are incised with marine-shell edge decoration. It would seem most likely that one or two Beakers were either used in the structure or were present in connection with the later cist burial and the construction of its cairn, but due to the lack

FIGURE 6.10: Artist's impression of the temporary structure.

of information it is not possible to investigate this further.

The comparison of this structure with other early Bronze Age examples is slightly easier to achieve than those of the late Neolithic because of their higher numbers and better preservation away from the machair zones of erosion. Sharples' unpublished work on Dalmore, Lewis (Figure 6.2), indicated there were a number of non-permanent superimposed structures with hearths of the early Bronze Age, partly dated through the typology of their associated lithic artefacts (Ballin 2002b), but further information is awaiting publication.

In machair areas, the preservation of structures that are considered temporary is not uncommon. Simpson (1971:138) described two Beaker structures that were oval in shape from Northton. Their walls were of drystone construction but instead of taking the weight of a roof they were thought to be revetments to the sides of the hollow the structures were built into. Like the temporary structure in Phase B at RUX6, the Northton buildings had a number of postholes in an irregular pattern in the floor, which could have held up an animal skin covering to the roof. At Kilellan, Islay, two circular to sub-rectangular early Bronze Age structures were built with a stone revetment to hold back the sand into which they were constructed (Ritchie 2005: 177). A subsidiary or secondary building at Barabhas in Lewis is also described as being of tent-like construction with a revetment against the sand. Cowie and MacLeod Rivett (2010) considered it could be a seasonally occupied hut of Beaker date. Other tent-like shelters have been noted in connection with settlement close to Bharpa Langais chambered tomb in North Uist (Badcock 2008: 16) (Figure 6.4) but these could be late Neolithic in date. The early Bronze Age buildings at Cladh Hallan (Parker Pearson 2015), when fully published should shed more light on the types of buildings and their functions during this period and later. The survival of the single temporary structure at RUX6 is rare but it is seemingly not alone, and indicates that people with permanent dwellings were probably located further inland.

The cist burial

The final recorded human intervention on the site until relatively modern times was the construction of a burial cist (BM) into the northern part of the abandoned temporary structure. With the cairn complex occupying the western part of the site, the standing stone (BD) surviving a few metres away, and the late Neolithic shaft and platform visible as a low cairn to the immediate north, they reinforced the ritual aspects of the landscape and it appropriateness for further burials. The temporary structure may have collapsed, naturally filled in and been grassed over, but enough of it survived for the cist to be constructed into its floor using one of its stones on edge as part of its west side. Through the preparation and the construction of the cist the floors of the building were disturbed and a saddle quern was removed to be used as a supporting stone for the eastern side of the cist. The availability of random stones for its ends and its eastern side could have come from any of the earlier stone-built structures nearby. The cist was not particularly well made, as its sides did not fit tightly together and its ends were not well positioned. The appearance of this cist implied it was very quickly assembled with probably a large stone used for its lid. The capping stone could, for example, have been robbed from the base of the standing stone, where larger flat stones had been placed during its erection.

An elderly male individual was placed into the cist with a whole cooking pot (SF 26259). Also accompanying the human remains were two or three small bone points and the remains of a calf. The cist was not large and its contents must have filled the available space.

Once the cist had been capped it was covered over with larger loose stones and a curved kerb was built around it and across the old building, with the gap from the kerb to the cist infilled with stone, sand and turf. There is no firm evidence, due to coastal erosion, to suggest the cairn continued around the cist in a circle but that was probably the intention. When the kerbed cairn was complete it had integrated the Phase D mound over the shaft and platform into its construction as possibly an expedient pile of stone, and created a low burial mound of c. 3 m diameter.

The location of the cist between two earlier monuments was probably more important than the uneven ground it was built upon as the standing stone (BD) survived to mark its position. Like the cairn complex to the west, this small cairn was not entirely a new construction, and by bringing in surviving ritual structures from the past, it gave extra meaning to the burial. The joining of the past with the present also implied that knowledge or memories existed of the earlier ritual monuments in this area. By embracing elements of an older religion into a new burial possibly imbued the deceased person with extra prestige, status or merit during his passage into the afterlife. Interestingly, this is the only burial on the excavated area where grave goods survived. The pot for liquids or dried food, and a beast to provide sustenance were necessary accompaniments for his journey and the burial in the ground was therefore not viewed as the end of a life.

The concluding marine incursion

The early and middle Bronze Age ritual monuments were subject to another flooding episode on the west side of the site that seemed to mirror the previous, but it was not as devastating. It was recorded as extending inland beyond the Phase B cairn complex where it removed organic layers that had built up around the structures. Interestingly, the cairn was not damaged in the flood, and the water left behind an enriched mixed horizon of fine residues. If any anthropogenic activity had occurred on the site after the construction of the cairns, it was removed by this incursion, which appears to be the last until 1974 (see PART 2). It is thought the marine flooding occurred sometime during the middle part of the Bronze Age.

The people

The remains of three early Bronze Age males, comprising two skeletons and one of disarticulated bone, were preserved at the site through burial (see Bohling and Buckberry, PART 3). The earliest SF 17436, which predated the complex of cairns, was placed in an informal pit and consisted of disarticulated bone. The middle interment SF 17642 was positioned in a high status cist beneath an elaborate cairn. The remaining burial SF 26319 was put in poorly constructed cist beneath a small cairn. All three were inhumations, and two were crouched burials, though the last was highly flexed as if it had been placed in a squatting position with its knees near its shoulders and its arms in front of its legs or its hands placed on its knees.

The diseases the two articulated individuals suffered are detailed in Part 3, as well as incidences of nutritional stress (starvation) and changes in diet in childhood and at other times in their early lives. Marine food was certainly a main part of the diet of SF 26319, although both individuals had terrestrial food markers. It was considered (Beaumont et al. in PART 3) that SF 26319 may have migrated (moved into the area) due to the changes in his diet from land-based foodstuffs to marine-based ones. Although this is a possibility that should not be discounted, other factors might also explain this change in diet. During periods where severe environmental events devastated crops and herds, it may seem more logical that the local population, out of necessity, would have turned to the sea for food. Scavenging for shell fish along the shore and fish from inlets and inshore waters might have been essential for survival, and either nutritional stress continued for prolonged periods or SF 26319 preferred to continue to eat marine foods as his main source of nutrition.

The earliest individual, SF 17642, died as a young adult probably between 2119–1892 cal BC (see Table 3.1) at the start of Phase B. SF 26319 died later, probably between 1877–1658 cal BC or c. 240 years later and buried on the east side of the site. However, the animal (a calf) buried with him died c. 1518–1409 cal BC, some 250 to 360 years after the human individual. There are two plausible interpretations of this data: that the cist was reopened a long time after it had been sealed and a calf was placed in the burial chamber retrospectively, or that the human individual had been curated and preserved before he was buried along with the calf and a pot. It was mentioned previously that the cist was small and it would have been a tight fit for the individual but also for the accompanying grave goods. The difference in age between the individual and the calf could have been resolved as a sample of carbonised food residue was submitted for radiocarbon dating from the pot found in the cist. However, it failed due to it not being a suitable deposit for dating.

Two questions remain - as to whether the individual after death had been tightly bound or wrapped and then curated for a considerable time before he was buried, and if so, what caused him to be removed from the curated place to a small cist in a small cairn? Excavations at Cladh Hallan, South Uist by Mike Parker Pearson and his team unearthed four mummified skeletons beneath the floors of later Bronze Age houses. One male individual had died c. 1500 BC and a female c. 1300 BC (Lobell and Patel 2010) and there was in addition a young child and a teenager. Scientific analysis indicated that the bodies had been mummified intact in a peat bog which had aided preservation before they were retrieved and curated for 300 to 500 years below the house floor (Parker Pearson 2015). One body was also tightly wrapped to maintain its crouched form.

This unusual and certainly rare form of dealing

with deceased individuals in north-western Europe has direct resonance with SF 26319 at Udal. The tightly flexed skeleton is similar to one found at Cladh Hallan, but its relatively poor preservation in comparison with the older SF 17642 skeleton could be accounted for by the mummification process as well as taphonomic conditions in a small and poorly sealed cist. The Cladh Hallan skeletons were not removed for reburial and this suggests there were other issues at play in the removal of the RUX6 skeleton from where it had previously been to finally rest in a cist. The threat to settlement buildings of further natural devastation by sand, wind or sea could have been one reason that led to a permanent solution to the burial of this individual. The mummified (although not proven scientifically) ancestor was obviously important but the poorly constructed cist suggests there was a certain urgency in his reburial. However, the accompanying grave goods, the potential use of Beaker pottery in the internment rites and the construction of a cairn, indicate that the individual was considered of high enough status to warrant a formal closure and a lasting monument. Crawford was lucky that this cist, its contents and the cairn, survived further erosion after 1974, as its preservation has added a little more to the complex belief systems operating in the Uists, during the early and middle Bronze Age, and the importance of the physical presence of ancestors to their communities.

PART 7 Conclusion - Life on the edge

By Beverley Ballin Smith

Iain Crawford had great vision and it is due to his perseverance in the field that the late Neolithic and early to middle Bronze Age structures were excavated and preserved by record, as imperfect as that record is. Without his attention to detail we would not have understood the processes involved in the interplay between natural environmental factors and human activities. Crawford was very lucky, as he was in the right place at the right time. The 1974 marine erosion allowed him that window into the past that he had not had in other parts of the project area – stratigraphy from the present day to the Holocene clays sitting on bedrock with intervening manmade structures. Even six years later when he began to dig at RUX6 again, coastal erosion had made a further impact on the site and the loss of archaeological evidence to the sea had accelerated. The site was literally falling over the shore face edge into the sea with every high tide.

Crawford's developing understanding of natural processes regarding the accumulation of sand, and the changing natural environment and landscape, were in the infancy of machair studies. Without his interest, and that of other researchers, the story of this particular site would have been considerably thinner. As it is, we have been able to document changes in people's lives from nearly 5000 years ago in the late Neolithic to the middle of the Bronze Age about 3500 years ago. Crawford had hoped his project would encompass the earliest settlement on the Hebrides, but that was not possible. Any earlier evidence close to this site had already been swept away by rising sea levels.

RUX6 has shown us that evidence for settlement prior to the late Neolithic in the Hebrides will only survive in rare coastal locations or possibly inland. Sites close to the present day sea level and especially on the machair will have been lost, unless the location is particularly exceptional, such as at An Doirlinn (Garrow and Sturt 2017), and at Barabhas, Lewis, where evidence of early Neolithic settlement is nearing publication (Cowie and MacLeod-Rivett forthcoming). More recent researchers into prehistoric settlement in the machair came later onto the scene than Crawford. They have mainly dealt with sparse evidence of later Neolithic date, or later structures, as the evidence for most of the Neolithic along the soft coastlines of the Uists will have disappeared. The implications of future research in the machair for the Neolithic in the Hebrides are bleak, as the resource is diminishing fast, if it has not already gone.

Through the post-excavation programme, and with considerable assistance from all the specialists that have contributed to this work, we have managed to tease out the information from the finds and samples to understand the activities of peoples from the distant past. If Crawford had published this site 10, 20 or even 30 years ago, he would not have gained the information that is presented here and the story will have been subtly different in many ways. We have built on a slightly deeper understanding of material culture and environmental evidence, and by using new and updated scientific techniques, brought the story of the site up to date. Some of that has been limited by loss of the record and the collection, but perhaps the greater loss has been that of Iain Crawford himself. He was not able to contribute to this volume apart from the words he wrote prior to the year 2000. It is doubtful that he would have recognised some of the story presented here and he may not necessarily have approved of, or agreed with, some of its interpretation. But the interpretation has been evidenced-based and this is the story that is told.

One of the aspects of the site that is most important is the compressed time scales of activities and events recorded. Within a period of 1500 to 2000 years much happened. The accumulation of windblown sand across a large extent of the coast, and in probably a short space of time, transformed the landscape completely and altered the lives of people living there. To have had to move house and resettle and to know that your territory and land has been changed beyond all recognition must have been traumatic beyond our understanding. We can only compare the experiences of this past community to those of the modern world, where the threats posed by tsunamis, sea level rise, war and devastation that we see and hear about daily in news broadcasts are pertinent. In the late Neolithic, people would have only had their selves to rely on, and situations they had to deal with, such as having no home, the loss of livestock and possibly imminent famine, were stark and pressing matters of survival or death. In some respects we can get closer to these people, some of whom would have been our ancestors, through empathy and understanding of modern population stresses and situations.

If one event was not enough for the community living at RUX6, another in the shape of what would have seemed like a mini tsunami transformed the landscape again and their lives and the society would have had to change correspondingly. How does a community deal with the physical and psychological changes of these threats, and realisation of dramatic landscape changes? This is truly living on the edge. We can perhaps understand some of the importance and necessities of ritual or religious activities to the communities that survived these changes, in caring for their ancestors, in finding suitable receptacles for them and building monuments to them, monuments that survived yet another sea inundation. By their actions, the community preserved its links to the past and looked forward to the future.

Archaeology should ultimately be about people and in RUX6 we have had the good fortune to not only analyse objects people made and used in the past but we have met some of them in the buried remains on the site. It is not always easy to understand all that they did and why they did it, but surprisingly as this story draws to an end, aspects of their lives on the edge between the land and the sea has been fuller, richer and more dramatic than we ever thought.

Afterword

By Iain Crawford and Imogen Crawford

The information retrieved from RUX6 has far transcended the initial assessment and its importance can scarcely be overstated. By amplifying the extent and timescale of the existing long term chronological continuity RUX6 has added significantly to the regional identification typology and dating sequence already constructed from the main Udal focus. Of equal consequence is the intimate association of this 'sub-site' with sea level change. This has provided the rare, if not unique, opportunity to date not only marine advance very closely, but large, adjacent inter-tidal deposits. This wealth of inter-related data provides the basis for a detailed study of landscape genesis and evolution. The complex history of man-environment interaction in this area, the way humans have used the available resources and their subsequent effect on the landscape, can now be fully illustrated. The data that can be used for geomorphological interpretation is far beyond what archaeology can usually be expected to provide.

It was also possible in these favourable circumstances to develop techniques for handling the problems of archaeological sites eroding from the shore face, and for short to medium term conservation which should prove of future value to any such research.

Appendices

Appendix 1: Marine Shell

Small Find Numbers from which shell was quantified by species.

Phase A	Phase B cont.	Phase D	Phase D cont.
23635	26476	23623	26017
26141	26477	23656	26018
	26478	23680	26332
	26485	23692	26333
Phase B	26521	23698	26407
25814	26522	23703	26423
26180	26523	23757	26431
26196	26537	23767	26532
23505	26552	23785	26454
23289	26553	23816	26455
23101	26554	23828	26724
23400	26561	25140	26725
23482	26562	25142	26726
23770	26563	25444	26727
25350	26573	25486	26728
25413	26574	25524	26729
25542	26579	25535	26730
25722	26916	25579	26731
25949	26917	25602	26744
25964	27189	25603	26745
26203	27198	25663	26746
26214	26486	25674	26747
26260	26498	25680	26757
26266	26499	25681	26758
26301	26510	25682	26759
26302		25683	26760

Phase D cont.	Phase C		
26303	26540	25734	26761
26304	26592	25752	26762
26323		25760	26763
26324		25789	26766
26325		25791	26767
26349		25792	26768
26350		25793	26769
26355		25927	26827
26356		25942	26828
26357		25952	26829
26372		25978	26847
26373		26012	26848
26374		26013	26849
26471		26014	26850
26472		26015	27213
26473		26016	27222

Appendix 2: Pottery catalogue

Vessel code	Sherd nrs	Sherds	Average sherd thickness (mm)	Weight (mm)	Coil joins present	Firing (Oxidised/ Reduced)	Description	Decorated	Rim	Carination/ cordon	Base
A01	9	25059/1-5	9.3	28	✓	O & R	Rim, inturned with flattened external bevel. Possibly one shoulder sherd and seven body sherds. Surfaces smoothed and burnished, possibly some slipping. The vessel is likely to have been a cooking vessel due to the survival of carbonised deposits on the external surface between the rim and the shoulder		✓		Missing
A02	6	25059/6-10, /12	15	23	✓		The vessel is has two rim sherds and four body sherds, their external surfaces were probably smoothed. The rim is simple but with convex surfaces. The function of the vessel is unclear.		✓		Missing
A03	1	24976/1	8	12	✓	O & R	The single, everted and irregularly profiled rim sherd has a flattened rim surface. Its exterior surface was smoothed. Exterior fireclouding suggests an open firing. Heavy sooting or carbonised food residues on both surfaces indicates a cooking pot.		✓		Missing
A04	1	23598/2	n/a	n/a	?	O & R	A single body sherd has an external ridge although the external surface was smooth. Possible external sooting suggests it was a cooking pot.				Missing

Vessel code	Sherd nrs	Sherds	Average sherd thickness (mm)	Weight (mm)	Coil joins present	Firing (Oxidised/ Reduced)	Description	Decorated	Rim	Carination/ cordon	Base
A05	1	23598/9	6	2	✓	O & R	A solitary body sherd had a smoothed external surface into which a decoration of horizontal parallel linear grooves was incised. The function of the vessel was not determined.				Missing
A06	2	23534/1, /3	6	2	✓	O & R	The vessel is represented by a decorated and everted rim sherd and a body sherd. The grooved decoration, is in the form of horizontal parallel linear lines.	✓	✓		Missing
A07	2	23518/1-2	12	6	✓	O & R	Two plain body sherds have exterior surfaces that were smoothed.				Missing
A08	1	23453/1	9	1	?	O	The external surface of the single rim sherd was smooth but its shape was incomplete. Possible carbonised food resides suggest it derived from a cooking pot		✓		Missing
A09	1	23312/1	5	n/a	✓	O	A single solitary body sherd has decoration, in the form of grooved parallel lines.	✓			Missing
A10	2	23246/1-2	n/a	4	✓	O & R	The vessel comprises two body sherds only.				Missing
A11	13	23246/3-5	5	16	✓	O & R	The vessel comprises two rim sherds and 11 body sherds, The rim has a convex surface. The external carbonised food remains suggest the vessel was for cooking.				Missing
A12	3	23035/1, 23048/2-3	9	15	✓	O & R	Two rim sherds and body sherds form this vessel. The rim combines a simple form with a convex rim surface. Internal carbonised food remains are noted on one sherd.				Missing
A13	2	23048/5-6	9	4	✓	O & R	Two plain body sherds, one of which has food residues.				Missing
A14	1	23025/1	9	n/a	✓	O & R	The vessel has a single, everted rim sherd suggesting a necked pot.				Missing
A15	1	17274/1	10	7	✓	O & R	A body sherd is decorated with parallel linear, shallow grooves.	✓			Missing
A16	1	16234/1	9	4	✓	O & R	The rim is everted. Grooved decoration occurs around the rim on both the interior and exterior surfaces. Horizontal parallel linear lines decorate the exterior; diagonal parallel linear lines, juxtaposed to form a herringbone motif, decorate the interior side of the everted rim.	✓	✓		Missing

Vessel code	Sherd nrs	Sherds	Average sherd thickness (mm)	Weight (mm)	Coil joins present	Firing (oxidised/ reduced)	Description	Decorated	Rim	Carination/ cordon	Base
B01	9	25324/6, 25373/1, 25522/3, 26285/2, 26231/2, 26231/4	12	53	✓	O & r	A possible burnished shoulder sherd and several body sherds with an amorphous depression which is possibly the vestiges of impressed decoration. Abrasion of the interior is probably use related external sooting or carbonised food residues suggest a cooking vessel.				
B02	11	25421/1, 25438/7, 25438/15, 25501/2	13	36	?	O & r	A burnished and possibly slipped, small rim sherd and body sherds. A shell inclusion is noted in the fabric.		✓		
B03	11	23942/1, 25199/4-5, 25257/5, 25303/6, 25303/8	13	97	✓	O & r	Body sherds. Thick carbonised food residues on the vessel indicate it was a cooking pot.				
B04	33	25385/1, 23391, 23403/1-9, 23403/11-13, 23403/15-20, 23403/22, 23403/11, 23404/3, 23404/15, 23420/3, 25112/1-3	9	120	✓	?	The vessel is represented by everted rim sherds, possible neck sherds, carinated sherds and body sherds, which have possibly been slipped and burnished. A slight horizontal ridge below the carination may be a product of moulding. The external surface is decorated with impressed cardium shell. The rim is decorated on both surfaces: below the rim externally with diagonal, parallel lines; the rim edge has diagonal, parallel lines; and internally below the rim are vertical, parallel lines. The neck of the vessel had diagonal parallel lines forming a herringbone motif, in repetitive horizontal bands. Immediately above and below the carination, were both vertical, parallel lines and diagonal lines, presumably in bands down the body of the vessel. There may also have been undecorated areas. The vessel exterior is abraded and has both sooting and carbonised food residues (also internal) indicating it was a cooking pot.	✓	✓	✓	
B05	1	25565/1	N/a	3	?	O & r	Small rim sherd with a flattened top, wiped and possibly slipped.		✓		
B06	22	25784/1-3, 25784/8-10, 25784/13	8	49	✓	?	Two flat base sherds and several body sherds. The base is more heavily tempered than the body sherds but is also abraded and possibly burnt. External sooting and carbonised food residues indicate the vessel was a cooking pot.				✓

Vessel code	Sherd nrs	Sherds	Average sherd thickness (mm)	Weight (mm)	Coil joins present	Firing (oxidised/ reduced)	Description	Decorated	Rim	Carination/ cordon	Base
B07	5	26290/1-2, 26231/6-7	10	11	✓	O & r	The vessel is represented by a shoulder sherd, a possible carinated sherd and body sherds. The sherds may be burnt and carbonised food residues indicate it was possibly cooking pot.			✓	
B08	2	25945/9, 25961/6	6	2	?	R	One carinated sherd another indeterminate body sherd.			✓	
B09	10	25041/1, 25153/4, 25257/2, 25438/6, 25802/1-2, 26192/3, 26220/1-2	11	75	✓	O & r	Burnished and possibly slipped body sherds. Carbonised food residues on the vessel interior (indicate a cooking pot.				
B10	696	26259/30, 26259/1-16, 26259/26-28, 26259/35-38, 26259/40-41, 26259/43-44, 26259/46, 26259/61	15	2222	✓	O & r	The vessel is represented by 11 everted rim sherds some bevelled externally, a neck sherd, body sherds and two fragmentary base sherds. Exterior sooting and carbonised food residues internally suggests the vessel was used on the hearth. The vessel is in a very fragmentary condition and was found in the bg24 cist.		✓		✓
B11	14	23649/4-7	12	42	✓	?	Body sherds with only slight evidence for incised decoration, in the form of diagonal, parallel lines.				
B12	26	23782/1, 25083/1, 25199/1, 25257/3, 25315/4, /6-7, 25324/3-5, 25470/1, 25558/1, 25649/1-5, 25928/8, 25961/1, 26142/3	11	251	✓	O & r open firing	The vessel is represented rim sherds, neck sherds, flat base sherds and body sherds. The rim combines a simple form with a flattened rim surface but is uneven in profile. A horizontal cordon is situated immediately below the rim. The estimated rim diameter is c. 160 Mm. Sooting is noted below the cordon. The abrasion and residues suggest the vessel was probably a cooking pot.		✓	✓	✓
B13	1	26143/1	N/a	1	?	?	A body sherd with well preserved incised decoration, in the form of parallel lines.	✓			
B14	3	26584/1, 26566/1	10	8	?	?	Burnished body sherds with evidence of grass tempering.				
B15	24	25749/1, 26495/1-6, 26504/1-2	13	141	✓	O & r	Body sherds with carbonised food residues suggesting a cooking pot.				
B16	2	26376/1-2	N/a	11	✓	O & r	Body sherds				
B17	1	26285/3	9	1	✓	O & r	A body sherd with incised decoration in the form of horizontal, parallel lines. Same as vessel b51 and b52.	✓			

Vessel code	Sherd nrs	Sherds	Average sherd thickness (mm)	Weight (mm)	Coil joins present	Firing (oxidised/ reduced)	Description	Decorated	Rim	Carination/ cordon	Base
B18	2	25819/1, 26232/1	5	13	✓	?	Two burnished and everted rim sherds.		✓		
B19	3	25447/1, 25558/5, 26231/1	11	29	✓	O & r	Body sherds with carbonised food residues on the internal surface suggest a cooking pot.				
B20	8	25928/10, 25938/10, 25925, 26143/1	14	51	✓	O & r	A decorated body sherd, three possible neck sherds, and several body sherds comprise the vessel. Incised decoration, in the form of a herringbone motif aligned horizontally. A single diagonal line on the vessel exterior neck is probably accidental.	✓			
B21	2	26285/1, 26176/2	14	25	✓	O & r	A possible shoulder sherd and a thick-walled body sherd.				
B22	11	26176/3, /5-6	7	18	?	O & r	Several body sherds, burnished externally. Sooting suggests the vessel was a cooking pot.				
B23	1	26176/9	6	1	?	R	A rim sherd has an internal bevel and carbonised food residues.		✓		
B24	10	25324/8, 25257/1, 25778/2-4, 25945/2, 25961/2	13	145	✓	?	A possible shoulder sherd, a neck sherd, a carination and several body sherds. Internal abrasion suggests heavy use. Thick carbonised food residues suggest the vessel was a cooking pot.			✓	
B25	1	25961/3	16	16	✓	O & r	A possible neck sherd, with finger moulding causing an indentation on the external surface.				
B26	6	25802/4, 25802/5	12	10	?	O & r	Body sherds with possible sooting suggest the vessel was probably a cooking pot.				
B27	1	25778/1	11	10	?	O & r	A straight rim sherd with a flattened top.		✓		
B28	1	25778/6	N/a	?	?	O & r	A possible, abraded, sooted and everted rim sherd from a cooking pot.		✓		
B29	7	21809/1, 24809/1, 25303/9-11, 25778/1, /7	9	37	✓	O & r	Two burnished, abraded carinated sherds and several body sherds. Sooting suggests a cooking pot.			✓	
B30	5	25501/1, 25515/4-6, 25558/2	9	20	✓	O & r	Two possible shoulder sherds and some body sherds, possibly slipped had carbonised food residues on the vessel exterior suggesting it was a cooking pot				
B31	1	25515/3	9	3	✓	O & r	The vessel is represented by a possible rim inturned sherd a flattened rim top.		✓		
B32	7	25396/1	13	26	✓		The everted rim was made by folding a flattened coil back upon itself, but several body sherds are also present.		✓		

Vessel code	Sherd nrs	Sherds	Average sherd thickness (mm)	Weight (mm)	Coil joins present	Firing (oxidised/ reduced)	Description	Decorated	Rim	Carination/ corcon	Base
B33	5	25470/2, /4, 25558/4, 25928/7	11	48	✓	O & r	A burnished shoulder sherd and several body sherds indicate that the vessel had a constricted neck. Sooting and carbonised food remains suggest the use of the pot on the hearth.				
B34	8	25438/2-3, 25438/10, 25438/14	N/a	26	✓	?	The vessel is represented by two hard fired, possible rim sherds and several body sherds.		✓		
B35	1	25438/5	N/a	1	?	?	The everted rim sherd has a flat top and has laminated. Carbonised residues are present on the rim surface suggesting the vessel's use as a cooking pot.		✓		
B36	3	25421/2, 25396/6-7	9	21	✓	?	Two rim sherds and a body sherd possibly burnished externally. The rim is broken at the junction of the rim coil with the body of the vessel. Uncertain whether the rim is everted or has an internal bevel.		✓		
B37	1	25396/8	N/a	2	?	?	A carinated sherd.			✓	
B38	2	25324/1, 25276/1	13	28	✓	O & r	The rim sherd is everted with an internal bevel sherd, the other sherd is a body sherd. The exterior is abraded. Sooting or carbonised food residues both surfaces of the sherds suggest the vessel was a cooking pot.		✓		
B39	4	25315/1, /3, /5	16	178	?	O & r	The vessel comprises a possible rim sherd with a convex top, a carinated sherd and two body sherds. Sooting below the rim infers its use as a cooking pot. The vessel was wiped with much abrasion internally through use.		✓	✓	
B40	3	25303/4-5, 25288/1	14	26	✓	?	Two, probably conjoinable rim sherds and a body sherd form the vessel. The rim is simple with a rounded top.				
B41	1	25288/2	9	11	?	O & r	This body sherd has two concentric grooves and two marked depressions which may represent deliberate decoration, rather than incidental manufacturing marks. The interior of the sherd is abraded.				
B42	3	25276/3-4	18	26	✓	O	Fireclouding suggests the body sherds were fired on with an open fire.				
B43	1	25153/8	8	1	?	R	The body sherd was burnished and is decorated with an incised line decoration	✓			
B44	4	25131/1-2	N/a	19	✓	O & r	Body sherds only				

Vessel code	Sherd nrs	Sherds	Average sherd thickness (mm)	Weight (mm)	Coil joins present	Firing (oxidised/ reduced)	Description	Decorated	Rim	Carination/ cordon	Base
B45	7	25131/3-7, 25131/11	5	14	✓	?	This is a simple rim sherd with a flattened top and several body sherds. External abrasion is probably use related. Sooting and food residues on both surfaces indicate the vessel was used for cooking.		✓		
B46	3	25438/1, 23942/3, 26176/1	11	32	✓	?	Two probable shoulder sherds and a body sherd, with burnished external surfaces and heavy external sooting/ carbonised food residues indicate the use of the vessel for food preparation.				
B47	10	23654/1-5	10	45	✓	O	Flecks of shell within the clay matrix give the vessel a distinctive appearance. It comprises a fragmentary rim sherd and several body sherds. The rim is simple with a flat top. A thickened upper part to the vessel may lie immediately below the rim to embellish the rim moulding. The external surfaces are abraded and sooted suggesting the vessel's use for food preparation.				
B48	27	23391/3, 23391/3-5, 23395/5-6, 23404/1-3, 23426/1, 23436/1, 23440/1	10	55	✓	O	This vessel comprises a rim sherd, a possible shoulder sherd, a possible carinated sherd and numerous body sherds, many of which are only provisionally identified with this vessel. The rim is simple with a rounded top. The shoulder is abrupt and the carination is also acute. Some burnishing is noted. Internal vessel abrasion is related to its use as a cooking pot, and the exterior of the pot is also sooted.		✓	✓	
B49	1	23440/3	6	2	✓	O & r	A small body sherd is possibly a carination.			✓	
B50	3	23403/25-27	5	5	?	R	The body sherds may have had external burnishing with incised decoration in the form of parallel lines, in a diagonal alignment. External sooting suggests the use of the pot on the hearth.	✓			
B51	1	23391/2	N/a	2	?	?	The single smoothed body sherd is decorated with horizontal parallel, lines. Same as vessels b17 and b52.	✓			
B52	9	23395/1-4	9	12	?	?	The externally smoothed body sherds carry incised decoration, in the form of horizontal parallel lines. Food residues on the vessel interior suggest it was a cooking vessel. Same as vessels b17 and b51.	✓			

Vessel code	Sherd nrs	Sherds	Average sherd thickness (mm)	Weight (mm)	Coil joins present	Firing (oxidised/ reduced)	Description	Decorated	Rim	Carination/ cordon	Base
B53	8	23381/1-3	8	15	?	O & r	The vessel is represented exclusively by body sherds where their external surfaces were probably smoothed. Food residues on the interior surfaces indicate the vessel was a cooking pot.				
B54	12	23620/1-3, 23620/5	8	54	✓	O & r	A possible neck sherd and body sherds were probably smoothed and possibly slipped. Heavy food residues particularly the shoulder sherd suggest its use as a cooking pot.				
B55	2	23187/1, 17611/1	9	31	✓	O & r	The necked vessel is represented by two burnished carinated sherds with a vertical profile above the carination. The thickness of the wall is noticeably greater below (10 mm) than above (8 mm) the carination. The external surface is profusely decorated with incised motifs separated by the carination. The neck of the vessel has parallel, lines in a herringbone or lattice design. Below the carination, the decoration comprises diagonal lines. The interior of the vessel is heavily abraded or worn.	✓		✓	
B56	1	23182/1	10	6	?	?	A rim sherd is everted with an internal bevel. Immediately below the rim it has incised decoration comprising short, vertical, parallel lines, arranged in horizontal bands. It has severe post-depositional abrasion.	✓	✓		
B57	1	23074/1	9	6	✓	O & r	A flat base sherd with fine horizontal striations on the external vessel wall is a result of rotating the vessel during use.				✓
B58	1	23372/1	8	2	✓	O	A slightly everted rim fragment.		✓		
B59	1	23372/2	9	1	?	O & r	The single body sherd has lost most of its external surface but carries a solitary groove. B59-62 same vessel.	✓			
B60	2	23372/3, /4	7	2	✓	?	Two tiny body sherds have incised decoration?, In the form of incised dashed lines, or a single line on their external surfaces. B59-62 same vessel.	✓			
B61	1	23372/8	N/a	1	✓	?	A very tiny sherd with a smoothed exterior surface with a possible incised line. B59-62 same vessel.	✓	✓		
B62	2	23372/5-6	N/a	2	✓	?	Two tiny body sherds survive and a fragment, with grooved parallel line decoration over a possible burnished finish. Possible sooting indicates its use near or on the hearth. B59-62 same vessel.	✓			

Vessel code	Sherd nrs	Sherds	Average sherd thickness (mm)	Weight (mm)	Coil joins present	Firing (oxidised/ reduced)	Description	Decorated	Rim	Carination/ cordon	Base
B63	1	23372/7	8	2	✓	O & r	A very small, possibly carinated sherd.				
B64	2	23372/9-10	9	6	✓	R & o	The vessel is represented by two conjoinable body sherds, which were smoothed externally.				
B65	2	26192/1-2	11	16	✓	?	Two body sherds were burnished and contained decoration in the form of a solitary incised line. Fireclouding suggests an open firing.	✓			
B66	1	25315/2	10	4	?	O & r	A solitary fragmentary rim sherd has an external surface that was probably smoothed. The rim form was indeterminate. Fireclouding confirms an open firing.		✓		
B67	1	25303/1	12	5	?	O & r	The neck sherd has a laminated and porous texture and its exterior was probably burnished. It has a vertical groove, with a curious circular expansion at one terminal, where the implement used to apply the decoration was stabbed into the clay, and is underlined by a single incised horizontal linear line.	✓			
B68	5	23273/1, 23273/4-5	13	20	?	O & r	The vessel is represented exclusively by body sherds, which were probably smoothed externally. A single external, horizontal, incised line is probably an accidental mark rather than deliberate decoration.	✓			
B69	1	26526/3	9	9	✓	O & r	A body sherd thought to have decoration in the form of horizontal parallel lines. These are widely spaced and faint and give a corrugated or ribbed appearance. Probably forming marks.	✓			
B70	3	26504/4	N/a	?	?	O	The vessel is represented by a small fragment of a rim sherd which may have been everted but with an internal bevel.		✓		
B71	2	26142/1-2	10	44	✓	O & r	The vessel is represented by a carinated sherd and a body sherd. A fragmentary cordon, added onto the carination, emphasises the change of vessel wall alignment. The external surface was probably smoothed and possibly slipped. The vessel, probably a cooking pot due to sooting, probably had a bipartite profile.			✓	
B72	9	25164/1, 25298/1, 25778/3, 25515/1-2, 25522/1, 25558/3	12	153	✓	O & r	A flat base sherd and several body sherds represent this vessel. Their exterior surfaces were smoothed and possibly burnished. Abrasion is perhaps related to use. Carbonised food residues, on both surfaces, suggest its use as a cooking pot.				

Vessel code	Sherd nrs	Sherds	Average sherd thickness (mm)	Weight (mm)	Coil joins present	Firing (oxidised/ reduced)	Description	Decorated	Rim	Carination/ cordon	Base
B73	1	23901/1	14	26	✓	O & r	A solitary body sherd with sooting on its external surface suggesting it was part of a cooking pot.				
B74	2	23420/2, 23366/1	N/a	9	?	R	The exterior surface was probably burnished. The vessel is represented by two conjoinable flat base sherds. External horizontal striations represent smoothing marks or attrition derived from rotating the vessel in an upright position during use.	✓			

Most sherds are smoothed and some with wiped marks. Many are abraded.
Could burnishing be due to sand blasting/food residues?
Fire clouding and o & r suggest an 'open' firing?

Vessel code	Sherd nrs	Sherds	Average sherd thickness (mm)	Weight (mm)	Coil joins present	Firing (Oxidised/ Reduced)	Description	Decorated	Rim	Carination/ cordon	Base
C01	35	23294/1-6, 23294/8, 23294/9/1-2, 23294/12-13, 23294/15, 23294/19, 23294/20, 23294/22, 23305/1-2, 23464/6	9	831	✓	?	The vessel is represented by several conjoinable base sherds and numerous body sherds. The basal diameter is c. 180 mm; the thickness of the base is c. 15 mm. Abraded internally and external sooting. A possible cooking pot.				✓
C02	12	23464/1-3	7	17	✓	O & R	Body sherds. Sooting or carbonised food residues on the exterior indicate a cooking pot.				
C03	4	23464/5, 23477/1	5	5	?	R	Body sherds contain deliberate shell inclusions which might be for effect and a distinctive appearance. Thick carbonised food residues indicate a cooking pot.				
C04	23	23414/1, 25214/3, 25356/1-2, 26205/6	10	49	✓	O	A rim sherd with a rounded top, a possible flat base sherd and several body sherds. Possible internal sooting indicates a cooking pot.		✓		✓
C05	11	23294/11, 23294/18, 23294/23, 23305/3, 23305/6	6	21	✓	O & R	An inturned rim sherd with a rounded top and several body sherds. Sooting or carbonised food residues on surfaces indicates a cooking pot.		✓		
C06	1	26186/1	8	62	✓	O & R	A burnished carinated sherd with the wall thickness measuring 8 mm above and 14 mm below. The lower part of the sherd is abraded.			✓	
C07	1	25370/1	11	11	?	O & R	A flat base sherd. Abrasion on the base indicates that the vessel was placed or rotated in an upright position during use.				✓

Vessel code	Sherd nrs	Sherds	Average sherd thickness (mm)	Weight (mm)	Coil joins present	Firing (Oxidised/ Reduced)	Description	Decorated	Rim	Carination/ cordon	Base
C08	12	25356/5, 25356/8	7	19	?	?	Body sherds. Some sooting indicates they derive from a cooking pot.				
C09	4	25379/1-3	7	6	?	O & R	A rim sherd with a flat top and body sherds. Sooting indicates its use on the hearth.		✓		
C10	1	26206/1	n/a	7	✓	O & R	A rim sherd with a rounded top. Thick carbonised food residues indicate a cooking pot.		✓		
C11	1	26206/2	n/a	9	?	O & R	An everted rim sherd with broad flat top and a wiped surface.		✓		
C12	3	23475/1, 25550/1, 26205/4	9	26	✓	?	Body sherds. Internal carbonised food residues indicates its possible use as a cooking pot.				
C13	1	23475/3	n/a	1	✓	O & R	A possibly burnished carinated sherd.			✓	
C14	1	25967/1	10	11	✓	R & O	A burnished body sherd. Possible sooting on the exterior suggests a cooking pot.				
C15	1	25967/2	13	12	✓	O & R	A thick body sherd with very coarse temper.				
C16	7	26590/1-3, 26590/16-17	13	12	?	O & R	Body sherds with smooth surfaces.				
C17	5	26590/4-5	9	9	?	?	Body sherds. Internal carbonised food residues indicate a cooking pot.				
C18	1	26205/1	n/a	1	?	R	A small, burnished rim sherd. The rim profile is ambiguous.		✓		
C19	5	26205/2, 26590/7-8	9	12	?	?	Three rim sherds with a flattened top and two body sherds. The vessel had a neutral profile. Thick carbonised food residues indicate a cooking pot.		✓		
C20	18	26205/8-10, 26205/13	8	33	?	O & R	Body sherds. Carbonised food residues indicate a cooking pot.				
C21	1	26205/23	9	2	?	O & R	A rim sherd with a rounded top.		✓		
C22	1	25411/1	5	1	?	O	A slightly everted rim sherd with a flat top. Fluted decoration externally, in the form of a curvilinear pattern lies below the rim.	✓	✓		
C23	1	26205/19	n/a	2	?	O & R	This includes a body sherd and a cordon applied to the vessel body. Sooting indicates the pot may have been a cooking vessel.			✓	

Vessel code	Sherd nrs	Sherds	Average sherd thickness (mm)	Weight (mm)	Coil joins present	Firing (Oxidised/ Reduced)	Description	Decorated	Rim	Carination/ cordon	Base
D01	25	26830/1, 26830/3/3	9	38	✓	O & R	5 small rim sherds and numerous body sherds were wiped and smoothed. A possible finger depression on the vessel exterior is a result of moulding the vessel.		✓		
D02	21	25620/3, 25620/5, 25620/9-10, 25620/14, 25620/19-23	8	51	✓	?	Includes a sherd from just above the base edge with moulding marks otherwise mostly tiny body sherds. Carbonised food residues indicate it was a cooking vessel.				
D03	76	25491/4, 25491/8, 25491/14, 25491/17-21, 25491/32	9	494	✓	O - O & R	Included are a possible neck sherd, possible shoulder sherds and body sherds where exterior surfaces were possibly slipped. A possible horizontal cordon at the base of the neck is questionable. A substantial linear groove on the vessel exterior is an isolated feature. Thick carbonised food residues on both surfaces indicate a cooking pot.	✓			
D04	10	25443/46, 25443/53, 25744/5, 25771/5, 25774/5, 26819/1-2, 26820/1, 26820/3-4	7	10	✓	O	Rim sherds may have been slipped, but are variable, and everted with a flat or rounded top. Thick carbonised food residues on both surfaces below the rim suggest a cooking pot.		✓		
D05	29	25447/7, 25620/1, 25994/3-5, 26027/1-6, 26027/9, 26027/13	9	123	✓	?	Flat base sherds, a possible neck sherd and several body sherds form the vessel with some body sherds possibly burnished. The base is abraded, probably through use on the hearth, as there are also carbonised food residues on both surfaces.				✓
D06	70	25442/9-10, 25442/17-19, 25442/21-23, 25442/26, 25443/22, 25443/50-52	8	179	✓	O & R	Shoulder sherds, some with a noticeable shoulder, and body sherds. Carbonised food residues are present on the internal surfaces of sherds.				
D07	39	25443/26, 25712/9, 25712/15-17, 25712/19, 25712/23, 25755/13-14, 26617/17	8	77	✓	O & R	A rim sherd, two rim fragments, a possible shoulder sherd and body sherds. The rim has an external bevel which carries diagonal, parallel lines incised decoration with horizontal and diagonal lines, on the interior surface. Possible exterior sooting indicates the pot was placed on the hearth.	✓	✓		

Vessel code	Sherd nrs	Sherds	Average sherd thickness (mm)	Weight (mm)	Coil joins present	Firing (Oxidised/ Reduced)	Description	Decorated	Rim	Carination/ cordon	Base
D08	69	25755/2-5, 26617/13	8	147	✓	?	A possible shoulder sherd and body sherds. Carbonised food residues and possible sooting indicate that the vessel was probably a cooking pot.				
D09	62	25984/11, 25984/21, 26818/1-5, 26819/3-5, 26819/7-9, 26819/11	11	267	✓	?	Includes a neck sherd, a shoulder sherd and body sherds. The vessel exterior is decorated with fingertip impressions. Possible sooting and carbonised food residues indicate the vessel may be a cooking pot.	✓			
D10	1	23571/5	7	1	?	O	A slightly everted rim sherd with a rounded top.		✓		
D11	71	25447/10-11, 25578/2-4, 25578/6-8, 25578/11, 25578/14, 25578/17, 25588/2, 25588/6-8, 25588/12, 25588/14, 25588/16, 25672/1-5, 25672/7, 25672/13, 25672/15, 25672/17, 25673/1-6, 25951/68, 26447/11, 26617/11	7	160	✓	?	This Grooved Ware vessel comprises rim sherds, probable rim, neck and shoulder sherds and body sherds, with exterior burnishing. The rim form is slightly everted, with either round or flat rim top. The shoulder sherds have a gradual change of angle, but the tight curvature of many sherds suggests the vessel is cylindrical with a long neck. The exterior surface is decorated with all-over incised geometric motifs, including lozenges divided into four quadrants and either a quarter, half or fully infilled with parallel lines. Some lozenges are enclosed by a larger lozenge, the intervening space infilled with parallel linear creating an elaborate border, and including a smaller infilled lozenge at its centre - the focus to the design. Some sherds, provisionally attributable to this vessel, have stabbed and impressed designs infilling larger motifs. Concentric curvilinear lines, forming indeterminate motifs, are also present on some sherds. Due to post-depositional abrasion the overall design remains obscure. Thick carbonised food residues on both surfaces suggest the vessel or vessels were used on the hearth.	✓	✓		
D12	1	26689/1	7	3	✓	?	A simple rim sherd has a rounded top and external sooting.		✓		
D13	1	26689/2	6	2	?	O	A possibly burnished rim sherd with an internal bevel and decorated externally with incised horizontal parallel lines. It also has carbonised food residues on its exterior surface indicating a cooking pot.	✓	✓		
D14	2	17623/1, 26722/2	7	3	?	O	Two straight rim sherds have rounded tops with sooting on the external surface.		✓		

Vessel code	Sherd nrs	Sherds	Average sherd thickness (mm)	Weight (mm)	Coil joins present	Firing (Oxidised/ Reduced)	Description	Decorated	Rim	Carination/ cordon	Base
D15	4	23571/1-2	8	14	?	O & R	Sizeable body sherds, with one with decoration of a single incised diagonal line and carbonised food residues.	✓			
D16	10	23608/1, 23624/1, 25574/1-2, 25644/1, 25644/4,	5	16	?	R	The clay contains pale shell inclusions which contrast with the dark pottery surface, which may indicate they were deliberately added. The vessel was also burnished. Simple rim sherds with flat tops or an internal bevel, carinations and body sherds. Faint incised cross-hatch decoration is noted below one rim sherd, suggesting other decoration on other sherds could have been removed by post-depositional abrasion. The presence of carbonised food residues on both surfaces indicates the vessel may have been a cooking pot.	✓	✓		
D17	8	23608/3-4, 25644/3	7	16	?	O & R	Two carinated sherds and body sherds. The angle of carination is acute and indicates a bipartite rather than biconical vessel, with carbonised food residues.			✓	
D18	10	23566/1, 23829/1, 23856/2, 25634/1-3	10	18	✓	?	Body sherds with carbonised food residues on both surfaces suggests a culinary purpose				
D19	1	26164/4	n/m	n/m	?	O	A rim sherd with an internal bevel.		✓		
D20	8	23702/2-4, 26817/1, 23856/1	10	147	✓	?	A flat base sherd and body sherds. The base is 14 mm thick.				✓
D21	3	26188/1-2, 26679/1	6	12	✓	O & R	A possible shoulder sherd and two body sherds. An incised decoration of short, parallel, diagonal lines form a horizontal band around the exterior surface of the shoulder. Thick carbonised food residues indicate a cooking pot.	✓			
D22	2	25976/2, 25976/5-6	12	15	✓	O	A large body sherd and two fragments. A circular impression on the exterior surface is tentatively interpreted as decoration.	✓			
D23	5	25976/1, 25976/3-4	8	53	✓	?	Surface treatments was inconsistent and included smoothing, wiping and burnishing. Three flat base sherds and two body sherds. The lower surface of the base is abraded, through use on the hearth. Sooting on the body is noted.				✓
D24	14	26737/2-3	11	166	?	O	A probable flat base sherd with carbonised food residues indicating a cooking pot.				✓

Vessel code	Sherd nrs	Sherds	Average sherd thickness (mm)	Weight (mm)	Coil joins present	Firing (Oxidised/Reduced)	Description	Decorated	Rim	Carination/cordon	Base
D25	9	23685/1, 23685/3-5	9	41	?	?	An abraded, flat base sherd and body sherds with thick carbonised food residues indicating its use as a cooking pot.				✓
D26	9	23664/1-3, 23665/5, 23685/2	8	66	✓	?	External voids could be the result of organic tempering. Body sherds with a possible horizontal cordon. Thick carbonised food residues are confined to the area above the cordon.			✓	
D27	3	26751/1-2	14	93	?	?	Flat base sherd with profuse post-depositional concretion and damage due to root action.				✓
D28	21	23665/1-3, 23665/7	10	73	?	O	Abraded, flat base sherds and body sherds, with possible sooting.				✓
D29	30	23834/3-4, 23834/6-7, 26820/24	7	53	?	O & R	Pale shell inclusions contrast with the dark exterior surface. A probable shoulder sherd, three carinated sherds, and body sherds. The angle of the carination is acute. Abraded and sooted sherds.			✓	
D30	17	23834/2, 23834/9, 23834/11, 23834/14, 26617/8	5	27	✓	?	Two conjoinable simple rim sherds, possibly inturned but with a flattened top, one shoulder sherd and body sherds with thick carbonised food residues indicating a cooking pot.		✓		
D31	216	25951/35-36, 25951/39, 25951/41-42, 25951/52-53, 25951/56-58, 25951/60, 26603/1	9	422			An indeterminate rim sherd, two possible shoulder sherds and body sherds with sooting and carbonised food residues.		✓		
D32	4	25588/24, 25588/39	8	17	?	?	A shoulder sherd and three body sherds have carbonised food residues on the exterior.				
D33	23	25588/33-34, 25951/71-72	6	68	✓	O & R	A rim sherd, two base sherds, and several body sherds have internal carbonised food residues. The flattened rim is decorated with a single horizontal row of stabbed dots, applied by stabbing.	✓	✓		
D34	13	27214/6-7	7	22	?	O	Body sherds with possible sooting externally.				
D35	1	27214/42	4	2	?	R	A rim sherd with a flattened top.		✓		
D36	1	27214/43	6	2	?	O	A rim sherd with a flattened top and external sooting.		✓		
D37	2	27214/44, 27214/49	4	2	?	?	An abraded rim sherd and a body sherd, with cordon and carbonised food residues.		✓	✓	

Vessel code	Sherd nrs	Sherds	Average sherd thickness (mm)	Weight (mm)	Coil joins present	Firing (Oxidised/ Reduced)	Description	Decorated	Rim	Carination/ cordon	Base
D38	131	26424/11-20	7	189	✓	R	A shoulder sherd, a neck sherd and body sherds. Three laminated fragments may represent the remnants of a rim, carination or even a base. Moulding or wiping marks are visible as are carbonised food remains.				
D39	107	25491/11-13, 25491/16, 25491/22-23, 25491/26-31, 25491/33-39, 25491/43	10	593	✓	O & R	Abraded flat base sherds, shoulder sherds, possible neck sherds, and body sherds have incidental marks on external surfaces through forming, detached temper and abrasion marks. Wipe marks and surface voids suggest wiping with organic matter. There are also carbonised food residues.				✓
D40	30	25491/41-42, 25491/50	6	49	?	?	A possible carinated sherd and body sherds with carbonised food residues on both surfaces indicating a cooking pot.			✓	
D41	12	25994/1, 25994/6-8	6	18	✓	R	Two irregularly shaped rim sherds with a flat or rounded rim top, and body sherds have carbonised food residues on both surfaces indicating a cooking pot.			✓	
D42	11	25994/2, 25994/5, 25994/9, 25994/18, 25994/21	5	13	?	?	Several rim and body sherds. The irregularly formed rim is slightly everted with a flat or rounded top, and sooting.			✓	
D43	13	25443/49, 25712/22, 25994/13-15, 25994/25	4	14	?	O & R/R	Six everted rim sherds with a flat or rounded top, and body sherds have carbonised food residues.		✓		
D44	6	25994/1, 25994/11-12, 25994/26	5	7	?	?	Four rim sherds with flat tops, and two body sherds 25994/11.		✓		
D45	20	25335/4-6 25335/6	n/a	44	?	O	Body sherds with carbonised food residues				
D46	41	25189/3, 25588/26, 25588/43-4, 25673/9-11, 25673/13, 25673/15	9	179	✓	?	Body sherds with moulding marks (from the potter's fingers) and carbonised food residues are noted on the surface.				
D47	25	25620/1, 25620/7, 25620/12-13, 25620/15,	7	31	?	O	An everted rim sherd with flat top, a possible carinated sherd, and body sherds with carbonised food residues.		✓	✓	

Vessel code	Sherd nrs	Sherds	Average sherd thickness (mm)	Weight (mm)	Coil joins present	Firing (Oxidised/ Reduced)	Description	Decorated	Rim	Carination/ cordon	Base
D48	7	25712/6-7, 25712/10-14	7	12	?	?	A shoulder sherd and body sherds. The vessel was decorated with discrete groups of incised parallel lines, to create indeterminate geometric motifs, which run horizontally across the body and the shoulder. There is also a possible lozenge and carbonised food residues. Same as Grooved Ware Vessel D11.	✓			
D49	2	23684/1, 25712/12, 25712/24	5	2	✓	?	Two rim sherds with internal bevels and carbonised food residues.		✓		
D50	31	25443/6, 25534/2, 25534/4-7, 25534/9-10	7	44	?	O & R	Body sherds with linear voids on the external surface either due to wiping the pottery with organic matter or the inclusion of organic matter as temper within the clay matrix. Carbonised food residues are also present.				
D51	52	25442/3, 25442/11, 25442/16, 25442/20, 25442/25, 25443/11-13, 25443/32-33, 25443/35-39, 25443/41, 25447/1-3, 25447/6, 25447/18, 25447/21-22, 26617/3,	9	180	✓	?	A rim sherd with a rounded top, possible neck sherds, shoulder sherds and body sherds with visible moulding marks. Some spalling, much abrasion and carbonised food residues or heavy sooting on the interior.		✓		
D52	1	25447/9	n/a	1	?	O	A rim sherd with a flat top.		✓		
D53	11	25442/8, 25443/57-59	14	84	✓	O & R	Four base sherds and body sherds. Horizontal parallel linear grooves decorate the lower, surviving portion of the vessel.	✓			✓
D54	2	25443/44	6	2	?	O	Two rim sherds with a flat top and carbonised food residues		✓		
D55	1	25443/45	5	1	?	?	A rim sherd with a flat top with carbonised food residues.				
D56	2	25442/1, 25588/13	9	8	?	R	A decorated, everted Beaker rim sherd and a fragment detached. The rim is everted. Below the rim is cardium impressed diagonal and parallel lines. The rim top, and the interior of the rim are decorated with diagonal and vertical cardium impressed parallel lines. The motifs are arranged in horizontal bands.	✓			

Vessel code	Sherd nrs	Sherds	Average sherd thickness (mm)	Weight (mm)	Coil joins present	Firing (Oxidised/Reduced)	Description	Decorated	Rim	Carination/cordon	Base
D57	31	25984/1-10, 25984/12, 26818/5-10	12	192	✓	O & R	Two abraded, flat base sherds, two probable neck sherds, three probable shoulder sherds and body sherds. Finger moulding marks are noted as well as two accidental incised lines on the exterior. Carbonised food residues are also present.				✓
D58	100	23841/1-2, 23841/7, 23841/10, 23841/12, 25644/2, 26833/1	10	200	?	?	A probable carinated sherd and body sherds are distorted. An external incised line is probably a manufacturing mark. Carbonised food residues are present.			✓	
D59	6	23841/6, 23841/13	2	2	?	R	Six rim sherds had burnished exterior surfaces and carbonised food residues. The rim shape is simple with an internal bevel.		✓		
D60	11	26617/1-2, 26617/10, 26617/12	10	102	✓	O	A possible carinated sherd has smoothed surfaces, a shoulder sherd and body sherds. Surfaces were probably smoothed and have external sooting.			✓	
D61	25	23893/2, 23893/4, 23893/9-11	7	51	✓	O & R	A possible neck sherd, two probable carinated sherds and body sherds had carbonised food residues. The difference in wall thickness above and below the alleged carinated is considerable.			✓	
D62	21	23893/1, 23893/3, 23893/5-7, 23893/13	5	29	✓	?	Three rim sherds with rounded tops and body sherds with carbonised food residues.		✓		
D63	17	974-6, 978-980, 984, 974	7	51	✓	O	A neck sherd, two shoulder sherds and body sherds with carbonised food residues on the external surfaces.				
D64	9	211, 213, 202-3, 26150/1, 26164/1-2	8	17	?	?	Three rim sherds with a flat or rounded tops and several body sherds with carbonised food residues.		✓		
D65	3	23834/10, 23834/15	6	7	✓	O & R	Body sherds have elongated voids on surface from the use of organic matter when wiping the outside of the vessel.				
D66	3	25951/62, 27214/52	n/a	3	?	?	Body sherds with a soapy feel.				
D67	9	27214/41, 27214/45	9	31	✓	O & R	A shoulder sherd and body sherds with finger moulding and carbonised food residues present on the external surfaces.				
D68	14	25335/1-3	6	19	?	O & R	Body sherds with internal carbonised food residues.				
D69	1	24000/1	10	7	✓	O & R	A shoulder sherd with interior carbonised food residues.				

Vessel code	Sherd nrs	Sherds	Average sherd thickness (mm)	Weight (mm)	Coil joins present	Firing (Oxidised/Reduced)	Description	Decorated	Rim	Carination/cordon	Base
D70	1	25712/8	4	1	?	R	A very small rim sherd expanded on one side with a flat top.		✓		
D71	38	25588/27-31, 25588/48, 25588/51, 25588/56,	9	94	✓	O & R	Two neck sherds and body sherds were smoothed externally, and with possible sooting and internal carbonised food residues suggesting a cooking pot.				
D72	1	25443/17	n/a	1	?	?	A probable everted rim sherd with an internal bevel has exterior sooting.		✓		
D73	1	25442/27	n/a	1	?	?	A body sherd.				
D74	1	26424/22	n/a	1	?	?	A small, abraded body sherd.				
D75	1	27214/46	7	3	?	O & R	A shoulder sherd decorated with a single horizontal row of stabbed dots and external sooting.	✓			
D76	1	27214/47	n/a	2	?	?	A shoulder sherd with two faint marks on the exterior surface made during the manufacture of the vessel.				
D77	2	25348/5, 26435/3	n/a	5	✓	?	Two body sherds with some internal carbonised food residues.				
D78	5	26603/3, 26435/2	8	6	?	?	Body sherds with carbonised food residues suggesting they are from a cooking pot.				
D79	2	25588/1, 25588/3	6	4	?	O	Two rim sherds with slightly pinched or everted profiles with flattened tops and sooting. Striations from moulding are visible beneath the rim.		✓		
D80	1	25588/4	6	1	?	O	A simple rim sherd with a flattened top was probably burnished and later sooted. An incised diagonal line immediately below the rim may be a result of forming.		✓		
D81	2	27214/48	n/a	2	?	?	Two small body sherds with decoration consisting of an exterior incised line.	✓			
D82	1	26424/21	6	2	✓	R	A slightly everted rim sherd.		✓		
D83	1	26820/2	6	1	?	O	A small rim sherd with external sooting.		✓		
D84	1	26869/2	n/a	1	?	?	A rim sherd with a rounded top and possible exernal sooting.		✓		
D85	1	26339/31	7	2	?	R	A simple rim sherd with a rounded shape has voids on its external surface suggest organic tempering or wiping of the surface with grasses before firing. Carbonised food residues on both surfaces suggest it may have been part of a cooking pot.		✓		
D86	2	26820/20		4	?	?	Two body sherds decorated with sets of incised parallel lines creating an indeterminate geometric motifs.	✓			

Vessel code	Sherd nrs	Sherds	Average sherd thickness (mm)	Weight (mm)	Coil joins present	Firing (Oxidised/ Reduced)	Description	Decorated	Rim	Carination/ cordon	Base
D87	1	25588/60	6	1	?	O	A rim sherd with carbonised food residues indicating a cooking pot.		✓		
D88	1	26452/3	6	1	?	O	An irregular rim sherd with a rounded top has surface voids suggesting the use of organic matter for tempering or wiping the vessel surface.		✓		
D89	1	26820/5	6	1	✓	O	A probable rim sherd.		✓		
D90	1	26820/6	6	1	✓	?	A possible neck sherd decorated with a single incised line.				
D91	1	26830/2	n/a	n/a	?	O	A rim sherd with a rounded top.		✓		
D92	1	25443/42	n/a	n/a	?	?	A flat base sherd.				✓
D93	47	27214/1-5, 27214/9-10, 27214/51	6	82	✓	?	A possible rim sherd with a flat top, possible neck sherds, flat base sherds and body sherds with carbonised food residues.		✓		✓
D94	13	27214/14-15	12	28	?	O	A body sherd and several indeterminate sherds. Fingertip impressions are noted on the external surface of sherds. Considered to be decoration.	✓			
D95	48	25443/2-4, 25442/6-7, 25443/9, 25442/12-15, 25443/21, 25443/56	8	175	?	?	Simple, slightly inturned rim sherds with rounded tops, notable shoulder sherds and body sherds with single incised lines decoration. Carbonised food residues on both surfaces suggest a cooking pot.	✓	✓		
D96	1	26820/23	6	1	?	O	Small rim sherd with an internal bevel.		✓		
D97	7	25712/1-5	8	19	✓	?	A possible shoulder sherd and several body sherds. Severe exterior abrasion due to use? Carbonised food residues indicate a cooking pot.				
D98	2	25588/35	n/a	7	?	?	Two burnished body sherds with incised parallel lines. External sooting indicates placement over an open fire.	✓			
E01	5	23891/1-2, 25578/13, 25673/8	9	8	?	?	Two abraded body sherds and three possible carinated sherds. Mostly burnished and wiped, evidence of the use of organic matter either in the clay or from wiping the surfaces with grass or straw. Carbonised food remains indicate a possible cooking pot.			✓	

Vessel code	Sherd nrs	Sherds	Average sherd thickness (mm)	Weight (mm)	Coil joins present	Firing (Oxidised/ Reduced)	Description	Decorated	Rim	Carination/ cordon	Base
E02	39	24000/2, 24000/10, 24000/15-16	7	72	✓	?	Four abraded rim sherds, four shoulder sherds and the rest are body sherds. The rim has a rounded top and is slightly intumed. A single diagonal line is incised across the shoulder on the exterior on one shoulder sherd. External sooting and carbonised food residues indicate a cooking vessel.	?	✓		
E03	20	24000/3, 24000/5-6, 24000/8-9	6	31	?	?	Two slightly everted rim sherds with rounded tops and body sherds indicate a small vessel. Three body sherds have circular or oval motifs, perhaps bone impressed, on their exterior surfaces. Carbonised food residues indicate its use on a hearth.		✓		
E04	1	23977/1	6	2	?	?	A body sherd				
E05	1	24000/13	6	1	?	?	A likely rim sherd with a straight to rounded top affected by adhering residues with a depression below the rim on the external surface possibly due to loss of residues - this is not decoration. Sooting is present.		✓		

Bibliography

Scottish Record Office documents

GD221/105 Documents of 1666

CC3/9/30 Folio 24 Document of 1666

C2/13/1 Charter dated 1469

Site records

Calendar or Day Books for the years 1974, 1980, 1981, 1983, 1984 and 1985.

Unpublished Interim reports

Crawford, I. A. 1963. Preliminary Interim Report: Excavations Coileagan an Udail, North Uist.

Crawford. I. A. 1964. 1st Interim Report: Excavations Coileagan an Udail, North Uist.

Crawford. I. A. 1968. 5th Interim Report: Excavations Coileagan an Udail, North Uist.

Crawford, I. A. 1970. 7th Interim Report: Excavations Coileagan an Udail, North Uist.

Crawford, I. A. 1974. 11th Interim Report: Excavations Coileagan an Udail, North Uist.

Crawford, I. A. 1980. 17th Interim Report: Excavations Coileagan an Udail, North Uist.

Crawford, I. A. 1981. 18th Interim Report: Excavations Coileagan an Udail, North Uist.

Crawford, I. A. 1983. 20th Interim Report: Excavations Coileagan an Udail, North Uist.

Crawford, I. A. 1984. 21th Interim Report: Excavations Coileagan an Udail, North Uist.

Other Udal unpublished works

Crawford, I. A. and Crawford, I. 1977. Udal Small Finds: notes and information.

Ballin Smith, B. 2013. The Udal. Post Excavation Research Design. Dated 31.03.2013.

Ballin Smith, B. 2015. The Udal, North Uist. Post Excavation Research Design: RUX6.

Armit, I. 1988. *Excavations at Loch Olabhat, North Uist, 3rd interim report*. Edinburgh: Department of Archaeology, University of Edinburgh Project Paper No 10.

Armit, I. 1990. *Excavations at Loch Olabhat, North Uist 1989, 4th interim report*. Edinburgh: Department of Archaeology, University of Edinburgh Project Paper No 12.

Armit, I. 1992. The Hebridean Neolithic, in N. Sharples and A. Sheridan (eds.) *Vessels for the Ancestors. Essays on the Neolithic of Britain and Ireland in honour of Audrey Henshall*: 307-321. Edinburgh: Edinburgh University Press.

Arneborg, J., Heinemeier, J., Lynnerup, N., Nielsen, H. L., Rud, N., Sveinbjörnsdóttir, Á. E. 1999. Change of diet of the Greenland Vikings determined from stable carbon isotope analysis and 14C dating of their bones. *Radiocarbon* 41(2): 157-168.

Ashmore, P. 1975. *Calanais: The Standing Stones*. Stornoway: Urras nan Tursachan Ltd.

Ashmore, P. 2016. *Calanais Survey and Excavation 1979-88*. Edinburgh: Historic Environment Scotland. Viewed 01 November 2017, <https://www.

historicenvironment.scot/archives-and-research/publications/publication/?publicationId=b6aee5fd-5980-4872-a2e0-a63c00cc7b68>.

Ashton, N., Dean, P. and McNabb, J. 1991. Flaked flakes: what, when and why? *Lithics* 12: 1-11.

Badcock, A. 2008. *Ancient Uists: exploring the archaeology of the Outer Hebrides*. Fort William: Comhairle nan Eilean Siar.

Ballin, T. B. 1996. *Klassifikationssystem for Stenartefakter*. Universitetets Oldsaksamling, Varia 36. Oslo: Universitetets Oldsaksamling.

Ballin, T. B. 1999. Bipolar Cores in Southern Norway - Classification, Chronology and Geography. *Lithics* 20: 13-22.

Ballin, T. B. 2000a. Classification and description of lithic artefacts. A discussion of the basic lithic terminology. *Lithics* 21: 9-15.

Ballin, T. B. 2000b. The lithic artefacts from Rosinish, Benbecula, Western Isles. Unpublished report commissioned by Historic Scotland.

Ballin, T. B. 2001. Shieldaig, Wester Ross: The quartz assemblage. Unpublished report commissioned by Historic Scotland/National Museums Scotland.

Ballin, T. B. 2002a. Later Bronze Age flint technology: a presentation and discussion of post-barrow debitage from monuments in the Raunds area, Northamptonshire. *Lithics* 23: 3-28.

Ballin, T. B. 2002b. The lithic assemblage from Dalmore, Isle of Lewis, Western Isles. Unpublished specialist report commissioned by Historic Scotland.

Ballin, T. B. 2002c. Shieldaig, Wester Ross: The flint and bloodstone assemblage. Unpublished report commissioned by Historic Scotland/National Museums Scotland.

Ballin, T. B. 2006. The plano-convex knife, in I.Suddaby and A. Sheridan, A pit containing an undecorated Beaker and associated artefacts from Beechwood Park, Raigmore, Inverness. *Proceedings of the Society of Antiquaries of Scotland* 136: 77-88.

Ballin, T. B. 2008a. The lithic artefacts from Barpha Langais, North Uist, Western Isles. Unpublished report commissioned by ARCUS Archaeology.

Ballin, T. B. 2008b. Quartz Technology in Scottish Prehistory. *Scottish Archaeological Internet Report (SAIR)* 26. Viewed 01 November 2017, <http://archaeologydataservice.ac.uk/archives/view/sair/volumes.cfm>.

Ballin, T. B. 2010a. The Lithic Assemblage from Barabhas 3, Lewis, Western Isles. Unpublished report commissioned by Historic Scotland.

Ballin, T. B. 2010b. The Lithic Assemblage from Barabhas 2001, Lewis, Western Isles. Unpublished report commissioned by Historic Scotland.

Ballin, T. B. 2010c. Barabhas surface finds, the Curtis Collection – Isle of Lewis, Western Isles. The lithic assemblage. Unpublished report commissioned by Historic Scotland.

Ballin, T. B. 2010d. Barabhas surface finds, the Murray Collection – Isle of Lewis, Western Isles. The lithic assemblage. Unpublished report commissioned by Historic Scotland.

Ballin, T. B. 2010e. The lithic and stone assemblages from the Barabhas Machair – summary and comparative analysis. Unpublished report commissioned by Historic Scotland.

Ballin, T. B. 2011a. Lithics, in C Nesbitt, M. J. Church and S. M. D. Gilmour, Domestic, industrial, (en)closed? Survey and excavation of a Late Bronze Age/Early Iron Age promontory enclosure at Gob Eirer, Lewis, Western Isles. *Proceedings of the Society of Antiquaries of Scotland* 141: 31-74.

Ballin, T. B. 2011b. Struck flint from West Cotton, Irthlingborough and Stanwick (SS 3.7.6), and Overview of the Lithic Evidence (SS 3.7.7), in J. Harding and F. Healy (eds.) *The Raunds Area Project. A Neolithic and Bronze Age Landscape in Northamptonshire*. Volume 2. Supplementary Studies: 433-506, 506-527. Swindon: English Heritage. Viewed 01 November 2017, <http://www.english-heritage.org.uk/publications/neolithic-and-bronze-age-landscape-vol2/raundsareaproj2-ss3.pdf>.

Ballin, T. B. 2012. Lithic artefacts, in M. Johnson and K. Cameron, An Early Bronze Age unenclosed cremation cemetery and Mesolithic pit at Skilmafilly, near Maud, Aberdeenshire. *Scottish Archaeological Internet Report (SAIR)* 53: 23-26. Viewed 01 November 2017, <http://archaeologydataservice.

ac.uk/archives/view/sair/volumes.cfm>.

Ballin, T. B. 2013. Lithics, in M. J. Church, C. Nesbitt and S. M. D. Gilmour, A special place in the saltings? Survey and excavation of an Iron Age estuarine islet at An Dunan, Lewis, Western Isles. *Proceedings of the Society of Antiquaries of Scotland* 143: 183-185.

Ballin, T. B. 2014a. The Lithic material, in I. Arabaolaza, The cliff hanging cists: Sannox Quarry, Isle of Arran. *Archaeology Reports Online (ARO)* 10: 8-9. Viewed 01 November 2017, <http://www.archaeologyreportsonline.com/PDF/ARO10_Sannox.pdf>.

Ballin, T. B. 2014b. The provenance of some Scottish lithic raw materials – identification, terminology and interpretation. Stonechat 1: 4-7. Viewed 01 November 2017, <http://implementpetrology.org/wp-content/uploads/2014/04/2014-03-stonechat.pdf>.

Ballin, T. 2014c. The stone ball, in B. Ballin Smith, Between Tomb and Cist: the funerary monuments of Crantit, Kewing and Nether Onston, Orkney: 51-53. Kirkwall: The Orcadian.

Ballin, T. B. 2015a. The quartz assemblage from Point Braighe, Isle of Lewis, Western Isles. Unpublished report commissioned by ARCHAS Cultural Heritage Ltd.

Ballin, T.B. 2015b. Making an Island World: Neolithic Shetland. 2013: Felsite polished axeheads/adzes and Shetland knives in Shetland Museum – recording, characterisation and interpretation of the collection. North Roe Felsite Project Report 2. Dublin: University College Dublin, School of Archaeology.

Ballin, T. B. 2016. The Lithic Assemblage, in P. Ashmore, Calanais, Isle of Lewis, Western Isles. Edinburgh: Historic Environment Scotland. Viewed 01 November 2017, <https://www.historicenvironment.scot/archives-and-research/publications/publication/?publicationId=b6aee5fd-5980-4872-a2e0-a63c00cc7b68>.

Ballin, T. B. forthcoming a. The lithic and stone assemblage, in M. MacLeod Rivett, The Elliott Collection from the Barabhas Machair. *Scottish Archaeological Internet Reports (SAIR)*.

Ballin, T. B. forthcoming b. The lithic assemblage from Guinnerso, Uig, Lewis (Uig Landscape Project).

Ballin Smith, B. (ed.) 1994. *Howe. Four Millennia of Orkney Prehistory*. Edinburgh: Society of Antiquaries of Scotland, Monograph Series: 9.

Ballin Smith, B. 2005. Catpund: a prehistoric house in Shetland. *Scottish Archaeological Internet Report (SAIR)* 7. Viewed 01 November 2017, <http://archaeologydataservice.ac.uk/archives/view/sair/volumes.cfm>.

Ballin Smith, B. 2014. Iain Crawford's Udal: the key to ceramic traditions of the western seaboard? in H. Mytum and B. Davey (eds.) Ceramics on the Edge: 39-44. *Medieval Ceramics* 34 (2013).

Ballin Smith, B. forthcoming. The coarse pottery, in B. Will, Excavations on Iona 2014-2015. *Archaeology Reports Online (ARO)*.

Barber, J. 2011. Characterizing archaeology in machair, in D. Griffiths and P. Ashmore (eds.) Aeolian Archaeology: the archaeology of sand landscapes in Scotland: 37-53. *Scottish Archaeological Internet Report (SAIR)* 48. Viewed 01 November 2017, <http://archaeologydataservice.ac.uk/archives/view/sair/volumes.cfm>.

Barclay, G. J. and Russell-White, C. J. 1993. Excavations in the ceremonial complex of the fourth to second millennium at Balfarg/Balbirnie, Glenrothes, Fife. *Proceedings of the Society of Antiquaries of Scotland* 123: 43-210.

Bass, W. 2005. *Human osteology: a laboratory and field manual*. 5th edition. Colombia, Missouri: Missouri Archeological Society.

Berry, A. and Berry, R. 1967. Epigenetic variation in the human cranium. *Journal of Anatomy* 101 (2): 361-369.

Beaumont, J., Gledhill, A., Lee-Thorp, J. and Montgomery, J. A. 2013. Protocol for Sectioning Human Dentine: Expanded from Methods 1 and 2. Unpublished guidelines for the Biological Anthropology Research Centre, University of Bradford.

Beaumont, J., Montgomery, J., Buckberry, J. and Jay, M. 2015. Infant mortality and isotopic complexity: new approaches to stress, maternal health and weaning. *American Journal of Physical Anthropology* 157 (3): 441-457.

Beaumont, J. and Montgomery, J. 2016. The Great Irish Famine: identifying starvation in the tissues of victims using stable isotope analysis of bone and incremental dentine collagen, Plos One. Viewed 01 November 2017, <http://journals.plos.org/plosone/article?id=10.1371/journal.pone.0160065>.

Bennett, K. D., Fossitt, J. A., Sharp, M. J. and Switsur, V. R. 1990. Holocene vegetational and environmental history at Loch Lang, South Uist, Western Isles, Scotland. New Phytologist 114 (2): 281-298.

Beveridge, E. 1911. reprinted 2001. North Uist: its archaeology and topography. Edinburgh: Birlinn Ltd.

Binns, R. E. 1972. Pumice on post-glacial strandlines and in prehistoric sites in the British Isles. Scottish Journal of Geology 8: 105-114.

Binford, L. R. 1981. Bones: Ancient Men and Modern Myths. San Diego, California: Academic Press, Inc.

Binford, L. R. 1983. In Pursuit of the Past. Decoding the Archaeological Record. London: Thames and Hudson.

Boessneck, J. 1971. Osteological differences between sheep and goats, in D. Brothwell and E. Higgs. Science in Archaeology: 1-58. London: Thames and Hudson.

Bohling, S. and Buckberry, J. 2016. Osteological analysis of the human skeletal remains from RUX6, Udal North Uist. Unpublished MSc thesis, Biological Anthropology Research Centre, University of Bradford.

Bogaard, A., Heaton, T. H. E., Poulton, P. and Merbach, I. 2007 The impact of manuring on nitrogen isotope ratios in cereals: archaeological implications for reconstruction of diet and crop management practices. Journal of Archaeological Science 34: 335-343.

Boldsen, J., Milner, G., Konigsberg, L. and Wood, J. 2002. Transitional analysis: a new method for estimating age from skeletons, in R. Hoppa and J. Vaupel (eds.) Paleodemography: age distributions from skeletal samples. Cambridge Studies in Biological and Evolutionary Anthropology 31: 73-106. Cambridge: Cambridge University Press.

Boocock, P., Roberts, C. and Manchester, K. 1995. Maxillary sinusitis in medieval Chichester, England. American Journal of Physical Anthropology 98 (4): 483-495.

Boyd, J. M. and Boyd, I. L. 1990. The Hebrides. London: Collins and Sons New Naturalist Series.

Branigan, K., Newton, A. J. and Dugmore, A. J. 1995. Pumice, in K. Branigan and P. Foster. Barra: Archaeological Research on Ben Tangaval: 144-148. Sheffield: Sheffield Academic Press Ltd.

Brickley, M. and McKinley, J. 2004. Guidelines to the standards for recording human remains. Vol. 19. BABAO and Institute of Field Archaeologists.

Britton, K., Muldner, G. and Bell, M. 2008. Stable isotope evidence for salt-marsh grazing in the Bronze Age Severn Estuary, UK: implications for palaeodietary analysis at coastal sites. Journal of Archaeological Science 35: 2111-2118.

Bronk Ramsey, C. 1995. Radiocarbon calibration and analysis of stratigraphy: the OxCal program. Radiocarbon 37: 425-30.

Bronk Ramsey, C. 1998. Probability and dating. Radiocarbon 40: 461-74.

Bronk Ramsey, C. 2001. Development of the radiocarbon calibration program OxCal, Radiocarbon 43: 355-63.

Bronk Ramsey, C. 2009. Bayesian analysis of radiocarbon dates. Radiocarbon 51(1): 337-360.

Brooks, S. and Suchey, J. 1990. Skeletal age determination based on the os pubis: a comparison of the Acsádi-Nemeskér and Suchey-Brooks methods. Human Evolution 5 (3): 227-238.

Brophy, K. and Sheridan, A. 2012. Neolithic Scotland. ScARF Panel Report. ScARF Summary Neolithic Panel Document. Edinburgh: ScARF. Viewed 01 November 2017, <http://www.scottishheritagehub.com/content/scarf-neolithic-panel-report>.

Brothwell, D. 1981. *Digging up Bones: the excavation, treatment and study of human skeletal remains.* Ithaca, New York: Cornell University Press.

Brown, N. (nd.) *Eilean Domhnuill Loch Olabhat Pottery*. Unpublished specialist report.

Brown, T. A., Nelson, D. E., Vogel, J. S. and Southon, J. R. 1988. Improved collagen extraction by modified Longin method. *Radiocarbon* 30: 171-177.

Buck, C. E., Cavanagh, W. G., and Litton, C. D. 1996. *Bayesian approach to interpreting archaeological data*. Chichester: John Wiley & Sons, Ltd.

Buckberry, J. and Chamberlain, A. 2002. Age estimation from the auricular surface of the ilium: a revised method. *American Journal of Physical Anthropology* 119 (3): 231-239.

Buikstra, J. and Ubelaker, D. (eds.) 1994. *Standards for data collection from human skeletal remains*. Vol. 44. Fayetteville, Arkansas: Arkansas Archeological Survey.

Butler, C. 2005. *Prehistoric Flintwork*. Stroud: Tempus Publishing Ltd.

Callahan, E. 1987. *An Evaluation of the Lithic Technology in Middle Sweden during the Mesolithic and Neolithic*. Aun 8. Uppsala: Archaeological Studies, Uppsala University Institute of North European Archaeology.

Campbell, E. 1991. Excavation of a wheelhouse and other Iron Age structures at Sollas, North Uist by R. J. Atkinson. *Proceedings of the Society of Antiquaries of Scotland* 121: 117-73.

Cappers, R. T. J., Bekker, R. M. and Jans, J. E. A. 2006. *Digital Seed Atlas of the Netherlands*. Groningen Archaeological Studies 4. Eelde, The Netherlands: Barkhuis Publishing.

Cardoso, F. and Henderson, C. 2010 Enthesopathy formation in the humerus: data from the known age-at-death and known occupation skeletal collections. *American Journal of Physical Anthropology* 141 (4): 550-560.

Carriker, M. C. 1981. Shell penetration and feeding by Naticacean and Muricacean predatory gastropods: a synthesis. *Malacologia* 20 (2): 403-422.

Childe, V. G. 1931. *Skara Brae: a Pictish village in Orkney*. London: Kegan Paul, Trench, Trubner and Co. Ltd.

Church, M. 2006. Plant Macrofossils, in D. D. A. Simpson, E. M. Murphy and R. A. Gregory *Excavations at Northton, Isle of Harris*. Oxford: British Archaeological Reports, British Series 408: 35.

Cibulka, M. 2004. Determination and significance of femoral neck anteversion. *Physical Therapy* 84 (6): 550-558.

Claassen, C. 1998. *Shells*. Cambridge: Cambridge Manuals in Archaeology.

Clarke, A. 1997. *A Report on the Quartz and Stone Assemblages from the Udal, RUX6, North Uist*. Unpublished preliminary report commissioned by Iain Crawford.

Clarke, A. 2005. Coarse stone artefacts, in A. Richie (ed.) *Kilellan Farm, Ardnave, Islay: excavations of a prehistoric to early medieval site by Colin Burgess and others 1954-1976* : 133-41. Edinburgh: Society of Antiquaries of Scotland.

Clarke, A. 2014. The pumice, in H. Moore and G. Wilson *Ebbing Shores: Survey and Excavation of Coastal Archaeology in Shetland 1995-2008*: 181-183. Edinburgh: Historic Scotland Archaeology Report 8.

Clarke, D. 1970. *Beaker Pottery of Great Britain and Ireland*. Cambridge: Cambridge University Press.

Clarke, D.V., Cowie, T.G. and Foxon, A. 1985. *Symbols of Power at the Time of Stonehenge*. Edinburgh: National Museum of Antiquities of Scotland.

Close-Brooks, J. 1995. Excavation of a cairn at Cnip, Uig, Isle of Lewis. *Proceedings of the Society of Antiquaries of Scotland* 125: 253-277.

Clough, T. H. McK. and Cummins, W. A. 1988. *Stone Axe Studies Volume 2: the petrology of prehistoric stone implements from the British Isles*. London: Council for British Archaeology.

Cook, G. T., Ascough, P. L., Bonsall, C., Hamilton, D., Russell, N., Sayle, K. L., Scott, M., and Bownes, J. 2015. Best practice methodology for 14C calibration of marine and mixed terrestrial/marine samples. *Quaternary Geochronology* 27: 164-71.

Copper, M. 2016. Prehistoric pottery from An Doirlinn, in D. Garrow and F. Sturt (eds.) 2016 An Doirlinn, South Uist, Outer Hebrides, Preliminary draft report on the 2012 excavations: 11-27. Unpublished University of Reading and University of Southampton report.

Cowie, R. 2005. The pottery, in A. Richie (ed.) *Kilellan Farm, Ardnave, Islay: excavations of a prehistoric to early medieval site by Colin Burgess and others 1954-1976*: 49-96. Edinburgh: Society of Antiquaries of Scotland.

Cowie, T. and MacSween, A. 1999. A Grooved Ware

from Scotland: a review, in R. Cleal and A. MacSween *Grooved Ware in Britain and Ireland*: 48-56. Oxford: Oxbow Books.

Cowie, T. and MacLeod Rivett, M. 2010. Barabhas 3: An early Bronze Age settlement at Barvas (Barabhas) Machair, Isle of Lewis. Unpublished data structure report.

Crawford, I. A. 1962. Kelp Burning, *Scottish Studies* 6/1: 104-7.

Crawford, I. A. 1965. Contributions to a History of Domestic Settlement in North Uist, *Scottish Studies* 9/1: 34-63.

Crawford, I. A. 1977: A corbelled Bronze-Age burial chamber and beaker and beaker evidence from the Rosinish machair, Benbecular. *Proceedings of the Society of Antiquaries of Scotland* 108: 94-107.

Crawford, I. A. 1986. *The West Highlands and Islands: a review of 50 centuries*. Cambridge: privately published. Produced for the British Museum Special Exhibition 1986-7.

Crawford, I. A. and Switsur, R. 1977. Sandscaping and C14: The Udal, North Uist. *Antiquity* 51: 124-36.

Crothers, J. H. 1985. Dog whelks: an introduction to the ecology of *Nucella lapillus* (L.). *Field Studies* 6: 291-360.

Curtis, M. R. and Curtis, G. R. 1999. Olcote, Breasclete Park, Callanish (Uig parish), burial cairn. *Discovery and Excavation in Scotland*: 94, Fig 25.

Dias, G. and Tayles, N. 1997. Abscess cavity - a misnomer. *International Journal of Osteoarchaeology* 7 (5): 548-554.

Dickens, A. 1990. A Study of the lithic assemblage from Allt Chrisal, Barra, Outer Hebrides. Unpublished dissertation submitted for the degree of BA in the Department of Archaeology and Prehistory, University of Sheffield.

Dickinson, G. and Randall, R. E. 1979. An interpretation of machair vegetation, in J. Morton Boyd (ed.) Natural Environment of the Outer Hebrides, *Proceedings of the Royal Society of Edinburgh* 77: 267-278.

Dickson, C. and Dickson, J. 2000. *Plants and People in Ancient Scotland*. Stroud: Tempus Publishing Ltd.

Dickson, J. H. 1992. North American driftwood, especially *Picea* (spruce), from archaeological sites in the Hebrides and Northern Isles of Scotland, in J. P. Pals, J. Buurman and M. van der Veen (eds.) Festschrift for Professor van Zeist. *Review of Palaeobotany and Palynology* 73: 49-56.

Djukic, K., Milenkovic, P., Milcvanovic, P., Dakic, M. and Djuric, M. 2014. The increased femoral neck anteversion in medieval cemetery of of Pecenjevce, aetiology and differential diagnosis in archaeological context. *Chungara Revista de Antropología Chilena* 46 (2): 295-303.

Downes, J. and Richards, C. 2005. The dwellings at Barnhouse, in C. Richards (ed.) *Dwelling among the Monuments: the Neolithic village of Barnhouse, Maeshowe passage grave and surrounding monuments at Stenness, Orkney: 57-127*. Cambridge: McDonald Institute for Archaeological Research.

Downes, J. 2012. *Chalcolithic and Bronze Age Scotland. ScARF Panel Report. ScARF Summary Bronze Age Panel Document*. Edinburgh: ScARF. Viewed 01 November 2017, <http://www.scottishheritagehub.com/content/scarf-bronze-age-panel-report>.

Dunbar, E., Cook. G. T., Naysmith, P., Tripney, B. G. and Xu, S. 2016. AMS 14C dating at the Scottish Universities Environmental Research Centre (SUERC) Radiocarbon Dating Laboratory. *Radiocarbon* 58 (1) March 2016: 9-23.

Dunwell, A. J., Neighbour, T. and Cowie, T. G. 1995. A cist burial adjacent to the Bronze Age cairn at Cnip, Uig, Isle of Lewis. *Proceedings of the Society of Antiquaries of Scotland 125: 279-288*.

Duška, F., Tůma, P., Mokrejš, P., Kuběna, A. and Anděla, M. 2007. Analysis of factors influencing nitrogen balance during acute starvation in obese subject with and without type 2 diabetes. *Clinical Nutrition* 26: 552-558.

Evans, G. J. 1972. *Land Snails in Archaeology*. London: London University Press.

Evans, S. J. 1897. *The Ancient Stone Implements, Weapons and Ornaments of Great Britain*. Second Edition. London: Longmans, Green, and Co.

Fenton, A. 1978. *The Northern Isles: Orkney and Shetland*. Edinburgh: John Donald.

Fenton, A. 1981. Early Manuring techniques, in R. Mercer (ed.) *Farming Practice in British Prehistory*. Edinburgh: Edinburgh University Press.

Fettes, D. J., Mendum, J. R., Smith, D. I. and Watson, J. V. 1992. *Geology of the Outer Hebrides*. British Geological Survey. London: Her Majesty's Stationery Office.

Finlay, J. I. 1981. The faunal remains from Cnoc Sligeach, Sollas, North Uist: A study in economic reconstruction. Unpublished MA dissertation. University of Edinburgh.

Finlay, J. I. 1984. Faunal evidence for prehistoric economy and settlement in the Outer Hebrides to c. 400 AD. Unpublished Ph.D. dissertation. University of Edinburgh.

Finlay, J. 2006. The Faunal Remains. in D.D.A. Simpson, E. Murphy and R. A. Gregory (eds.) *Excavations at Northton, Isle of Harris*: 147-149. Oxford: British Archaeological Reports, British Series 48.

Finlayson, B. 1990. Lithic exploitation during the Mesolithic in Scotland. *Scottish Archaeological Review* 1990: 41-58.

Finlayson, B. 2000. Chipped Stone Assemblage, in J. Downes and R. Lamb *Prehistoric Houses at Sumburgh in Shetland. Excavations at Sumburgh Airport 1967-74*. Oxford: Oxbow Books.

Finlayson, B. 1997. The plano-convex knife, in R. J. Mercer and M. S. Midgley The Early Bronze Age cairn at Sketewan, Balnaguard, Perth and Kinross, *Proceedings of the Society of Antiquaries of Scotland* 127: 281-338.

Finnegan, M. 1978. Non-metric variation of the infracranial skeleton. *Journal of Anatomy* 125 (1): 23-57.

Garrow, D. and Sturt, F. (eds.) 2016 An Doirlinn, South Uist, Outer Hebrides, Preliminary draft report on the 2012 excavations. Unpublished University of Reading and University of Southampton report.

Garrow, D. and Sturt, F. (eds.) 2017 *Neolithic Stepping Stones: excavation and survey within the Western Seaways of Britain 2008-2014*. Oxford: Oxbow books.

Geological Survey of Great Britain 2017 Geology of Britain viewer. Viewed 01 November 2017, <http://mapapps.bgs.ac.uk/geologyofbritain/home.html>.

Gibson, A. M. 1995. The Neolithic Pottery from Allt Chrisal, in K. Branigan and P. Foster. *Barra: archaeological research on Ben Tangaval*: 100-115. Sheffield: Sheffield Academic Press.

Gibson, A. M. 2006. Pottery, in D. D. A. Simpson, E. M. Murphy and R. A. Gregory *Excavations at Northton, Isle of Harris*: 90-133. Oxford: British Archaeological Reports, British Series 408.

Goodman, A. and Rose, J. 1991. Dental enamel hypoplasias as indicators of nutritional stress, in M. Kelley and C. Larsen (eds.) *Advances in dental anthropology*: 279-293. New York: Wiley Liss.

Graham-Campbell, G. and Batey, C. E. 1998. (reprinted 2005). *Vikings in Scotland*, Edinburgh: Edinburgh University Press.

Grant, A. 1975. Appendix B: the use of tooth wear as a guide to the age of domestic animals – a brief explanation, in B. Cunliffe *Excavations at Portchester Castle: I. Roman*. London: Society of Antiquaries of London.

Grant, A. 1982. The use of tooth wear as a guide to the age of domestic ungulates, in R. Wilson, C. Grigson and S. Payne *Ageing and sexing animal bones from archaeological sites*. 91-108. Oxford: British Archaeological Reports, British Series 109.

Green, H. S. 1980. *The Flint Arrowheads of the British Isles. A detailed study of material from England and Wales with comparanda from Scotland and Ireland*. Oxford: British Archaeological Reports, British Series 75 (i).

Gregory, R. 2006a. Coarse stone artefacts, in D. D. A. Simpson, E. M. Murphy and R. A. Gregory *Excavations at Northton, Isle of Harris*: 25-26. Oxford: British Archaeological Reports, British Series 408.

Gregory, R. 2006b. Drift pumice, in D. D. A. Simpson, E. M. Murphy and R. A. Gregory *Excavations at Northton, Isle of Harris*: 69,133. Oxford: British Archaeological Reports, British Series 408.

Gregory, A. 2006c. Radiocarbon dating, in D. D. A. Simpson, E. M. Murphy and R. A. Gregory *Excavations at Northton, Isle of Harris*: 89-90. Oxford: British Archaeological Reports, British Series 408.

Gregory, R. and Simpson, D. A. A. 2006. Discussion,

in D. D. A. Simpson, E. M. Murphy and R. A. Gregory *Excavations at Northton, Isle of Harris: 78-84 and 149-154.* Oxford: BAR British Series 408.

Guthrie, H. A. and Picciano, M. F. 1995. *Human Nutrition*. Missouri: Mosby.

Haggarty, A. 1991. Machrie Moor, Arran: recent excavations at two stone circles. *Proceedings of the Society of Antiquaries of Scotland* 121, 51-94.

Hallén, Y. 1994. The use of bone and antler at Foshigarry and Bac Mhic Connain, two Iron Age sites on North Uist, Western Isles. *Proceedings of the Society of Antiquaries of Scotland* 124, 189-231.

Hallsworth, C. R. and Knox, R. W. O'B. 1999. *BGS Rock Classification Scheme. Volume 3. Classification of sediments and sedimentary rocks*: 99-103. British Geological Survey Research Report.

Hamilton, M. and Sharples, N. 2012 Early Bronze Age settlements at Machair Mheadhanach and Cill Donnain, in M. Parker Pearson *From Machair to Mountains*: 199-214. Oxford and Oakville: Oxbow Books.

Hansom, J. D. 2003. Machair, in V. J. May and J. D. Hansom (eds.) *Coastal Geomorphology of Great Britain*: 472-514. Peterborough: Geological Conservation Review Series No. 28. Joint Nature Conservation Committee.

Hansom, J. D. and Angus, S. 2006 Machair nan Eilean Siar (Machair of the Western Isles). *Scottish Geographical Journal* 121 (4): 401-412.

Hardy, K. 2009. Worked and modified shell, in K. Hardy and C.Wickham-Jones (eds.) Mesolithic and later sites around the Inner Sound: the work of the Scotland's First Settlers project 1998-2004. *Scottish Archaeological Internet Reports (SAIR)* 31, Section 3.5. Viewed 01 November 2017, <http://archaeologydataservice.ac.uk/archives/view/sair/volumes.cfm>.

Harman, M. 1983. Animal remains, in G. Graham Ritchie and H. Welfare Excavations at Ardnave, Islay: 343-350. *Proceedings of the Society of Antiquaries of Scotland* 113: 302-366.

Harman, M. 2010. Appendix G: Animal bone, in T. Cowie and M. MacLeod Rivett, Barabhas 1: Data Structure Report: A Bronze Age settlement at Barvas (Barabhas) Machair, Isle of Lewis: 44-46. Unpublished report.

Hatch, K. A., Crawford, M. A., Kunz, A. W., Thomsen, S. R., Eggett, D. L., Nelson, S. T. and Roeder, B. L. 2006. An objective means of diagnosing anorexia nervosa and bulimia nervosa using 15N/14N and 13C/12C ratios in hair. *Rapid Communications in Mass Spectrometry* 20: 3367-3373.

Hayward, P., Nelson-Smith, T. and Shields, C. 1996. *Sea Shore of Britain and Europe*. London. Collins Pocket Guide.

Henley, C, 2012, Loch a' Choire Neolithic Settlement – Finds, in M. Parker Pearson (ed.) *From Machair to Mountains: Archaeological Survey and Excavation in South Uist*: 189-178. Oxford and Oakville: Oxbow Books.

Henshall, A. S. 1983. Catalogue of Artefacts, in A. Ritchie 1983 Excavation of a Neolithic farmstead at Knap of Howar, Papa Westray, Orkney: 75-78. *Proceedings of the Society of Antiquaries of Scotland* 113: 40-121.

Higgins, M. W. 1971. *Cataclastic Rocks*. Geological Survey Professional Paper 687. Washington: United States Government Printing Office.

Hobson, K. A., Alisauska, R. T. and Clark, R. G. 1993. Stable nitrogen isotope enrichment in avian tissues due to fasting and nutritional stress: implications for isotopic analyses of diet. *The Condor* 95: 388-394.

Inizan, M-L., Roche, H. and Tixier, J. 1992. *Technology of Knapped Stone*. Meudon: Cercle de Recherches et d'Etudes Préhistoriques.

Jones, A. 2005. The Grooved-Ware from Barnhouse, in C. Richards *Dwelling among the Monuments: the Neolithic village at Barnhouse, Maeshowe passage grave and surrounding monuments at Stenness, Orkney*. Cambridge: McDonald Institute for Archaeological Research: 261-282.

Juel Jensen, H. 1994. *Flint Tools and Plant Working. Hidden Traces of Stone Age Technology. A use wear study of some Danish Mesolithic and TRB implements*. Aarhus: Aarhus University Press.

Katzenberg, M. A. and Lovell, N. C. 1999. Stable isotope variation in pathological bone. *International Journal of Osteoarchaeology* 9: 316-324.

Kerney, M. P. and Cameron, R. A. D. 1996. *Land Snails*

of Britain and North West Europe. London: Collins.

Klales, A., Ousley, S. and Vollner, J. 2012. A revised method of sexing the human innominate using Phenice's nonmetric traits and statistical methods. *American Journal of Physical Anthropology* 149 (1): 104-114.

Lacaille, A. D. 1937. A stone industry, potsherds, and a bronze pin from Valtos, Uig, Lewis, *Proceedings of the Society of Antiquaries of Scotland* 71: 279-296.

Larsen, G., Newton, A. J., Dugmore, A. J. and Vilmundardóttir, E. 2001. Geochemistry, dispersal, volumes and chronology of Holocene silicic tephra layers from the Katla volcanic system, Iceland. *Journal of Quaternary Science* 16 (2): 119-132.

Lauer, F., Prost, K., Gerlach, R., Pätzold, S., Wolf, M., Urmersbach, S. et al. 2014. Organic fertilization and sufficient nutrient status in prehistoric agriculture? – Indications from multi-proxy analyses of archaeological topsoil relicts, *PLoS ONE 9(9): e106244*. Viewed 01 November 2017, <http://journals.plos.org/plosone/article?id=10.1371/journal.pone.0106244>.

Law, M. and Thew, N. 2015. Land snails, sand dunes and archaeology in the Outer Hebrides. *Journal of the North Atlantic* 9: 125-133.

Lightfoot, E., O'Connell, T. C., Stevens, R. E., Hamilton, J., Hey, G. and Hedges, R. 2009. An investigation into diet at the site of Yarnton, Oxfordshire, using stable carbon and nitrogen isotopes. *Oxford Journal of Archaeology* 28: 301-322.

Linklater, E. 2002. Udal Law – Past, Present and Future? Unpublished dissertation for LLB (Hons.) European Law, University of Strathclyde.

Lobell, A. and Patel, S. S. 2010. Cladh Hallan, *Archaeology Magazine* 63/3. Viewed 01 November 2017, <http://archive.archaeology.org/1005/bogbodies/cladh_hallan.html>.

Marshall, D. N. 1977. Carved stone balls. Proceedings of the Society of Antiquaries of Scotland 108: 40-72.

Mays, S. 2016. Estimation of stature in archaeological human skeletal remains from Britain. American Journal of Physical Anthropology 161 (4): 646-655.

Mays, S. and Cox, M. 2000. Sex determination in skeletal remains. in M. Cox and S. Mays (eds.) Human osteology in archaeology and forensic science: 117-130. London: Greenwich Medical Media.

McNeill, F. M. 1974. The Scots Kitchen: Its Lore and Recipes. London: Granada Publishing.

Mears, R. and Hillman, G. 2007 Wild Foods. London: Hodder and Stoughton.

Meindl, R. and Lovejoy, O. 1985. Ectocranial suture closure: a revised method for the determination of skeletal age at death based on the lateral-anterior sutures. American Journal of Physical Anthropology 68 (1): 57-66.

Mekota, A., Grupe, G., Ufer, S. and Cuntz, U. 2006. Serial analysis of stable nitrogen and carbon isotopes in hair: monitoring starvation and recovery phases of patients suffering from anorexia nervosa. Rapid Communications in Mass Spectrometry 20: 1604-1610.

Merrifield, R. 1987. The Archaeology of Ritual and Magic. London: B. T. Batsford Ltd.

Meyer, C., Nicklisch, N., Held, P., Fritsch, B. and Alt, K. 2011. Tracing patterns of activity in the human skeleton: an overview of methods, problems, and limits of interpretation. Journal of Comparative Human Biology 62: 202-217.

Mills, C. M., Crone, A., Edwards, K. J. and Whittington, G. 1994. The excavation and environmental investigation of a sub-peat stone bank near Loch Portain, North Uist, Outer Hebrides, Proceedings of the Society of Antiquaries of Scotland 124: 155-171.

Milner, N. 2004. An analysis of the marine molluscs from the Mesolithic site of Sand, Scotland, in K. Hardy and C. Wickham-Jones (eds.) Mesolithic and later sites around the Inner Sound: the work of the Scotland's First Settlers project 1998-2004. Scottish Archaeological Internet Reports (SAIR) 31, Viewed 01 November 2017, <http://archaeologydataservice.ac.uk/archives/view/sair/volumes.cfm>.

Milner, N. 2007. Sea Loch Survey: an analysis of the marine molluscs, in K. Hardy and C. Wickham-Jones (eds.) Mesolithic and later sites around the Inner Sound: the work of the Scotland's First Settlers project 1998-2004. *Scottish Archaeological Internet Reports (SAIR)* 31, Viewed 01 November 2017. <http://archaeologydataservice.ac.uk/archives/view/sair/volumes.cfm>.

Milner, N. 2009. Consumption of crabs in the Mesolithic: side stepping the evidence? in K. Hardy and C. Wickham-Jones (eds.) Mesolithic and later sites around the Inner Sound: the work of the Scotland's First Settlers project 1998-2004. *Scottish Archaeological Internet Reports (SAIR)* 31. Viewed 01 November 2017, <http://archaeologydataservice.ac.uk/archives/view/sair/volumes.cfm>.

Montgomery, J., Evans, J. A. and Neighbour, T. 2003. Sr isotope evidence for population movement within the Hebridean Norse community of NW Scotland. *Journal of the Geological Society* 160 (5): 649-653.

Montgomery, J., Evans, J. A. and Cooper, R. E. 2007. Resolving archaeological populations with Sr-Isotope mixing models. *Applied Geochemistry* 22 (7): 1502-1514.

Montgomery, J., Beaumont, J., Jay, M., Keefe, K., Gledhill, A. R., Cook, G. T., Dockrill, S. J. and Melton, N. D. 2013. Strategic and sporadic marine consumption at the onset of the Neolithic: increasing temporal resolution in the isotope evidence. *Antiquity* 87 (338): 1060-1072.

Moore, M. 2013. Sex estimation and assessment, in E. Digangi and M. Moore (eds.) *Research methods in human skeletal biology*: 91-116. Oxford: Academic Press.

Müldner, G., Chenery, C. and Eckardt, H. 2011. The Headless Romans: multi-isotope investigations of an unusual burial ground from Roman Britain. *Journal of Archaeological Science* 38 (2): 280-290.

Murphy, E. and Simpson D. 2006a. Worked bone, in D. D. A. Simpson, E. M. Murphy and R.A. Gregory *Excavations at Northton, Isle of Harris*: 72-75. Oxford: British Archaeological Reports, British Series 408.

Murphy, E. and Simpson D. 2006b. Worked bone and antler, in D. D. A. Simpson, E. M. Murphy and R.A. Gregory *Excavations at Northton, Isle of Harris*: 140-147. Oxford: British Archaeological Reports, British Series 408.

(NCCA) National Coastal Change Assessment. 2017. Viewed 01 November 2017, <http://www.dynamiccoast.com/webmap.html>.

Neighbour, T. 2005. Excavation of a Bronze Age kerbed cairn at Olcote, Breasclete, near Calanais, Isle of Lewis. Scottish Archaeological Internet Report 13. Viewed 01 November 2017, <http://archaeologydataservice.ac.uk/archives/view/sair/volumes.cfm>.

Nelis, E. 2006a. Various entries, 23-25; 69-72; 133-139 and 170, in D. D. A. Simpson, E. M. Murphy and R.A. Gregory *Excavations at Northton, Isle of Harris*. Oxford: British Archaeological Reports, British Series 408.

Nelis, E. 2006b. Chipped polished and coarse stone artefacts, in D. D. A. Simpson, E. M. Murphy and R.A. Gregory *Excavations at Northton, Isle of Harris*: 133-139. Oxford: British Archaeological Reports, British Series 408.

Newton, A. J. 1999a. *Ocean-transported pumice in the North Atlantic*. Unpublished PhD dissertation, University of Edinburgh.

Newton, A. J. 1999b. Report on the pumice, in B. E. Crawford and B. Ballin Smith *The Biggings, Papa Stour, Shetland: the history and archaeology of a royal Norwegian farm*. Edinburgh: Society of Antiquaries of Scotland Monograph Series No 13.

Newton, A. J. 2001. The pumice, in S. Mithen (ed.) *Hunter-gather landscape archaeology: The Southern Hebrides Project 1988-98*: 403-405. Cambridge: McDonald Institute Monographs.

Newton, A. J. 2004. Pumice, in C. Wickham-Jones and K. Hardy, Camas Daraich: a Mesolithic site at the Point of Sleat, Skye. Scottish Archaeological Internet Report *(SAIR)* 12, 47-49. Viewed 01 November 2017, <http://archaeologydataservice.ac.uk/archives/view/sair/volumes.cfm>.

Newton, A. J. 2006. Pumice, in I. Armit *Anatomy of an Iron Age roundhouse: the Cnip wheelhouse excavations, Lewis*: 153-154. Edinburgh, Society of Antiquaries of Scotland.

Newton, A. J. 2014. Pumice, in J. Coolen and N. Mehler *Excavations and Surveys at the Law Ting Holm, Tingwall, Shetland*: 73-74. Oxford: British Archaeological Reports, British Series 592.

Newton, A. J. and Dugmore, A. J. 1995. The Pumice: analytical report, in K. Branigan and P. Foster *Barra: Archaeological Research on Ben Tangaval*: 145-148. Sheffield: Sheffield Academic Press.

Newton, A. J. and Dugmore. A. J. 2003. Analysis of pumice from Baleshare, in J. Barber (ed.) Bronze Age farms and Iron Age farm mounds of the Outer

Hebrides. *Scottish Archaeologial Internet Report (SAIR)* 3: 135-138. Viewed 01 November 2017, <http://archaeologydataservice.ac.uk/archives/view/sair/volumes.cfm>.

Nicolson, J. R. 1990. *Traditional Life in Shetland.* London: Robert Hale.

Ogden, A., Pinhasi, R. and White, W. 2007. Gross enamel hypoplasia in molars from subadults in a 16th-18th century London graveyard. *American Journal of Physical Anthropology* 133 (3): 957-966.

Parker Pearson, M. 2012. Settlement, agriculture and society in South Uist before the clearances, in M. Parker Pearson *From Machair to Mountains: Archaeological Survey and Excavation in South Uist*: 401-425. Oxford: Oxbow Books.

Parker Pearson, M. 2014. Bone and stone tools, in M. Parker Pearson and M. Zvelebil *Excavations at Cill Donnain: A Bronze Age Settlement and Iron Age Wheelhouse in South Uist* : 35-37. Oxford and Philadelphia: Oxbow Books.

Parker Pearson, M. 2015. The Prehistoric Village at Cladh Hallan - Part III. Viewed 01 November 2017, <https://www.shef.ac.uk/archaeology/research/cladh-hallan/cladh-hallan03>.

Parker Pearson, M. and Zvelebil, M. 2014. Excavations at Cill Donnain: A Bronze Age Settlement and Iron Age Wheelhouse in South Uist. Oxford: Oxbow Books.

Parker Pearson, M., Chamberlain, A., Craig, O., Marshall, P., Mulville, J., Smith, H., Chenery, C., Collins, M., Cook, G., Craig, G. and Evans, J. A. 2005. Evidence for mummification in Bronze Age Britain. Antiquity 79 (305): 529-546.

Parker Pearson, M. et al. forthcoming Cadh Hallan.

Payne, S. 1972. On the interpretation of bone samples from archaeological sites, in E. Higgs (ed.) Papers in economic prehistory: 65-82. Cambridge: Cambridge University Press.

Pellant, C. 1992. Rocks and Minerals. The visual guide to over 500 rock and mineral specimens from around the world. Eyewitness Handbooks. London: Dorling Kindersley.

Pollard, T. 1997. Excavation of a Neolithic settlement and ritual complex at Beckton Farm, Lockerbie, Dumfries and Galloway. Proceedings of the Society of Antiquaries of Scotland 127: 113-117.

Powanda, M. C. and Beisel, W. R. 2003. Metabolic effects of infection on protein and energy status. Journal of Nutrition 133: 332s-327s.

Reimer, P. J., Bard, E., Bayliss, A., Beck, J. W., Blackwell, P., Bronk Ramsey, C. et.al. 2013. IntCal13 and Marine13 radiocarbon age calibration curves 0-50,000 years cal BP. Radiocarbon 55: 1869-87.

Renfrew, C. 1976. Megaliths, territories and populations, in S. J. De Laet Acculturation and continuity in Atlantic Europe: mainly in the Neolithic period and the Bronze Age: 198-220. Bruge: De Temple/International Union of Prehistoric and Protohistoric Sciences.

Richards, M. P., Fuller, B. T. and Molleson, T. I. 2006. Stable isotope palaeodietary study of humans and fauna from the multi-period (Iron Age, Viking and late Medieval) site of Newark Bay, Orkney. Journal of Archaeological Science 33 (1): 122-131.

Ritchie, A. 1983. Excavation of a Neolithic farmstead at Knap of Howar, Papa Westray, Orkney. Proceedings of the Society of Antiquaries of Scotland 113: 40-121.

Ritchie, A. (ed.) 2005. Kilellan Farm, Ardnave, Islay: Excavations of a prehistoric to early medieval site by Colin Burgess and others1954-76. Edinburgh: Society of Antiquaries of Scotland.

Ritchie, W. 1979. Machair development and chronology in the Uists and adjacent Islands, in J. Morton Boyd (ed.) The Natural Environment of the Outer Hebrides: 107-122. Edinburgh: The Royal Society of Edinburgh.

Roberts, C. 2007. A bioarchaeological study of maxillary sinusitis. American Journal of Physical Anthropology 133 (2): 792-807.

Rokade, S. and Mane, A. 2008. Femoral anteversion: comparison by two methods. The Internet Journal of Biological Anthropology 3 (1).

Roper, K. and Murray, R. 2017. Historic Scotland Human Remains Call Off Contract: Horgabost, Harris. Archaeological Excavation. Unpublished AOC Data Structure Report 22470-18.

Rose, J. and Ungar, P. 1998. Gross dental wear and

dental microwear in historical perspective, in K. Alt, W. Rösing and M. Teschler-Nicola (eds.) Dental anthropology: fundamentals, limits, and prospects: 349-386. New York: Springer-Verlag/Wein.

(RCAHMS) Royal Commission on the Ancient and Historical Monuments of Scotland. 1928. Outer Hebrides, Skye and the Small Isles. Edinburgh: HMSO.

Russell, N., Cook, G. T., Ascough, P. L., Scott, E. M. 2015. A period of calm in Scottish seas: A comprehensive study of ΔR values for the northern British Isles coast and the consequent implications for archaeology and oceanography. Quaternary Geochronology 30: 34-41.

Saville, A. and Ballin, T. B. 2009. Upper Palaeolithic evidence from Kilmelfort Cave, Argyll: a re-evaluation of the lithic assemblage. Proceedings of the Society of Antiquaries of Scotland 139: 9-45.

Scheuer, L. and Black, S. 2000. Developmental juvenile osteology. San Diego, California: Academic Press.

Schoeninger, M. J. and DeNiro, M. J. 1984. Nitrogen and carbon isotopic composition of bone collagen from marine and terrestrial animals. Geochimica et Cosmochimica Acta 48: 625-639.

Schulting, R., Sheridan, A., Crozier, R. and Murphy, E. 2010. Revisiting Quarterness: new AMS dates and stable isotope data from an Orcadian chamber tomb. Proceedings of the Society of Antiquaries of Scotland 140, 1-50.

Schweingruber, F. H. 1990. Anatomy of European Woods. Berne and Stuttgart: Haupt.

Scott, W. L. 1951. Eilean an Tighe: a pottery workshop of the second millennium BC. Proceedings of the Society of Antiquaries of Scotland 85: 1-37.

Serjeantson, D. 2013. Farming and Fishing in the Outer Hebrides AD 600 to 1700: The Udal, North Uist. Southampton: The Highfield Press.

Serjeantson, D., Smithson, V. and Waldron, T. 2005. Animal husbandry and the environmental context, in A. Ritchie (ed.) Kilellan Farm, Ardnave, Islay: Excavations of a prehistoric to early medieval site by Colin Burgess and others, 1954-76: 151-167. Edinburgh: Society of Antiquaries of Scotland.

Sharples, N. 2005. A Norse Farmstead in the Outer Hebrides: Excavations at Mound 3, Bornais, South Uist. Oxford: Oxbow Books.

Sharples, N. 2009. Beaker settlement in the Western Isles, in M. J. Allen, N. Sharples and T. O'Connor Land and People: papers in memory of John G Evans: 147-158. Oxford: Oxbow Books.

Sharples, N. 2012. The Beaker-period and early Bronze Age settlement at Sl geanach, Cill Donnain, in P. Parker Pearson (ed.) From Machair to Moorlands: archaeological survey and excavation in South Uist: 215-258. Oxford and Oakville: Oxbow Books.

Sharples, N. 2015. A short history of archaeology in the Uists, Outer Hebrides, Journal of the North Atlantic. Special Volume 9: 1-15.

Shepherd, I. A. G. 1976. Preliminary results from the Beaker settlement at Rosinish, Benbecula, in C. B. Burgess and R. Miket (eds.) Settlement and Economy in the Third and Second Millennia BC: 209-217. Oxford: BAR British Series 33.

Shepherd, I. A. G. and Tuckwell, A. N. 1977. Traces of Beaker-period cultivation at Rosinish, Benbecula. Proceedings of the Society of Antiquaries of Scotland 108: 108-13.

Shepherd, W. 1972. Flint: its origin, properties and uses. London: Faber and Faber.

Sheridan, A., Henshall, A.S. and Johnson, M. 2016. The pottery assemblage, in P. Ashmore Calanais Survey and Excavation 1979-88: 573-622. Edinburgh: Historic Environment Scotland.

Silver, I. A. 1969 The ageing of domestic animals, in D. R. Brothwell and E. S. Higgs (eds.) Science in Archaeology: 283-302. London: Thames and Hudson.

Simpson, D. D. A. 1971. Beaker houses and settlements in Britain, in D. D A. Simpson Economy and Settlement in Neolithic and Early Bronze Age Britain and Europe: 131-152. Leicester: Leicester University Press.

Sinclair, A. and Finlayson, B. 1989. Report on the Southern Hebrides lithic raw material survey, in S. Mithen and B. Finlayson The Southern Hebrides Mesolithic Project: Second Interim Report. (Unpublished).

Smith, H 2012 Plant remains, in N. Sharples The Beaker-period and early Bronze Age settlement at Sligeanach, Cill Donnain: 241-246, in P. Parker Peason (ed.) From Machair to Moorlands: archaeological survey and excavation in South Uist: 215-258. Oxford and Oakville: Oxbow Books.

Stace, C. 1997. New Flora of the British Isles. 2nd Edition. Cambridge: Cambridge University Press.

Stevenson, R. B. K. 1953. Prehistoric pot-Building in Europe. Man 53: 65-68.

Steyn, M. and Patriquin, M. 2009. Osteometric sex determination from the pelvis - does population specificity matter? Forensic Science International 191: 1-3.

Stirland, A. 1994. The angle of femoral torsion: an impossible measurement? International Journal of Osteoarchaeology 4 (1): 31-35.

Stuiver, M. and Reimer, P. J. 1986. A computer program for radiocarbon age calibration, in Stuiver, M. and Kra, R. S. (eds.) Proceedings of the 12th International 14C Conference. Radiocarbon 28(2B): 1022-1030.

Stuiver, M. and Reimer, P. J. 1993. Extended 14C data base and revised CALIB 3.0 14C calibration program. Radiocarbon 35(1): 215-230.

Stuiver, M. and Kra, R. S. 1986. Editorial comment. Radiocarbon 28(2B): ii.

Stuiver, M. and Polach, H. A. 1977. Reporting of 14C data. Radiocarbon 19: 355-363.

Szpak, P., Orchard, T. J., McKechnie, I. and Gröcke, D. R. 2012. Historical ecology of late Holocene sea otters (Enhydra lutris) from northern British Columbia: isotopic and zooarchaeological perspectives. Journal of Archaeological Science 39 (5): 1553-1571.

Thew, M. 2003. The molluscan assemblage, in J. Barber, Bronze Age Farms and Iron Age Farm Mounds in the Outer Hebrides: 163-173. Scottish Archaeological Internet Report (SAIR) 3. Viewed 01 November 2017, <http://archaeologydataservice.ac.uk/archives/view/sair/volumes.cfm>.

Trotter, M. 1970. Estimation of stature from intact limb bones, in T. Stewart (ed.) Personal Identification in Mass Disasters: 71-83. Washington DC: National Museum of Natural History, Smithsonian Institute.

Vickers, K., ul Haq, S. and Hamshaw-Thomas, J. 2014. Chapter 13: The faunal remains, in M. Parker Pearson and M. Zvelebil, Excavations at Cill Donnain: A Bronze Age Settlement and Iron Age Wheelhouse in South Uist. Oxford: Oxbow Books.

Villotte, S., Castex, D., Couallier, V., Dutour, O. and Knüsel, C-G. 2010. Enthesopathies as occupational stress markers: evidence from the upper limb. American Journal of Physical Anthropology 142 (2): 224-234.

Villotte, S. and Knüsel, C. 2013. Understanding entheseal changes: definition and life course changes. International Journal of Osteoarchaeology 23 (2): 135-146.

von den Driesch, A. 1976. A Guide to the Measurement of Animal Bones from Archaeological Sites (Peabody Museum Bulletins). Cambridge, Massachusetts: Peabody Museum Press, Harvard University.

Waldron, T. 2009. Palaeopathology. Cambridge Manuals in Archaeology. Cambridge: Cambridge University Press.

Walker, P. 2008. Sexing skulls using discriminant function analysis of visually assessed traits. American Journal of Physical Anthropology 136 (1): 39-50.

Warren, G. 2005. Chipped stone, in T. Neighbour, Excavation of a Bronze Age kerbed cairn at Olcote, Breasclete, near Calanais, Isle of Lewis: 34-46. Scottish Archaeological Internet Report (SAIR) 13. Viewed 01 November 2017, <http://archaeologydataservice.ac.uk/archives/view/sair/contents.cfm>.

Wickham-Jones, C. R. 1990. Rhum. Mesolithic and Later Sites at Kinloch. Excavations 1984-86. Edinburgh: Society of Antiquaries of Scotland, Monograph 7.

Wickham-Jones, C. R. and Collins, G. 1978. The Sources of flint and chert in Northern Britain. Proceedings of the Society of Antiquaries of Scotland 109: 7-21.

Young, A, 1966. The sequence of Hebridean pottery, in A. L. F. Rivet (ed.) The Iron Age of North Britain: 45-58. Edinburgh: Edinburgh University Press.

Index

A' Croig Bheag: 1, 4, 9, 20, 24, 25, 28, 41, 58, 203, 214, 217, Figure 1.1, 6.1

Allt Chrisal, Barra: 163, 165, 173, 174, 193, Figure 6.2

An Doirlinn, South Uist: 92, 196, 197, 204, 212, 227, Figure 6.2

An Dunan, Uig, Lewis: 163, 164

Antrim, Ireland: 123

Ard (see Ploughing)

Ard a' Mhorrain (a'Bhorain) escarpment: 1, Figure 1.1, 4, 5, 6, 7, 8, 19, 20, 60

Ardnave, Islay: 110, Figure 6.2

Atkinson, R J C: 8

Auchrynie, Aberdeen: 197

Axe /axehead: 34, 176, 181, 182, 183, 209, 212-213, Figure 2.15, 2.16, 5.51

Bac Mhic Connain, North Uist: 202, Figure 6.4

Bailie: 5

Baleshare / Baile Sear, North Uist: 24, 165, Figure 6.4

Balfarg/Balnirnie, Fife: 206, Figure 6.2

Barabhas / Barvas, Lewis: 110, 127, 129, 130, 140, 142, 146, 147, 155, 161-164, 174, 224, 227, Figure 6.2

Barnhouse, Orkney: 196, 211, Figure 6.2

Barns Farm, Dalgety: 72

Barpha Langais, North Uist: 163, Figure 6.4

Barra: 62, 123, 174, 177

Beaker (see also Pottery): 6, 10, 14, 110, 174, 182, 215,

    Cemetery complex: 8

Beckton Farm, Dumfries and Galloway: 206, Figure 6.2

Benbecula: 5, 8, 213

Beveridge, Erskine: 4, 6, 9

Bharpa Carinish, North Uist: 211, Figure 6.4

Bone, worked / tools: 197-202, 206, 210, 223

Boreray: 1

Bornais, South Uist: 113, Figure 6.2

Botanical remains (see also Machair, Wood): 1, Figure 1.4, 88-94, 210

    Cereals/ cereal crops: 89, 90, 92, 94, 109, 209, 211, 212, 216, 222, Figure 6.6

Boundary post-setting: 22-23, 39-41, 43, 44, Figure 2.24

Bowl pits: 22-24, 36, 39-41, 43, 44, 51, 53, 108, 119, 159, 214, 215, 221, Figures 2.25, 2.26, 6.1

Bowler, David: 114

Breasclete, Lewis: 220, 221, Figure 6.2

British Association for Biological Anthropology and Osteoarchaeology: 70

British Museum: 174

Burials and human remains: 46, 216, Figures 2.29

    Early burial (disarticulated): 46, 47, 60, 70, 78-80, 217, 225, Figures 2.28-2.30, 3.18-3.20

    Cist burial with skeleton (west): 47-49, 64, 66, 67, 70-76, 79, 80, 83, 117-118, 196, 217, 219, 220, 225-226, Figures 2.29, 2.30, 2.32-2.36, 3.10-3.17, 3.21, 3.22, 3.24, 6.8

    Cist burial with skeleton (east): 55, 64, 66, 67, 70-76, 79, 80-81, 83, 114, 159, 185, 194, 224, 225, 226, Figures 2.44-2.46, 3.5-3.9, 3.21-3.23

Calanais, Lewis: 162, 163, 164, 165, 196, 212, 215, 220, Figure 6.2

Camas Daraich, Skye: 165, Figure 6.2

Cambridge: 1, 64, 67, 86, 95, 122
Catpund, Shetland: 44
Ceardagh Ruad: 25, Figure 6.2, 6.4
Celtic Ox: 102
Cereals (see Botanical Remains)
Chartered Institute for Archaeologists: 70
Christ's College, Cambridge: 4
Cill Donnain (Kildonan), South Uist: 110, 165, 183, 215, Figure 6.2
Cladh Hallan, South Uist: 81, 174, 197, 224, 225, 226, Figures 3.22, 6.2
Clarke, Ann: 122, 174
Cnip, Lewis: 165, 215, 216, 220, 221, Figure 6.2
Coastal (marine) change /erosion / flooding / incursion: 6, 8, 9, 10, 16, 22-27, 29, 31, 36, 39, 44, 46, 47, 49, 55, 57, 183, 203, 204, 211, 213, 214, 216, 217, 221, 225, Figures 2.3-2.5
Coilegean an Udail, North Uist: 4, 6, 8, Figure 1.3
Comhairle nan Eilean Siar: 17
Conjested Districts Board: 60
Cottenham, Cambridgeshire: 176
Craonabhal: 220, Figure 6.4
Crantit tomb, Orkney: 177
Crofting-township: 5
Crawford, Barbara: xxvi, 4
Crawford, Harriet: 6, 15
Crawford, Iain A: xxvi, xxvii, 1, 4-10, 12-15, 17, 19, 20, 24, 27, 29-32, 35-37, 40, 43, 44, 46, 48, 51, 53, 58-62, 64, 67, 86, 87, 95, 119, 122, 123, 161, 174, 175, 176, 180, 181, 183, 197, 201, 203, 204, 206, 208, 213, 214, 217, 220, 221, 226, 227, 228
Crawford, Imogen: 5, 7, 87, 123, 197
Crawford's Chrystal Palace: 16, 176, 183, 197, 201, Figure 1.11
Crustacea (see Marine Shell): 116-118

Dalmore, Lewis: 147, 162, 163, 164, 224, Figure 6.2
Denstone, Bernard: 67
Diet (human): 64, 80, 81, 83
Duckworth Laboratory, Cambridge: 67
Dumfries and Galloway: 122
Dun Mor Vaul, Tiree: 5
Dun Skellor, North Uist: 8, Figure 1.1
Dun Toloman, North Uist: 8, Figure 1.1

Earth houses: 6
Edinburgh: 5
Eilean Domhnuill (Olabhat), North Uist: 163, 164, 182, 211, Figure 6.4
Eilean an Tighe: 208, 211, 213, Figure 6.4
Exchequer Rolls: 9

Faunal remains:
    Antler: 97, 98, 105, 110, 164, 197, 198, 201, 202, 209, 210, 212
    Birds / fowling: 87, 95, 96, 97, 105, 106, 109, 110, 210, 212, 222
    Cattle / calf: 46, 55, 62, 65, 80, 81, 82, 96, 97, 98, 99, 100, 101, 102, 104, 105, 108-110, 113, 120, 197-198, 201, 209, 210, 212, 222, 224, Figure 2.45, 3.21, 4.9
    Deer, Red: 97, 98, 99, 100, 105, 108, 109, 110, 198, 199, 209, 210, 212, 222
    Dog: 97, 98, 105, 108, 109, 110, 210, 222
    Fish: 87, 95, 96, 97, 105, 107, 108, 109, 110, 113, 198, 210, 212
    Horse: 97, 99, 110
    Pig: 97, 110
    Seal: 99, 105, 110
    Sheep, sheep/goat: 96, 97, 98-102, 103, 104, 105, 108-110, 198, 199, 201, 202, 209, 210, 212, 222, Figures 4.5-4.8
Fishing: 113, 115, 116, 175
Flint: 123-129, 175, 182, 210-212, 222
    Arrowhead: 126, 127
    Cores: 125, 127
    Retouched pieces: 127
    Scrapers: 126, 128,
    Scale-flaked knife: 126, 127
    Scraper - end: 126, 127, Figure 5.2
    Scraper - thumbnail: 126, 127, 128, Figure 5.2
    Scraper - truncated; 126, 127
Foshigarry, North Uist: 202, Figure 6.4
Fresh water spring (AY); 62, Figure 2.49
Frit (see Aeolianite)

Gaelic: 4, 5
Gob Eirer, Uig, Lewis: 163, 164
Graham, Professor G: 28
Grenetote: 1, 8, 9, 58-60, 62, Figure 6.4

Griminish: 8, Figure 6.4
Grinder: 179, 181, 222, Figure 5.53
GUARD, University of Glasgow: 17
Gunnerso, Uig, Lewis: 162, 163, 164

Hammerstone (see also Quartz): 178, 179, 180, 182
    Hammerstone/ anvil: 178, 179, 181, 222
    Hammerstone / grinder: 179
    Hammerstone / pounder: 179, 181
Hansom, Dr Jim: 7
Harris: 62, 174
Hebrides Thrust Zone: 124
Historic Environment Scotland: 18
Historic Scotland: 17, 122
Horgabost, Harris: 220, 221, Figure 6.2
Hougarry: 60

Iceland: 165, 166
Inner Hebrides: 4, 123
Islay: 123
Isotopic analysis: 80-85
Iron Age: 4, 5, 9

Jarlshof, Shetland: 6
Judith's pot pit: 194

Katla Volanic System: 166
Kelp:
    Kelp drying dyke: 22, 27, 51, 58, 61, 62, Figure 2.49
    Kelp drying kiln/ pit: 22, 58, 62, 63, Figure 2.49
Kildonan, South Uist (see Cill Donnain)
Kilellan, Ardnave, Islay: 110, 183, 197, 224, Figure 6.2
Kilmelfort, Argyll and Bute: 162
Kintyre: 5
Knap of Howar, Papa Stour, Orkney: 202, 204, 215, Figure 6.2

Lewis, Butt of: 8
Lewisian Complex: 1
Lian an Udail: 4
Links of Noltland, Orkney: 110, Figure 6.2
Loch a' Choire, South Uist: 174, Figure 6.2

Loch Laing, South Uist: 92
Lochmaddy: 1
Loch Portain, North Uist: 92

Machair / environment / meadows / systems: 1, 3-10, 12, 18, 109, 119, 120, 178, 185, 215, 216, 227
Machair Leathann: 8
Machair Mheadhanach: 133, figure 6.2
Machrie Moor, Arran: 196
Machair Research Group: 9, 62
MacKie, Euan: xxvi, xxvii, 5
McLeod's Stone, Harris: 221
MacInnes, Donald: 58, 59, 61
MacInnes, Norman: 61
Manuring: 85, 215
Marine shell /crustacea: 110-115, 116-118, 209, 212, 222
    Perforated: 115
    Shell for pottery decoration: 188, 194, 195, 197, Figure 5.59
Marine incursion (see Coastal change)
Mattock marks / mattock: 22, 23, 36, 39, 40, 185, 213, 216, Figure 2.21
Migration (human): 80, 85, 225
Museum nan Eilean: 18

National Library of Scotland: 9
National Museums Scotland: 102, 174
    Artefact Research Unit: 122
Newark Bay, Orkney: 81, Figure 3.22
North Uist: 1, 4-6, 8-10, 12, 18, 27, 20, 28, 60, 61, 123, 174, 176, 212
Northton, Harris: 72, 109, 110, 163, 174, 182, 197, 201, 211, 212, 215, 216, 224, Figure 6.2
Northern Isles: 116
North Roe, Shetland: 176
Norse: 4
Norway: 4, 165, 166

Odal (see Udal)
Olcote, Lewis: 163, 164
Orkney: 165, 174
Otairnis, Boreray: 220, Figure 6.2, 6.4
Outer Hebrides: 4, 5, 6, 7, 8, 9, 110, 165

Pastoralism: 109, 110, 212
Peat: 6, 88-94
Perihelion event / tide: 9, 10, 22
Plants (see Botanical Remains, Machair)
Ploughing / plough marks / land: 4, 21-24, 30, 36, 39-41, 44, 58, 183, 185, 204, 211, 214, 215, 216, Figures 2.22, 2.23, 6.5
    Ard: 41, 110, Figure 2.23
Pollen: 75
Point Braighe, Lewis: 163, 164, Figure 6.2
Polisher: 178, 180, 181
Pool, Sanday, Orkney: 215, Figure 6.2
Port nan Long: 8, Figure 6.4
Pottery / clay: 183-197, 206, 210, 211, 212, 215, 222-223
    Beaker: 188, 192, 193, 194, 195, 196, 197, 210, 217, 225, Figure 5.55
    Carbonised food residues: 64, 65, 66, 185, 187, 191, 192, 196, 197, 203, 210
    Decoration (see Marine Shell): 187-188, 194, Figure 5.56, 5.57, 5.59, 5.60
    Grooved Ware: 188, 193, 195, 196, 197, 206, 210, 212, Figure 5.55, 5.56, 5.57
    Heavy cooking pot: 197, 224, 225, Figure 5.55
    Organic temper (vegetable matter): 184, 185
    Shell temper: 184, 185, 188, Figure 5.57
Pounder (see also Quartz): 180
Preece, Dr Richard: 86
Pumice: 165-168, 175, 212
    Worked: 169-174, 209 210

Quanterness, Orkney: 81, Figure 3.32
Quarternary Research Group: 1
Quaternary Research Laboratory: 64
Quartz: 127, 128, 129-165, 175, 182, 183, 206, 210 212, 222
    Atypical scrapers: 132, 147, 150, 157, Figure 5.17
    Atypical cores: 132
    Backed knives: 146, 157
    Bipolar cores: 138, 139, 145, 157, Figure 5.28, Figure 5.29
    Blades: 135, 136, 157, Figure 5.9
    Blade-scrapers: 147, 148, 157, Figure 5.32
    Cores: 136, 137, 138, 139, 157
    Core fragments: 146, 157
    Cores with two platforms at an angle: 143, 157
    Crested pieces: 132, 134, 135, 137, 148, 155, 156, 157, 158, 159, 160, 162, Figure 5.11
    Denticulated pieces: 151, 157
    Discoidal scrapers: 147, 148, 157, Figure 5.33
    Double-scrapers: 147, 149, 157
    End-/side-scrapers: 147, 148, 149, 157
    Flaked flakes': 144, 157
    Hammerstones: 153, 157, Figure 5.41
    Hammerstones/anvils: 153, 157,
    Hammerstones/split pebbles: 153, 154, 157
    Irregular cores: 138, 143, 157, Figure 5.26, 5.27
    Microblades: 134, 135, 157, Figure 5.8
    Opposed-platform cores: 143, 157, Figure 5.24, 5.25
    Percussoirs: 152, 157
    Percussoirs/anvils: 152, 157, Figures 5.40, 5.51
    Pieces with edge-retouch: 152, 157, Figure 5.38
    Pieces with retouched notch(es): 151, 157
    Piercers : 146, 150, 151, 157, Figure 5.37
    Points: 152, 157, Figure 5.39
    Pounders: 154, 157, Figure 5.24
    Scale-flaked knives: 146, 147, 157, 161, 210, Figure 5.30
    Scraper-edge fragments: 147, 150, 157
    Short end-scrapers: 147, 148, 157, Figure 5.34. 5.35
    Side-scrapers: 147, 148, 149, 151, 157, Figure 5.36
    Single-platform cores: 138, 141, 142, 157, Figure 5.17, 5.18-5.21
    Single-platform/discoidal cores: 138, 142, 157
    Split pebbles: 138, 155, 157
    Tools: 132

Radiocarbon dates; 23, 25, 26, 46, 47, 64-67, 166, 196, 197, 201, 221, 224
Ritchie, Professor William: 7

Spring (fresh water): 62
Standing stones: 23, 28, 32, 43, 44, 46, 53, 55, 58, 59, 62, 159, 174, 180-181, 221-222, 224, Figures 2.8, 2.30, 2.40, 2.41, 2.44
Staosnaig, Colonsay: 165
Stone (see also Flint, Pumice, Quartz)
    Aeolianite (frit): 1
    Amphibolite: 133, 175, 179, 180, 181, 182
    Basalt: 175, 177, 178, 179, 181, 182, 222
    Chalcedony: 123, 127
    Diorite: 175, 176, 177, 181, 182
    Dolerite: 175, 176, 178, 179, 181, 182, 222
    Gneiss: 1, 22, 124, 133, 162, 175, 176, 178, 179, 180, 181, 182, 184, 185
    Granite: 1
    Greenstone: 182
    Mudstone: 182
    Mylonite: 123, 126, 147, 161, 162, 163, 164
    Pseudotachylite: 123, 124, 127, 176, 178, 180, 182, Figure 5.1
    Rhyolite: 1
    Serpentine: 182
    Tephra: 166, 168
Stone ball: 177, 181, 182, 213, Figure 5.51
Stone robbing / holes / pit: 22, 27, 36, 49, 57, 58, 60, 61, 120, 183
Sumburgh, Shetland: 81, Figure 3.22
Sutherland: 5
Svalbard: 165
Sweden: 165

Terrestrial snails / molluscs: 118-121, 209, 210
Tidal channels: 8
Till, glacial (subsoil): 20, 25, 28, 95, 184
Toftsness, Sanday, Orkney: 215, Figure 6.2
Traigh an Udail, North Uist: 4, Figure 1.1
Traigh Iar, North Uist: Figure 1.1
Treasure Trove Unit: 18
Treaty of Perth: 4
Trondheim Convention: 64
Turf: 1, 3, 9, 16, 22, 24, 26, 27, 29, 34, 35, 37, 39, 41, 43, 46-48, 51, 53, 57, 58, 60, 61, 88-94, 119, 203, 208-209, 211, 213-215, 217, 219-220, 222-224

Udal
    Peninsula or promontary: 1, 4, 5, 6
    Sites: RUX1, RUX2, RUX3, RUX6, UN and US 6, 8, 9, 13, 44, 87, 165, 213, Figures 1.1, 1.3, 1.6
    Fieldwork statistics: 12
    Place name: 4
    Research project: 12
    Udal (Odal) Law; 4
University of
    Aberdeen: 7
    Bradford, Biological Anthropology Research Centre: 67
    Edinburgh, School of Scottish Studies: 4, 9, 10, 122
    Glasgow: 122
    St Andrews: xxvi, 95
    Swansea; 28
Uists: 5, 6, 8, 174
Unival, North Uist: 196, 212, Figure 6.4
Unstan ware: 182

Vegetation (see Machair)
Valtos, Lewis: 163, 164, Figure 6.2

West, Professor Richard G: 95
West Highlands: 4, 5, 6
West of Scotland: 4
Western Europe: 5
Western Isles: 123, 128, 129, 161, 162, 164, 165, 170, 177, 185, 196
Whale bone: 29-31, 64, 65, 66, 105, 180, 195, 198, 201, 206, Figures 2.8-2.10, 2.30
Wood:
    Driftwood: 59, 89, 91, 93, 114, 209, 210, 212, 214, 222, 223
    Wooden posts: 31, 32, 35, 206, 208, 214, 221
    Wood species, woods, woodland: 88-91, 93, 114, 119, 210, 212, 221
Wrecks (maritime): 27, 58, 60

Yarnton, Oxford 81, Figure 3.22

Rosinish, Benbecula: 110, 162, 163, 174, 215, 220, Figure 6.2

Rubha an Udail: 1-4, 6, 9, 13, 28, 58, 62 Figure 1.1

Rubha Bheilis: 1, Figure 1.1

Rubha Huilis, Huilis, Uilish, Oinlish: 8, 25, 60 Figures 1.1, 1.3, 6.1

RUX1, RUX2, RUX3 (see Udal sites)

RUX6 - records:
- Database: 18
- Data structure reports: 17
- Site grid: 13, 14
- Site recording and records: 14, 16, 17
- Site seiving: 15, 86, 87, 96, 108, 110, 116, 123, 129, 209

RUX6 - features and structures:
- Ash mound DG: 30, 203, Figure 2.7
- Burnt area DK: 37, 159, 195, 203, 208, Figure 2.7
- Cairn complex (west): 20, 21, 23, 25-27, 44, 46-49, 51, 58, 60, 61, 67, 183, 194, 208, 217, 219, 220, 222, Figures 2.5, 2.30, 2.31, 2.38, 2.39, 6.9
- Clay pit: 28, 37
- Earliest structure (DA) ?building 3: 30, 159, 203, Figure 2.7
- Great Auk stone: 23, 30, 32, 37, 174, 179, 180, 181, 206-208, Figures 2.8, 2.10-2.14, 5.54, 6.3
- Hearths, hearth pits / fire pits: 30, 35-37, 54, 55, 66, 156, 159, 203, 209, 210, 211, Figures 2.7, 2.15, 2.17-2.19, 2.41
- Kerbed-cairn (east): 22, 23, 44, 55, 57, 108, 159, 174, 179, 208, 224, Figures 2.47, 2.48
- Neolithic Building 1 (DJ): 29, 30, 34, 36, 37, 40, 47, 65, 67, 95, 108, 109, 115, 119, 123, 127, 128, 129, 159, 160, 161, 162, 177, 181, 182, 193, 195, 203, 204, 208-211, Figures 2.7, 2.15
- Neolithic Building 2 (DH): 28, 29, 32, 36, 37, 66, 67, 86, 87, 95, 108, 109, 110, 115, 119, 127-129, 156, 159-162, 179, 181, 184, 193-195, 197, 199, 200, 201, 203, 204, 208-211, 212, Figures 2.7, 2.12, 2.15, 2.17-2.19
- Linear trench (saw-pit?) (BH): 57 Figure 2.44
- Othello stone: 31, Figures 2.11, 2.13
- Platforms (internal to structures): 37, 209, Figure 2.15
- Pot pit (CE): 44, Figure 2.24
- Shaft (libation pit), platform and stone settings ('garden gnomes') / alignments: 22, 23, 29, 30-32, 66, 109, 115, 159, 160, 161, 180, 194, 199, 206, 208, 221, 222, 224, Figures 2.8, 2.9, 2.11-2.13
- Stone boxes: 44, 55, 159, 160, Figures 2.24, 2.27, 2.41-2.43
- Temporary structure (BG24): 46, 53-55, 64, 66, 95, 108, 114, 127, 159, 161, 162, 174, 179, 180, 194, 197, 198, 222-224, Figures 2.41, 2.42, 6.10

Sand / blown sand / accumulation: 1, 20, 23, 24, 39, 40, 41, 46, 48, 58, 59, 61, 114, 115, 119, 204, 206, 210, 213, 215, 216, 221
- Sand hills/dunes/ hillocks: 4, 6, 8, 61
- Sand abrasion: 133, 183, 185, 187, 199

Sand, Inner Sound: 116

Saw-pit: 22, 27, 30, 36, 37, 44, 46, 53, 57, 58, 59, 60, 62, 120, 204, 223, Figures 2.11, 2.12

Saddle quern: 180, 183, 222

Scots Law: 4

Scottish Development Department (SDD): 7, 12, 14, 122

Scottish Records Office (National Archives of Scotland): 9

Scottish Universities Research Centre: 64

Sea level (see also Coastal change): 1, 7, 9, 10, 18, 19, 20, 24, 28, 39, 41, 46, 229, 203, 211, 217

Seaweed (tangle): 58, 62, 91, 92, 93, 94, 211

Serjeantson, Dr Dale: 87

Settlement (origin and history): 4, 5-10, 13, 18

Sgeirean /bile (see kelp drying dykes)

Shetland: 117, 165, 174

Shieldaig, Wester Ross: 162

Shingle: 9, 23, 24, 95, 217

Sioachadh Ghroaidh: 8

Skara Brae, Orkney: 44, 202, 204, 211, 213, 215, Figure 6.2

Skilmafilly, Aberdeenshire: 148, Figure 6.2

Sligeanach, Kildonnan, South Uist: 6, 215, Figure 6.2

Sound of Eriskay: 8

Sollas, North Uist: 27, Figure 1.1, Figure 6.4
- Wheelhouse: 8

Sound of Harris: 8

South Harris: 1

South Uist: 8